T0142960

The International Series in Video Computing

Series Editor: Mubarak Shah, Ph.D
University of Central Florida
Orlando, Florida

For further volumes:
http://www.springer.com/series/6673

Saad Ali • Ko Nishino • Dinesh Manocha
Mubarak Shah

Editors

Modeling, Simulation and Visual Analysis of Crowds

A Multidisciplinary Perspective

 Springer

Editors
Saad Ali
Center for Vision Technologies
SRI International
Princeton, NJ, USA

Dinesh Manocha
Department of Computer Science
University of North Carolina
Chapel Hill, NC, USA

Ko Nishino
Department of Computer Science
Drexel University
Philadelphia, PA, USA

Mubarak Shah
Center for Research in Computer Vision
University of Central Florida
Orlando, FL, USA

ISSN 1571-5205
ISBN 978-1-4939-4628-0   ISBN 978-1-4614-8483-7 (eBook)
DOI 10.1007/978-1-4614-8483-7
Springer New York Heidelberg Dordrecht London

Preface

Accurate analysis and synthesis of human behavior in crowds, a large and dense group of people with varying characteristics and goals, is a common requirement across a wide range of domains. If the human behavior, including those of individuals, small groups of people, and even the crowd as a whole – can be interpreted and anticipated in arbitrary real-world situations, a repertoire of important applications, many of which are societally important, can be realized: For example, perpetrators disguised in a busy street corner will be easily spotted and tracked in a surveillance video feed; new buildings, public places and outdoor environments will be designed to optimize the space use with the dynamically changing flow of people in mind, while minimizing the time need for evacuation whenever necessary; and the social psychology of people can be studied based on large-scale, longitudinal observations, and many more.

The goal of this book is to provide the readers a comprehensive map of the current state of the art in distinct but related fields, mainly in computer vision, graphics, and evacuation dynamics, towards the common goal of better analyzing and synthesizing the pedestrian movement in dense, heterogeneous crowds. The monograph is organized into different parts that consolidate various aspects of research towards this common goal, namely the modeling, simulation, and visual analysis of crowds. Many of the chapters in these parts extend the works that were presented at the first workshop on the same topic at International Conference on Computer Vision, 2011, and collectively cover the diverse challenges involved in better understanding of human crowds. Our hope is, through this book, the readers will see the common ideas and vision as well as the different challenges and techniques for modeling, analyzing, and simulating crowds, that will stimulate novel approaches to getting us a step closer to fully grasping "crowds."

This book grew out of the first IEEE Workshop on Modeling, Simulation and Visual Analysis of Large Crowds, that was held in conjunction with International Conference of Computer Vision 2011. Therefore, first of all we would like to acknowledge the workshop program committee who worked tirelessly for the success of the workshop and authors that contributed their valuable pieces of work. We would also like to thank Prof. Jie Yang and National Science Foundation (NSF)

v

funding based on grant IIS-1142382 to provide travel support for the workshop. We are also grateful to our host institutions (SRI International, Drexel University, University of North Carolina and University of Central Florida) for providing a highly stimulating research environment that enables pursuit of new research ideas and discoveries. Springer has provided excellent support throughout the preparation of the book, and we would like to specially thank their staff for their support and professionalism. Many people have helped proof reading draft material and providing comments and suggestions. We would like to thank all of them for their time and valuable contribution towards improving the quality of the book.

Princeton, NJ, USA   Saad Ali
Philadelphia, PA, USA   Ko Nishino
Chapel Hill, NC, USA   Dinesh Manocha
Orlando, FL, USA   Mubarak Shah

Contents

Contributors

Saad Ali Center for Vision Technologies, SRI International, 201 Washington Road, Princeton, NJ, USA

Norman I. Badler University of Pennsylvania, Philadelphia, USA

Stefania Bandini Department of Computer Science, Systems and Communication, Complex Systems and Artificial Intelligence (CSAI) Research Center, University of Milan – Bicocca, Milano, Italy

Maik Boltes Jülich Supercomputing Centre, Forschungszentrum Jülich GmbH, Jülich, Germany

Avishy Y. Carmi Department of Mechanical and Aerospace Engineering, Nanyang Technological University, Singapore

Antoni B. Chan Department of Computer Science, City University of Hong Kong, Hong Kong, China

Ke Chen Queen Mary University of London, London, UK

M. Chraibi Jülich Supercomputing Centre, Forschungszentrum Jülich, Jülich, Germany

M. Cristani Pattern Analysis and Computer Vision (PAVIS), Istituto Italiano di Tecnologia, Genova, Italy

Sean Curtis University of North Carolina at Chapel Hill, Chapel Hill, USA

A. Del Bue Pattern Analysis and Computer Vision (PAVIS), Istituto Italiano di Tecnologia, Genova, Italy

Funda Durupinar University of Pennsylvania, Philadelphia, USA

Simon J. Godsill Department of Engineering, University of Cambridge, Cambridge, UK

Shaogang Gong Queen Mary University of London, London, UK

Helena Grillon Centrale de Compensation, Geneva, Switzerland

Pini Gurfil Department of Aerospace Engineering, Technion Israel Institute of Technology, Kesalsaba, Israel

Stephen J. Guy University of North Carolina at Chapel Hill, Chapel Hill, USA

Mubbasir Kapadia Center for Human Modeling and Simulation, University of Pennsylvania, Philadelphia, USA

Louis Kratz Department of Computer Science, Drexel University, Philadelphia, USA

Laura Leal-Taixé Leibniz University Hannover, Hannover, Germany

Chen Change Loy Vision Semantics Limited, The Chinese University of Hong Kong, Hong Kong, China

Jonathan Maim Bengaluru Area, India

Dinesh Manocha Department of Computer Science, University of North Carolina, Chapel Hill, USA

Lyudmila Mihaylova School of Computing and Communications, Lancaster University, Lancaster, UK

V. Murino Pattern Analysis and Computer Vision (PAVIS), Istituto Italiano di Tecnologia, 16163 Genova, Italy

Ko Nishino Department of Computer Science, Drexel University, Philadelphia, USA

Sze Kim Pang DSO National Laboratories, Singapore

Jan-Frederik Pietschmann Center for Industrial and Applied Mathematics (CIAM), Royal Institute of Technology, Stockholm, Sweden

R. Raghavendra Pattern Analysis and Computer Vision (PAVIS), Istituto Italiano di Tecnologia, Genova, Italy

Bodo Rosenhahn Leibniz University Hannover, Hannover, Germany

E. Sangineto Pattern Analysis and Computer Vision (PAVIS), Istituto Italiano di Tecnologia, Genova, Italy

A. Schadschneider Institute for Theoretical Physics, Universität zu Köln, Köln, Germany

François Septier Signal Processing and Information Theory Group, TELECOM Lille 1, Villeneuve d'Ascq Cedex, France

Armin Seyfried Jülich Supercomputing Centre, Forschungszentrum Jülich, Jülich GmbH, Germany

Computer Simulation for Fire Safety and Pedestrian Traffic, Bergische Universität Wuppertal, Wuppertal, Germany

Mubarak Shah Center for Research in Computer Vision, Harris Corporation Engineering Center, University of Central Florida, Orlando, USA

Alexander Shoulson University of Pennsylvania, Philadelphia, USA

Daniel Thalmann Institute for Media Innovation, Nanyang Technological University, Singapore

Nuno Vasconcelos Department of Electrical and Computer Engineering, University of California, San Diego, USA

Giuseppe Vizzari Department of Computer Science, Systems and Communication, Complex Systems and Artificial Intelligence (CSAI) Research Center, University of Milan – Bicocca, Milano, Italy

Tao Xiang Queen Mary University of London, London, UK

Barbara Yersin Bengaluru Area, India

Basim Zafar Hajj Research Institute, Umm al-Qura University, Mecca, Saudi Arabia

Jun Zhang Jülich Supercomputing Centre, Forschungszentrum Jülich GmbH, Jülich, Germany

Chapter 1
Modeling, Simulation and Visual Analysis of Crowds: A Multidisciplinary Perspective

Saad Ali, Ko Nishino, Dinesh Manocha, and Mubarak Shah

Abstract Over the last several years there has been a growing interest in developing computational methodologies for modeling and analyzing movements and behaviors of 'crowds' of people. This interest spans several scientific areas that includes Computer Vision, Computer Graphics, and Pedestrian Evacuation Dynamics. Despite the fact that these different scientific fields are trying to model the same physical entity (i.e. crowd of people), research ideas have evolved independently. As a result each discipline has developed techniques and perspectives that are characteristically it's own. In this chapter we provide a brief overview of major research themes from these different scientific fields, discuss common challenges and point to problem areas that will benefit from common synthesis of perspectives from these fields. In addition we introduce various pieces of work that appear in this monograph as separate chapters.

S. Ali (✉)
Center for Vision Technologies, SRI International, 201 Washington Road, Princeton, NJ, USA
e-mail: saad.ali@sri.com

K. Nishino
Department of Computer Science, Drexel University, 3141 Chestnut Street,
Philadelphia, PA 19104, USA
e-mail: kon@drexel.edu

D. Manocha
University of North Carolina, Chapel Hill, NC, USA
e-mail: dm@cs.unc.edu

M. Shah
University of Central Florida, Orlando, FL, USA
e-mail: shah@crcv.ucf.edu

S. Ali et al. (eds.), *Modeling, Simulation and Visual Analysis of Crowds*, The International
Series in Video Computing 11, DOI 10.1007/978-1-4614-8483-7_1,
© Springer Science+Business Media New York 2013

1.1 Introduction

Over the last several years there has been a growing interest in developing computational methodologies for modeling and analyzing movements and behaviors of 'crowds' of people. This interest spans several scientific areas: in *Computer Vision* the need to carry out visual surveillance in crowded scenes is fueling research on topics related to visual representations of crowds [2], tracking of individuals and groups [3, 18, 32, 45], detection of normal and abnormal behaviors [44, 62], segmentation and classification of motion patterns [2, 63], and mathematical modeling of interactions among the pedestrians in the crowd [81]; in *Computer Graphics* the goal of modeling and simulating crowd behaviors in different real-world or synthetic environments, including models for homogeneous and heterogeneous crowd simulation [30, 66], is advancing the state of art in areas aggregate flow [30, 31, 91], agent-based motion simulation [49, 84, 96], motion planning for large scale crowds [36, 70, 71, 94], obstacle and collision avoidance [52, 82, 92], modeling group behaviors [57, 65], and representation of virtual humans at multiple levels of detail [16, 60, 61]; in *Evacuation Dynamics* a parallel effort is underway to develop motion, interaction and self-organization models for pedestrian simulation and evacuation analysis [24]. However, as opposed to computer graphics, the emphasis is more on empirical validation of simulated movements and collective behaviors in terms of fundamental diagrams (which captures relationship between crowd density and velocity) and flows [11, 82]. In addition to above mentioned scientific areas, the diversity of context in which 'crowds' are studied has a long history and includes studies from areas of Anthropology, Psychology, and Sociology [9, 53].

Despite the fact that these different scientific fields are trying to model the same physical entity (i.e. crowds of people), many research ideas have evolved independently, and as a result each discipline has developed techniques and perspectives that are characteristically it's own. However, we strongly believe that in order to make the next big leap in terms of solving the crowd modeling and related computational problems, there is a need to develop common insights and understanding of general principles that characterize various aspects of a crowd. This requires development of a common-platform for cross-disciplinary exchange of ideas and interaction that allows benefiting from each other's experience and scientific discoveries. Some of the recent research in computer vision [3, 62, 69, 99] points towards merits of such cross-disciplinary work, where pedestrian interaction pedestrian interaction models, originally developed in evacuation dynamics, have been successfully used to carry out visual tracking and abnormal behavior estimation. Similarly, recent research in data-driven crowd simulation in computer graphics makes use of crowd trajectories and behaviors that are extracted from videos using computer vision algorithms [52, 67, 72].

The central goal of this monograph is to facilitate a process of cross-disciplinary interaction among researchers from areas of compute vision, computer graphics and evacuation dynamics by providing a common platform. For this purpose, a number of peer-reviewed chapters from leading researchers in these fields are compiled.

These chapters provide an understanding of the state of the art and open problems related to crowd modeling in each scientific discipline.

The rest of the chapter is organized as follows: In Sect. 1.2 we discuss various aspects of crowd which make their modeling a challenging task. In Sect. 1.3 we introduce central themes of the book and provide an overview of the related literature. In Sect. 1.4 we provide an overview of the organization of the rest of the book.

1.2 Aspects of Crowds

In order to bring about a common understanding of concepts and approaches related to crowds of people, it is important to answer the question: how should we think about crowds? What are the particular characteristic of crowds which make its modeling a challenging task? As to the former, in the most basic sense, a crowd is any collection of individuals or pedestrians where behavior of one individual is influenced by the other. We believe the flexible nature of this definition makes it applicable to all scientific areas that are focus of this monograph. For instance, we have examples of computer vision techniques which represent this influence at a notional level through particle interactions [2] or through dynamic floor fields [3]. Similarly, in computer graphics and simulated environment, this influence is taken into consideration during design of algorithms for collision avoidance and local interaction [20, 92]. Finally, in evacuation dynamics, computational approaches for emergent behaviors tend to use force-based models and cellular automata [82]. Various chapters of the monograph will provide many more example of how influence among participants of crowds are represented in different settings.

Agreeing to this definition leads us to the next question. What are the particular aspects of crowds which make them really challenging to model. We list some of them next in no particular order:

- Human behavior is extremely complex and exhibit large variation based on situations and settings. It also depends on individual characteristics such as age, sex, height, and cultural background, to name the few. There is no existing mathematical model that can account of all these complexities in human behavior.
- Human behavior can vary drastically based on the given situation. For instance, transition from walking to panic can be instantaneous given a dangerous situation (e.g. stampede).
- When viewed from visual sensors, it is hard to discern individuals in dense crowds due to low resolution (i.e. few pixels per individual). This results in appearance ambiguity and severe occlusion which often results in breakdown of visual processing pipelines. For instance, detection of individual person in the crowd might not be possible.
- Inability to detect individuals in a crowds that are observed through visual sensors makes it difficult to explicitly model interactions among individuals.

- Crowds tend to be heterogeneous in nature and within the same scene some parts of the crowd may be behaving very differently. This makes it hard to represent the dynamics of the crowd using a single global model.
- Behavior and dynamics of an individual in a crowd are connected with other individuals, both at the level of structure as well as behavior. This means actions of individuals can not be modeled in isolation and the fact that crowd will react to it has to be taken into consideration.
- Lack of datasets representing the richness of crowd behaviors and associated ground truth (e.g. 3D scene layout, individual tracks, personal characteristics such as height, sex etc.) makes it difficult to verify and validate crowd modeling techniques.

1.3   Central Themes and Topics

With this set of ideas in mind, we now introduce some of the main themes and topics considered in this monograph and the ways in which they reinforce the underlying principles related to modeling crowds. We also provide pointers to the chapters that are related to each of these themes.

1.3.1   Visual Analysis of Crowds

Visual analysis of crowded scenes is an integral component of a wide array of applications that span a number of areas with direct social impact. For instance, the rising prevalence of video recording technology in crowded areas presents a dire need for automatic visual analysis that can operate on videos containing a large numbers of individuals. Due to a large number of pedestrians in close proximity, crowded areas are at a high risk for dangerous activities including crowd panic, stampedes, and accidents involving a large number of individuals. Crowded scenes, are one such scenario of high-density, cluttered scenes that contain a large number of individuals. The extent of activity within such scenes is difficult for even human observers to analyze, making crowded scenes perhaps in the most need of automatic video analysis. Analyzing the behavior of pedestrians in such crowded scenes is also essential to the understanding and prediction of human behavior in similar but different scene context. Video analysis of crowded scenes can thus directly serve as the means to obtain in situ measurements of human behavior for data-driven crowd simulation.

Despite these strong needs, crowded scenes pose unique challenges that severely impede the development of robust video analysis methods. The complexity of the scene, largely owing to the sheer number of people in the crowds, becomes a direct burden on the computational method for visually analyzing the scene. Such complexity manifests itself in frequent, partial or complete occlusions among the pedestrians; the fact that every individual is moving and also surrounded by other moving people blurs the boundary of foreground and background pixels in the scene; and the arbitrary directions pedestrians may take based on their personal goals, neighboring pedestrians, and the physical obstacles within the scene. These all combined result in a heterogeneous and dynamically evolving crowd motion that is often too complex to analyze with conventional computer vision methods.

Conventional video analysis methods learn the behavior of the scene in three steps: detecting objects, tracking objects, and compiling the tracked results into higher order models for individual or global crowd behavior modeling. The applicability of such object-centric methods is limited to scenes with relatively few objects. Discerning individuals in crowded scenes is difficult since they are typically surrounded by other moving pedestrians. Tracking is also difficult due to the frequent partial or complete occlusions in crowded scenes. Finally, such methods suffer from problems of scale: each new pedestrian that enters the scene increases the complexity of the model.

Next we briefly describe some popular approaches to modeling crowds and their behaviors in videos.

1.3.1.1 Object-Centric Visual Analysis

Conventional video analysis methods are mostly object-centric; they begin by analyzing each scene object. Such methods detect the scene objects (e.g., pedestrians or automobiles), track them, and then analyze the trajectories to model the behavior of the objects. These methods work well on scenes that are relatively sparse (roughly 5–20 pedestrians) and, as noted by Zhan et al. [102], are not appropriate for dense crowded scenes. We review related object-centric work that are designed for crowded scenes, but emphasize that there are many challenges in videos containing high density crowds.

Detecting the scene objects is often the first step in object-centric video analysis. Zhao et al. [103], for example, track pedestrians in videos of crowds by detecting each individual using a model of human shapes. Rodriguez and Shah [76] detect pedestrians using a voting scheme on the contours around each individual. The contours are computed by subtracting the background from each video frame. In high density crowded scenes, however, the background is rarely visible and pedestrians are often partially occluded, making the contours difficult to estimate. Leibe et al. [51] also segment pedestrians from the background, but use global image cues to add robustness to partial occlusions. Their method handles some partial occlusions well, but assumes that the torso of the pedestrian is visible. This is often not the case in near-view scenes where only the heads of most pedestrians are visible.

Other work detect pedestrians by assuming that they exhibit unique motion. Brostow and Cipolla [8] group short feature tracks (or "tracklets") to identify similarly moving pedestrians. They assume that the subjects move in distinct directions and thus disregard possible local motion inconsistencies between different body parts.

As noted by Sugimura et al. [90], such inconsistencies cause a single pedestrian to be detected as multiple targets. In addition, pedestrians that are moving in the same direction are identified as a single group. Crowded scenes, especially when captured in relatively near-field views, as is often the case in video surveillance, necessitate a method that represents the multiple motions of a single individual or similar motions of different pedestrians.

After detection, the objects are tracked as they move through the scene. Data association methods, such as that of Betke et al. [7] or Gilbert and Bowden [19], track multiple targets in cluttered scenes by associating detection results of consecutive frames. These techniques assume that the detection is always reliable, and thus degrade in very crowded scenes. Wu and Nevatia [97] are able to track partially occluded pedestrians by detecting body parts, rather than the full pedestrian. The data association problem itself is NP-hard, and thus becomes less tractable in scenes with a large number of pedestrians. Often, approximation techniques are used to estimate a solution such as the Bayesian framework of Li et al. [54].

Other data association methods do not rely on detecting individuals. Khan et al. [40] model the interaction among detected interest points to improve the tracking of each object. Hue et al. [29] use a Markovian model on each tracked point to augment data association in generic domains. As noted by Khan et al. [41], however, a single point may be shared between multiple targets and can result in ambiguities. Shared points are often the result of motion boundaries or clutter, both of which occur frequently in videos of crowded scenes.

After tracking, the trajectories are used to characterize behaviors of objects within the scene. Wang et al. [95], for example, cluster trajectories to learn the common routes taken by pedestrians and automobiles. Dee and Hogg [14] use the tracking information to identify pedestrians that deviate from a goal-specific behavior. Hu et al. [26] learn global motion patterns (i.e., that describe motion over the entire frame) and use them to detect anomalies and predict future behaviors. Johnson and Hogg [34] estimate different distributions of trajectories, and attach semantics to each in order to identify specific events within the scene. Such methods not only depend on reliable detection and tracking, which may not be available in videos of crowded scenes, but also face problems of scale, in terms of handling large, crowded scenes. As more pedestrians enter the scene, the complexity of these methods increases and may become intractable with even moderately dense crowds.

1.3.1.2 Crowd Motion Patterns

To address the complexity of real-world scenes containing crowds, many researchers propose holistic techniques that characterize the scene as a collection of local motion estimates rather than a collection of objects. Often, holistic methods aim to identify behaviors within the scene that are part of the same physical process [27]. Mahadevan et al. [58] describe the typical dynamics of the crowd with a mixture of dynamic textures (previously used for segmentation by Chan and Vasconcelos [10]).

Using dynamic textures, however, retains appearance variations which can introduce noise into the model and degrade results. This approach is further elaborated in Chap. 11.

Moore et al. [64], Mehran et al. [63], and Ali and Shah [3] model crowds based on a hydrodynamics model that essentially treats each pedestrian as a particle in a fluid. As noted by Still [88], however, specific behaviors that occur in crowds, such as lane formations or clustering, do not occur in fluids. While particles are affected only by the external forces around them (such as other particles or the environment), the motion of pedestrians is a result of both external forces and their individual desires. Such differences between individual pedestrians form dynamic space-time structures in the crowd motion that can not be represented with a hydrodynamics model.

Other work assumes that the crowd flow is constant over the entire video. Ali and Shah [2] average the optical flow over a video clip, and use it to model a Finite Time Lyapunov Exponent field for segmenting the motion of the crowd. Similarly, Mehran et al. [62] measure the "social force" [25] by comparing the instantaneous optical flow to the optical flow averaged over the video clip. Raghavendra et al. [73] also estimate the social force, but do so using a particle swarm method that clusters similar motion vectors. In many crowded scenes, especially those with unconstrained environments, the motion of pedestrians can change dramatically in a short period of time as individuals move towards different goals.

Some researchers assume the crowd exhibits homogeneous motion in each area of the scene. Hu et al. [28], for example, identify global motion patterns (i.e., ones that take up the entire frame) in crowded scenes by clustering optical flow vectors in similar spatial regions. Similarly, Cheriyadat and Radke [13] detect dominant motions in crowds by clustering low-level tracked features. Such methods can not handle dynamically varying crowds or those with heterogeneous motions in local areas. A crosswalk, for example, naturally has pedestrians moving in two directions who emerge together as they pass each other.

Other methods capture the multi-modal nature of the crowd, but ignore the important temporal relationship between sequentially occurring motions exhibited by pedestrians. Rodriguez et al. [77] use a topical model (similar to the bag-of-word models) over quantized optical flow directions to describe the crowd motion. They later improve their tracking using a crowd density estimate [78]. Though they model the heterogeneous nature of the crowd, they does not encode the relationship between temporally co-occurring motions. By disregarding the temporal variations in the motions exhibited by pedestrians, these approaches cannot represent the underlying temporal pattern within the crowd motion.

Andrade et al. [4, 5] captures the temporal structure of the crowd by training hidden Markov models on optical flow vectors. They demonstrate that their method is a good indicator of emergency situations in simulated crowd flow data. Real-world crowded scenes, however, were not evaluated. Kratz and Nishino [44] modeled the local motion patterns with the distribution of the spatio-temporal gradients and derived a distribution-based hidden Markov model to encode their spatio-temporal variations. They successfully demonstrated the use of this crowd motion model for anomaly detection [44] and tracking of individuals [45, 47] in very crowded scenes. They further extended this model with directional statistics distributions to more faithfully encode the local motion patterns and introduced the use of pedestrian efficiency for modeling the magnitude of deviation of a person from the crowd motion to better detect and track unusual activities [46]. Saleemi et al. [79] developed a statistical representation of motion patterns of pedestrians in a scene observed by a static camera. Motion patterns are learned in an unsupervised manner directly from the salient patterns of optical using mixture model representation.

1.3.1.3 Particle-Based Representation of Crowds

To overcome the difficulty of detection individual objects in dense crowded scene, a particle based representation of crowd motion has been used by many researchers. Ali and Shah [2] introduced this representation where crowd motion is represented in terms of trajectories of a dense grid on particles. They used this representation to carry out segmentation dynamically distinct crowd flows (further elaborated in Chap. 8). In following years, particle based representation has been successfully used for tracking individual pedestrians in crowds [3], abnormal event detection [62, 98], and semantic scene understanding [55]. The work of Moore et al. [64] provides an expanded explanation of the intuition behind the particle based representation and its applicability to various problem areas.

1.3.1.4 Abnormal Event Detection

Various approaches have been proposed to perform abnormal event detection in dense crowded scenes. These can be characterized based on whether abnormality is detected locally or globally in the scene. The local abnormality detection methods mostly employ local 2D or 3D motion or appearance features with some added information to capture the local context information (e.g. using co-occurrence matrices). For instance, Adam et al. [1] measure the probability of optical flow in a local patch using histograms for detecting abnormal patterns. Kratz and Nishino [44] fit a Gaussian model to spatio-temporal gradients and then use HMM to detect abnormal events. Kim and Grauman [42] model local optical flow and enforce the consistency using Markov Random Field for detection of abnormal motion patterns. Mahadevan et al. [58] model the normal crowd behavior by mixtures of dynamic textures.

The global abnormality detection approaches label the motion in the entire scene as abnormal. This often happens in cases of panic situations such as stampede. In this direction, Mehran et al. [62] proposed a formulation of abnormal crowd behavior by adopting the social force model and then using Latent Dirichlet Allocation (LDA) to detect abnormality. In [98], chaotic invariants of particle trajectories are used for detecting abnormal motion.

Next we describe related approaches from area of crowd simulation and pedestrian evacuation dynamics.

1.3.2 Crowd Simulation and Behavior Modeling

Techniques to simulate and understand crowd behaviors and motion have been studied in computer graphics, virtual reality, social science, statistical physics, robotics, pedestrian and evacuation dynamics, and other areas of science and engineering. One of the major goals is to develop appropriate models that can be used to simulate and predict crowd behaviors in different real-world or synthetic environments. A key computation within these models at the *microscopic* level is to simulate the trajectory of each agent that avoids the static or dynamic obstacles and other human agents in the environment. The microscopic models also take into account the psychological and social behavior of each individual and how they respond to external events or stress. At the same time, we are also interested in *macroscopic* techniques that focus on the tendency and movement pattern of the entire crowd or the aggregate flow. These models seek to predict plausible or likely motion of human-like agents in terms of paths as well as other characteristics such as densities, speeds, and emerging behaviors. Overall, the design and formulation of such crowd simulation models is a challenging and multi-faceted problem. The range of trajectories that the humans follow, the pattern of crowd behaviors that we observe, and the variety of situations which the humans encounter are almost endless. As a result, crowd simulation remains an active area of research in many disciplines.

Any model for crowd simulation needs to take into account several components. This includes specifying a computer-based geometric model of the environment; computing the movement and trajectories of various agents, taking into account the interactions amongst the agents and with the environment; model external and dynamic events that affect crowd behavior. Some of the widely used models are based on multi-agent simulation. A crowd is composed of human-like agents (i.e., individuals in the crowds), with a collection of goals and a set of obstacles that constitute the environment. The individuals constituting a crowd may have similar or distinct goals. In heterogeneous crowd simulation, each individual in the crowd is assumed to have a physically embodied goal. The representation of this goal can vary based on the simulation scenario; for example, a goal may correspond to a specific position or a certain region in the environment. These goals may be dynamic and may change over time, say following some other individual or dynamic obstacle in the crowd. In addition to specifying the region and goals for each agent, the model must take into account the environment, which consists of walls, obstacles or other regions that may not be accessible to human-agents. These obstacles may be static (e.g. buildings) or dynamic (e.g. moving vehicles). Given the description of the agents and the environment, a key component of crowd simulation is the computation of trajectories for each agent that adhere to environmental factors, avoid collisions with the obstacles and other agents, and guide each agent towards its immediate goal. In particular, Reynolds refers to this process of intermediate-level planning as *steering behaviors* [75]. These steering behaviors are largely independent of the particulars of the agent's means of locomotion, but are used to navigate around the environment in a life-like and improvisational manner, and also results in collision avoidance. The combinations of such steering behaviors can be used to achieve some higher level goals, like following a walkay or a corridor, joining some other group of agents, etc.

1.3.2.1 Models for Crowd Motion

There is extensive literature on simulating crowd movements and dynamics. Many techniques have been proposed to compute the motion of individuals in crowds. Their application depends on the type of crowd patterns or behaviors that we want to simulate and the surrounding environment. An important issue with respect to crowd simulation is whether the crowds being simulated are homogeneous or heterogeneous. Homogeneous crowds correspond to instances where each agent has very similar behavior or goal. The study of heterogeneous crowds dates back to at least the work of Le Bon more than a century ago, who analyzed how members of a crowd can have different races, genders, intents, and backgrounds [50]. In heterogeneous crowds, every individual agent in the crowd maintains a distinct, observable identity. This identity is observed in the goals, desired speeds, aggressiveness, cooperation, and many other factors which affect the motion and trajectory of each agent. In contrast, homogeneous crowds are observed when a clear unity of action leads to a "disappearance of conscious personality", which results in a homogeneity of motion [50]. In terms of simulating homogeneous crowds, it may be possible to exploit the coherence in individual motion to accelerate the overall simulation. These include flow-based models [30, 31] that are governed by differential equations that uniformly dictate the flow of crowds across space. Other examples include models based on continuum crowds [91], which allow a small, fixed number of goals, and aggregate dynamics for dense crowd simulation [66].

1.3.2.2 Agent-Based Crowd Simulations

In contrast to continuum methods, agent-based simulation methods allow for true heterogeneity in simulating the motion and trajectory of each individual. In these simulations, each human-like character in the crowd is represented as a simulated agent. Since the motion of each agent is computed separately, it is possible to simulate crowds with varying characteristics and personalities for each agent. One of the most popular agent-based approaches was proposed by Reynolds in the Boids algorithm [74], which can generate steering behaviors that resemble flocking, herding, and school behaviors commonly observed in animal motion. This algorithm has been widely used in games and generating special effects in movies.

There is considerable literature in robotics and related areas on computing the motion and trajectories of multiple agents in a shared environment. The underlying motion-planning problem is known to have exponential complexity in terms of number of agents or the degrees of freedom [49]. At a broad level, prior work on motion planning can be classified into two kinds of approaches. The *centralized* approaches [48,49] consider the sum of all the robots or agents and treat the resulting system as a single composite system. In these methods, the configuration spaces of individual robots are combined (using the Cartesian product) in a composite space, and the resulting algorithm searches for a solution in this combined configuration space. In contrast, the *decoupled* planners proceed in a distributed manner, and coordination among them is often handled by exploring a *coordination space*, which represents the parameters along each specific path or are computed some kind of local rules. Decoupled approaches [84,96] are much faster than centralized methods, but may not be able to guarantee theoretical completeness. Some of the techniques from robot motion planning have been used to generate group behaviors [6, 37] and real-time navigation of large numbers of agents amongst obstacles [17, 89].

Most widely-used techniques for handling a large number of human-like agents are based on decentralized methods. This is a challenging task, particularly in densely-packed, crowded scenarios with several hundreds or thousands of agents. Each agent essentially has to navigate through an unknown dynamic environment; it has no prior knowledge of how other agents or the dynamic obstacles will move in the future. The standard approach to this class of problems is to let each agent run a continuous cycle of sensing and acting. During each cycle, the agent observes its surroundings, acquires information about the positions and velocities of other agents and obstacles in the synthetic environment, and computes a local path towards a goal that avoids collisions. If this cycle is executed at a high frequency, the agent is able to react in a timely way to changes in its surroundings. The computation of an agent's motion breaks down into two tasks: global and local navigation. Global navigation aims at computing a collision-free path towards a goal position that only takes into account the static obstacles, while local navigation techniques are used to avoid collisions with other agents and dynamic obstacles and steer each agent towards its goal position.

1.3.2.3 Global Navigation

Global path computation is typically performed using a global data structures, such as roadmaps or navigation meshes. A roadmap is a graph consisting of a set of vertices positioned in freespace (i.e. not inside an obstacle) and a set of edges connecting these vertices. Two vertices are connected by an edge if and only if the direct path between the two nodes is collision-free (i.e. no obstacles block the direct path). Such roadmaps can be constructed by an artist, or can be automatically generated using visibility graphs [56], probabilistic methods [39], or other techniques [48, 49]. Each agent can compute its global path using such roadmaps and performing graph search, such as Dijkstra's algorithm [15] or A* search [23]. In games and related applications, navigation meshes [36, 86, 94] have begun to supplant roadmaps. A navigation mesh is a decomposition of the freespace of the environment into a mesh consisting of convex polygons. As in roadmaps, the connectivity of the mesh is stored as a graph; however, navigation meshes have advantages over roadmaps in that all edges of a polygon are implicitly connected to each other, i.e. because of the convexity there is a straight-line path from any point in the polygon to any boundary. In addition, a single navigation mesh can encode clearance for arbitrarily sized agents. Computing a global path with a navigation mesh simply requires searching the connectivity graph for the shortest path between two nodes. The cost of a graph edge between two polygons depends on the length of the shared edge of those two polygons. If the edge is not large enough to accommodate the agent, the cost is infinite.

1.3.2.4 Local Navigation and Crowd Simulation

Several techniques have been proposed to animate or simulate large groups of autonomous agents or crowds. Most of these methods use a rather simple representation for each agent – for example, a circular shape in a 2D plane or a cylindrical object in the 3D space – and compute a collision-free trajectory for each agent. After computing the trajectory using a simple representation, these techniques use either footstep planning or walking synthesis methods to compute a human-like motion for each agent along the given trajectory.

Local navigation computation takes into account the motion of dynamic obstacles and other agents in the environment. At a broad level, prior methods for local collision avoidance and navigation can be classified into the following categories:

- *Potential-based methods:* These algorithms focus on modeling agents as particles with potentials and forces [24].
- *Boid-like methods:* These approaches, based on the seminal work of Reynolds, create simple rules for computing the velocities [74, 75].
- *Geometric and velocity methods:* These algorithms compute collision-free paths using sampling in the velocity space obstacles [92] or by using optimization methods [20, 85, 93].
- *Field based methods:* These algorithms compute fields for agents to follow [12, 33, 72, 100], or generate navigation fields for different agents [67].
- *Least effort crowds:* These methods for modeling the paths of crowds are based on the classic principle of Least Effort proposed by Zipf [104], many researchers have used that formulation to model the paths of crowds [35, 38, 80, 87]. Recently, it has been combined with multi-agent collision avoidance algorithms [93] and used to efficiently and automatically generate emergent behaviors for a large number of agents [21].
- *Data-driven methods:* These methods use real-world or data-driven techniques to simulate realistic crowd simulation as well as evaluate their accuracy [52, 72, 82].

In addition to these broad classifications, there are many other specific approaches designed to simulate crowd behavior based on cognitive modeling and behavior [83, 101], sociological or psychological factors [68], personality models [22], and dynamic behaviors [43].

1.4 Looking Ahead

Through the collection of chapters presented in this book, we hope to provide reader with an insight that will ultimately lead to addressing some of the following questions:

- What are the general principles that characterize complex crowd behavior of heterogeneous individuals?
- How can verifiable mathematical models of crowd motion and interaction can be developed based on these principles?
- How these general principles can be used to enhance performance of low level vision tasks such as object detection, tracking, and activity analysis in crowds?
- What are the possible problem areas in visual analysis of crowds that will benefit from crowd simulation and behavior models (e.g. tracking, target acquisition across sensor gaps, and sensor hand-off techniques etc.) and vice versa.

The rest of the book is organized into two parts. The first part presents a collection of chapters that focus on experimental validation of various pedestrian motion and interaction models (Chaps. 2 and 4), crowd simulation and behavior modeling (Chaps. 5–8), and relationship between micro and macroscopic models (Chap. 3). The second part of the book focus on approaches of visual analysis of crowded scenes. It covers topics of modeling crowd flows (Chaps. 9, 10, and 12), interaction among crowd participants (Chaps. 11 and 13), crowd counting (Chap. 14) and abnormal event detection (Chap. 15).

Acknowledgements We like to acknowledge Army Research Office, Nippon Telegraph and Telephone, National Science Foundation, and Office of Naval Research for providing support for writing this chapter. We also like to thank Louis Kratz for his contribution to this chapter.

References

1. Adam, A., Rivlin, E., Shimshoni, I., Reinitz, D.: Robust real-time unusual event detection using multiple fixed-location monitors. IEEE Trans. Pattern Anal. Mach. Intell. **30**(3), 555–560 (2008)
2. Ali, S., Shah, M.: A Lagrangian particle dynamics approach for crowd flow segmentation and stability analysis. In: Proceedings of the IEEE International Conference on Computer Vision and Pattern Recognition, 17–22 June 2007, pp. 1–6. http://ieeexplore.ieee.org/stamp/stamp.jsp?tp=&arnumber=4270002&isnumber=4269956
3. Ali, S., Shah, M.: Floor fields for tracking in high density crowd scenes. In: Proceedings of European Conference on Computer Vision. Lecture Notes in Computer Science, vol. 5303 (2008)
4. Andrade, E.L., Blunsden, S., Fisher, R.B.: Modelling crowd scenes for event detection. In: Proceedings of the 18th International Conference on Pattern Recognition, vol. 1, pp. 175–178 (2006). http://ieeexplore.ieee.org/stamp/stamp.jsp?tp=&arnumber=1698931&isnumber=35817

5. Andrade, E.L., Blunsden, S., Fisher, R.B.: Hidden Markov models for optical flow analysis in crowds. In: Proceedings of the 18th International Conference on Pattern Recognition, vol. 1, pp. 460–463 (2006). . http://ieeexplore.ieee.org/stamp/stamp.jsp?tp=&arnumber=1698931& isnumber=35817

6. Bayazit, O.B., Lien, J.-M., Amato, N.M.: Better group behaviors in complex environments with global roadmaps. In: Standish, R.K., Bedau, M.A., Abbass, H.A. (eds.) International Conference on the Simulation and Synthesis of Living Systems (ICAL 2003) (Alife), pp. 362–370. MIT, Cambridge (2002)

7. Betke, M., Hirsh, D.E., Bagchi, A., Hristov, N.I., Makris, N.C., Kunz, T.H.: Tracking large variable numbers of objects in clutter. In: Proceedings of IEEE International Conference on Computer Vision and Pattern Recognition (CVPR '07), 17–22 June 2007, pp. 1–8 (2007). http://ieeexplore.ieee.org/stamp/stamp.jsp?tp=&arnumber=4270019&isnumber=4269956

8. Brostow, G.J., Cipolla, R.: Unsupervised Bayesian detection of independent motion in crowds. In: Proceedings of IEEE International Conference on Computer Vision and Pattern Recognition, vol. 1, pp. 594–601, 17–22 June 2006. http://ieeexplore.ieee.org/stamp/stamp. jsp?tp=&arnumber=1640809&isnumber=34373

9. Cartwright, D., Zander, A.: Group Dynamics: Research and Theory, 3rd edn. Tavistock Publications, London (1968)

10. Chan, A.B., Vasconcelos, N.: Modeling, clustering, and segmenting video with mixtures of dynamic textures. IEEE Trans. Pattern Anal. Mach. Intell. **30**(5), 909–26 (2008)

11. Chattaraj, U., Seyfried, A., Chakroborty, P.: Comparison of pedestrian fundamental diagram across cultures. Adv. Complex Syst. **12**(03), 393–405 (2009)

12. Chenney, S.: Flow tiles. In: Proceedings 2004 ACM SIGGRAPH/Eurographics Symposium on Computer Animation (SCA '04), pp. 233–242. Eurographics Association, Aire-la-Ville (2004). http://dx.doi.org/10.1145/1028523.1028553

13. Cheriyadat, A., Radke, R.: Detecting dominant motions in dense crowds. IEEE J. Sel. Top. Signal Process. **2**(4), 568–581 (2008)

14. Dee, H., Hogg, D.: Detecting inexplicable behaviour. In: Proceedings of British Macine Vision Conference, The British Machine Vision Association, pp. 477–486 (2004)

15. Dijkstra, E.W.: A note on two problems in connexion with graphs. Numer. Math. **1**(1), 269–271 (1959)

16. Dobbyn, S., Hamill, J., O'Conor, K., O'Sullivan, C.: Geopostors: a realtime geometry/impostor crowd rendering system. In: Proceedings of the Symposium on Interactive 3D Graphics and Games, New York, pp. 95–102 (2005)

17. Gayle, R., Sud, A., Andersen, E., Guy, S., Lin, M., Manocha, D.: Interactive navigation of independent agents using adaptive roadmaps. IEEE Trans. Vis. Comput. Graph. **15**(1), 34–48 (2009)

18. Ge, W., Collins, R., Ruback, B.: Vision-based analysis of small groups in pedestrian crowds. IEEE Trans. Pattern Anal. Mach. Intell. **34**(5), 1003–1016 (2012)

19. Gilbert, A., Bowden, R.: Multi person tracking within crowded scenes. In: Elgammal, A., Rosenhahn, B., Klette R. (eds.) IEEE Workshop on Human Motion, pp. 166–179. Springer, Berlin/Heidelberg (2007)

20. Guy, S.J., Chhugani, J., Kim, C., Satish, N., Lin, M.C., Manocha, D., Dubey, P.: Clearpath: highly parallel collision avoidance for multi-agent simulation. In: Proceedings of ACM SIGGRAPH/Eurographics Symposium on Computer Animation, pp. 177–187 (2009)

21. Guy, S., Chuggani, J., Curtis, S., Dubey, P., Lin, M., Manocha, D.: Pledestrians: a least-effort approach to crowd simulation. In: Proceedings of Eurographics/ACM SIGGRAPH Symposium on Computer Animation, pp. 119–128 (2010)

22. Guy, S., Kim, S., Lin, M., Manocha, D.: Simulating heterogeneous crowd behaviors using personality trait theory. In: Proceedings of Eurographics/ACM SIGGRAPH Symposium on Computer Animation, pp. 43–52 (2011)

23. Hart, P.E., Nilsson, N.J., Raphael, B.: A formal basis for the heuristic determination of minimum cost paths. IEEE Trans. Syst. Sci. Cybern. **4**(2), 100–107 (1968)

24. Helbing, D., Molnar, P.: Social force model for pedestrian dynamics. Phys. Rev. E **51**, 4282 (1995)
25. Helbing, D., Molnar, P., Farkas, I.J., Bolay, K.:   Self-organizing pedestrian movement. Environ. Plan. B Plan. Des. **28**(3), 361–383 (2001)
26. Hu, W., Xiao, X., Fu, Z., Xie, D., Tan, T., Maybank, S.: A system for learning statistical motion patterns. IEEE Trans. Pattern Anal. Mach. Intell. **28**(9), 1450–1464 (2006)
27. Hu, M., Ali, S., Shah, M.: Detecting global motion patterns in complex videos. In: Proceedings of International Conference on Pattern Recognition (ICPR 2008), Dec 2008, pp. 1–5. http://ieeexplore.ieee.org/stamp/stamp.jsp?tp=&arnumber=4760950&isnumber=4760915
28. Hu, M., Ali, S., Shah, M.: Learning motion patterns in crowded scenes using motion flow field. In: Proceedings of International Conference on Pattern Recognition (ICPR 2008), pp. 1–5, 8–11 Dec (2008). http://ieeexplore.ieee.org/stamp/stamp.jsp?tp=&arnumber=4761183&isnumber=4760915
29. Hue, C., Le Cadre, J.-P., Perez, P.: Posterior Cramer-Rao bounds for multi-target tracking. IEEE Trans. Aerosp. Electron. Syst. **42**(1), 37–49 (2006)
30. Hughes, R.: A continuum theory for the flow of pedestrians. Transp. Res. B Methodol. **36**(6), 507–535 (2002)
31. Hughes, R.L.: The flow of human crowds. Ann. Rev. Fluid Mech. **35**, 169–182 (2003)
32. Izadinia, H., Saleemi, I., Li, W., Shah, M.: (MP)2T: multiple people multiple parts tracker. In: Computer Vision – ECCV. Lecture Notes in Computer Science, vol. 7577, pp. 100–114 (2012)
33. Jin, X., Xu, J., Wang, C.C.L., Huang, S., Zhang, J.: Interactive control of large crowd navigation in virtual environment using vector field. IEEE Comput. Graph. and Appl. **28**(6):37–46 (2008). http://ieeexplore.ieee.org/stamp/stamp.jsp?tp=&arnumber=4670099&isnumber=4670088
34. Johnson, N., Hogg, D.: Learning the distribution of object trajectories for event recognition. In: Proceedings of British Macine Vision Conference, pp. 583–592 (1995)
35. Kagarlis, M.: Method and apparatus of simulating movement of an autonomous entity through an environment. United States Patent No. US 7,188,056, Sept. 2002
36. Kallmann, M.:   Shortest paths with arbitrary clearance from navigation meshes.   In: Proceedings ACM SIGGRAPH Eurographics Symposium on Computer Animation (SCA '10), pp. 159–168. Eurographics Association, Aire-la-Ville (2010)
37. Kamphuis, A., Overmars, M.: Finding paths for coherent groups using clearance. In: Proceedings of ACM SIGGRAPH/Eurographics Symposium on Computer Animation (SCA '04), pp. 19–28. Eurographics Association, Aire-la-Ville (2004). http://dx.doi.org/10.1145/1028523.1028526
38. Karamouzas, I., Heil, P., Beek, P., Overmars, M.: A predictive collision avoidance model for pedestrian simulation. In: Proceedings of Motion in Games, pp. 41–52 (2009)
39. Kavraki, L., Svestka, P., Latombe, J., Overmars, M.: Probabilistic roadmaps for path planning in high-dimensional configuration spaces. IEEE Trans. Robot. Autom. **12**(4), 566–580 (1996)
40. Khan, Z., Balch, T., Dellaert, F.: MCMC-based particle filtering for tracking a variable number of interacting targets. IEEE Trans. Pattern Anal. Mach. Intell. **27**(11), 1805–1819 (2005)
41. Khan, Z., Balch, T., Dellaert, F.: MCMC data association and sparse factorization updating for real time multitarget tracking with merged and multiple measurements. IEEE Trans. Pattern Anal. Mach. Intell. **28**(12), 1960–1972 (2006)
42. Kim, J., Grauman, K.: Observe locally, infer globally, a spacetime MRF for detecting abnormal activities with incremental updates. In: IEEE Conference on Computer Vision and Pattern Recognition (CVPR2009), 20–25 June 2009, pp. 2921–2928. http://ieeexplore.ieee.org/stamp/stamp.jsp?tp=&arnumber=5206569&isnumber=5206488
43. Kim, S., Guy, S., Manocha, D., Lin, M.C.: Interactive simulation of dynamic crowd behaviors using general adaptation syndrome theory.   In: Proceedings of Interactive 3D Graphics Symposium (2012)

44. Kratz, L., Nishino, K.: Anomaly detection in extremely crowded scenes using spatio-temporal motion pattern models. In: Proceedings of IEEE International Conference on Computer Vision and Pattern Recognition(CVPR 2009), 20–25 June 2009, pp. 1446–1453. http://ieeexplore.ieee.org/stamp/stamp.jsp?tp=&arnumber=5206771&isnumber=5206488

45. Kratz, L., Nishino, K.: Tracking with local spatio-temporal motion patterns in extremely crowded scenes. In: Proceedings of IEEE International Conference on Computer Vision and Pattern Recognition (2010). http://ieeexplore.ieee.org/stamp/stamp.jsp?tp=&arnumber=5540149&isnumber=5539770

46. Kratz, L., Nishino, K.: Going with the flow: pedestrian efficiency in crowded scenes. In: Proceedings of European Conference on Computer Vision (ECCV 2012). Lecture Notes in Computer Science, vol. 7575, pp. 558–572 (2012)

47. Kratz, L., Nishino, K.: Tracking pedestrians using local spatio-temporal motion patterns in extremely crowded scenes. IEEE Trans. Pattern Anal. Mach. Intell. **34**(5), 987–1002 (2012)

48. Latombe, J.C.: Robot Motion Planning. Kluwer Academic Publishers, Boston (1991)

49. LaValle, S.M.: Planning Algorithms. Cambridge University Press (2006). Also available at http://msl.cs.uiuc.edu/planning/

50. Le Bon, G.: The Crowd: A Study of the Popular Mind Macmillan, New York (1896). Reprint available from Dover Publications

51. Leibe, B., Seemann, E., Schiele, B.: Pedestrian detection in crowded scenes. In: Proceedings of Conference on Computer Vision and Pattern Recognition (CVPR 2005), 20–25 June 2005, vol. 1. http://ieeexplore.ieee.org/stamp/stamp.jsp?tp=&arnumber=1467359&isnumber=31472

52. Lerner, A., Chrysanthou, Y., Shamir, A., Cohen-Or, D.: Data driven evaluation of crowds. In: Proceedings of the 2nd International Workshop on Motion in Games. Lecture Notes in Computer Science vol. 5884, pp. 75–83 (2009)

53. Lewin, K.: In: Cartwright, D. (ed.) Field Theory in Social Science; Selected Theoretical Papers. Harper & Row, New York (1951)

54. Li, Y., Huang, C., Nevatia, R.: Learning to associate: hybridboosted multi-target tracker for crowded scene. In: Proceedings of IEEE International Conference on Computer Vision and Pattern Recognition (CVPR 2009), 20–25 June 2009. http://ieeexplore.ieee.org/stamp/stamp.jsp?tp=&arnumber=5206735&isnumber=5206488

55. Lin, D., et al.: Modeling and estimating persistent motion with geometric flows. In: IEEE Conference on Computer Vision and Pattern Recognition (CVPR), 13–18 June 2010, pp.1–8. http://ieeexplore.ieee.org/stamp/stamp.jsp?tp=&arnumber=5539848&isnumber=5539770

56. Lozano-Pérez, T., Wesley, M.: An algorithm for planning collision-free paths among polyhedral obstacles. Commun. ACM **22**(10), 560–570 (1979)

57. Magnenat-Thalmann, N., Seo, H., Cordier, F.: Automatic modeling of virtual humans and body clothing. J. Comput. Sci. Technol. **19**(5), 575–584 (2004)

58. Mahadevan, V., Li, W., Bhalodia, V., Vasconcelos, N.: Anomaly detection in crowded scenes. In: Proceedings of IEEE International Conference on Computer Vision and Pattern Recognition (CVPR), 13–18 June 2010, pp. 1975–1981 (2010). http://ieeexplore.ieee.org/stamp/stamp.jsp?tp=&arnumber=5539872&isnumber=5539770

59. Mahadevan, V., Li, W., Bhalodia, V., Vasconcelos, N.: Anomaly detection in crowded scenes. In: CVPR (2010)

60. Maïm, J., Yersin, B., Pettré, J., Thalmann, D.: YaQ: an architecture for real-time navigation and rendering of varied crowds. IEEE Comput. Graph. Appl. **29**(4), 44–53 (2009)

61. McDonnell, R., Larkin, M., Dobbyn, S., Collins, S., O'Sullivan, C.: Clone attack! perception of crowd variety. ACM Trans. Graph. **27**(3), 1–8 (2008)

62. Mehran, R., Oyama, A., Shah, M.: Abnormal crowd behavior detection using social force model. In: Proceedings of Conference on Computer Vision and Pattern Recognition (CVPR 2009), 20–25 June 2009, pp. 935–942. http://ieeexplore.ieee.org/stamp/stamp.jsp?tp=&arnumber=5206641&isnumber=5206488

63. Mehran, R., Moore, B.E., Shah, M.: A streakline representation of flow in crowded scenes. In: European Conference on Computer Vision (ECCV 2010). Lecture Notes in Computer Science, vol. 6313, pp. 439–452 (2010)
64. Moore, B.E., Ali, S., Mehran, R., Shah, M.: Visual crowd surveillance through a hydrodynamics lens. Commun. ACM **54**, 64–73 (2011)
65. Musse, S.R., Thalmann, D.: A hierarchical model for real time simulation of virtual human crowds. IEEE Trans. Vis. Comput. Graph. **7**(2), 152–164 (2001)
66. Narain, R., Golas, A., Curtis, S., Lin, M.C.: Aggregate dynamics for dense crowd simulation. ACM Trans. Graph. **28**(5), 1–8 (2009)
67. Patil, S., van den Berg, J., Curtis, S., Lin, M.C., Manocha, D.: Directing crowd simulations using navigation fields. IEEE Trans. Vis. Comput. Graph. **17**(2), 244–254 (2011)
68. Pelechano, N., Allbeck, J.M., Badler, N.I.: Controlling individual agents in high-density crowd simulation. In: Proceedings of the 2007 ACM SIGGRAPH/Eurographics Symposium on Computer Animation (SCA '07), pp. 99–108. Eurographics Association, Aire-la-Ville (2007)
69. Pellegrini, S., Ess, A., Schindler, K., Van Gool, L.: You will never walk alone: modeling social behavior for multi-target tracking. In: Proceedings of IEEE International Conference on Computer Vision and Pattern, Sept. 29-Oct. 2 2009, pp. 261–268. http://ieeexplore.ieee.org/stamp/stamp.jsp?tp=&arnumber=5459260&isnumber=5459144
70. Pettré, J., de Heras Ciechomski, P., Maïm, J., Yersin, B., Laumond, J.-P., Thal-mann, D.: Real-time navigating crowds: scalable simulation and rendering. J. Vis. Comput. Animat. **17**(3–4), 445–455 (2006)
71. Pettre, J., Grillon, H., Thalmann, D.: Crowds of moving objects: navigation planning and simulation. In: Proceedings of IEEE International Conference on Robotics and Automation, 10–14 April 2007, pp. 3062–3067. http://ieeexplore.ieee.org/stamp/stamp.jsp?tp=&arnumber=4209555&isnumber=4209049
72. Pettre, J., Ondrej, J., Olivier, A., Cretual, A., Donikian, S.: Experiment-based modeling, simulation and validation of interactions between virtual walkers. In: Proceedings of the 2009 ACM SIGGRAPH/Eurographics Symposium on Computer Animation (SCA '09), pp. 189–198. ACM (2009). http://doi.acm.org/10.1145/1599470.1599495
73. Raghavendra, R., Del Bue, A., Cristani, M., Murino, V.: Optimizing interaction force for global anomaly detection in crowded scenes. In: Proceedings of IEEE International Conference on Computer Vision (ICCV Workshops), 6–13 Nov 2011, pp. 136–143. http://ieeexplore.ieee.org/stamp/stamp.jsp?tp=&arnumber=6130235&isnumber=6130192
74. Reynolds, C.W.: Flocks, herds and schools: a distributed behavioral model. Proc. ACM SIGGRAPH **21**, 25–34 (1987)
75. Reynolds, C.W.: Steering behaviors for autonomous characters. In: Game Developers Conference (1999)
76. Rodriguez, M.D., Shah, M.: Detecting and segmenting humans in crowded scenes. In: Proceedings of the 15th International Conference on Multimedia (MULTIMEDIA '07), pp. 353–356. ACM, New York (2007). http://doi.acm.org/10.1145/1291233.1291310
77. Rodriguez, M., Ali, S., Kanade, T.: Tracking in unstructured crowded scenes. In: Proceedings of IEEE International Conference on Computer Vision, 29 Sept-2 Oct 2009, pp. 1389–1396. http://ieeexplore.ieee.org/stamp/stamp.jsp?tp=&arnumber=5459301&isnumber=5459144
78. Rodriguez, M., Sivic, J., Laptev, I., Audibert, J.: Density-aware person detection and tracking in crowds. In: Proceedings of IEEE International Conference on Computer Vision (ICCV), 6–13 Nov 2011, pp. 2423–2430. http://ieeexplore.ieee.org/stamp/stamp.jsp?tp=&arnumber=6126526&isnumber=6126217
79. Saleemi, I., Hartung, L., Shah, M.: Scene understanding by statistical modeling of motion patterns. In: IEEE Conference on Computer Vision and Pattern Recognition (CVPR), 13–18 June 2010, pp. 2069–2076. http://ieeexplore.ieee.org/stamp/stamp.jsp?tp=&arnumber=5539884&isnumber=5539770

80. Sarmady, S., Haron, F., Hj, A.Z.: Modeling groups of pedestrians in least effort crowd movements using cellular automata. In: Proceedings of 3rd Asia International Conference on Modeling and Simulation (AMS '09), pp. 520–525. IEEE Computer Society, Washington, DC (2009). http://dx.doi.org/10.1109/AMS.2009.16

81. Scovanner, P., Tappen, M.: Learning pedestrian dynamics from the real world. In: Proceedings of IEEE International Conference on Computer Vision, 29 Sept-2 Oct 2009, pp. 381–388. http://ieeexplore.ieee.org/stamp/stamp.jsp?tp=&arnumber=5459224&isnumber=5459144

82. Seyfried, A., Boltes, M., Kähler, J., Klingsch, W., Portz, A., Rupprecht, T., Schadschneider, A., Steffen, B., Winkens, A.: Enhanced empirical data for the fundamental diagram and the flow through bottlenecks. In: Klingsch, W.W.F., Rogsch, C., Schadschneider, A., Schreckenberg, M. (eds.) Pedestrian and Evacuation Dynamics 2008, pp. 145–156. Springer, Berlin/Heidelberg (2010)

83. Shao, W., Terzopoulos, D.: Autonomous pedestrians. In: SCA '05: Proceedings of the 2005 ACM SIGGRAPH/Eurographics Symposium on Computer Animation (SCA '05), pp. 19–28. ACM, New York (2005). http://doi.acm.org/10.1145/1073368.1073371

84. Simeon, T., Leroy, S., Laumond, J.: Path coordination for multiple mobile robots: a geometric algorithm. In: Proceedings of the International Joint Conference on Artificial Intelligence (IJCAI), pp. 1118–1123 (1999)

85. Snape, J., van den Berg, J., Guy, S.J., Manocha, D.: Independent navigation of multiple mobile robots with hybrid reciprocal velocity obstacles. In: Proceedings of the IEEE/RSJ International Conference on Intelligent Robots and Systems, St. Louis, pp. 5917–5922 (2009)

86. Snook, G.: Simplified 3D movement and pathfinding using navigation meshes. In: DeLoura, M.A. (ed.) Game Programming Gems, pp. 288–304. Charles River, Hingham (2000). Chapter 3

87. Still, G.: Crowd dynamics. Ph.D. thesis, University of Warwick (2000)

88. Still, K.: Crowd dynamics. Ph.D. thesis, University of Warwick (2000)

89. Sud, A., Andersen, E., Curtis, S., Lin, M., Manocha, D.: Real-time path planning for virtual agents in dynamic environments. In: ACM SIGGRAPH 2008 classes (SIGGRAPH '08), Article 55 , 9pp. ACM, New York (2008)

90. Sugimura, D., Kitani, K., Okabe, T., Sato, Y., Sugimoto, A.: Using individuality to track individuals: clustering individual trajectories in crowds using local appearance and frequency trait. In: Proceedings of IEEE International Conference on Computer Vision, 29 Sept-2 Oct 2009, pp.1467–1474. http://ieeexplore.ieee.org/stamp/stamp.jsp?tp=&arnumber=5459286&isnumber=5459144

91. Treuille, A., Cooper, S., Popovic, Z.: Continuum crowds. ACM Trans. Graph. 25(3), 1160–1168 (2006)

92. van den Berg, J., Guy, S.J., Lin, M., Manocha, D.: Reciprocal n-body collision avoidance. In: International Symposium of Robotics Research. Robotics Research Springer Tracts in Advanced Robotics, vol. 70, pp. 3–19 (2009)

93. van den Berg, J., Seawall, J., Lin, M.C., Manocha, D.: Virtualized traffic: reconstructing traffic flows from discrete spatio-temporal data. Proc. IEEE Trans. Vis. Comput. Grap. 17(1), 26–37 (2009). IEEE Computer Society. http://doi.ieeecomputersociety.org/10.1109/TVCG.2010.27

94. van Toll, W., Cook, A.F., Geraerts, R.: Navigation meshes for realistic multi-layered environments. In: Proceedings of IEEE RSJ International Conference on Intelligent Robots and Systems, 25–30 Sept 2011, pp. 3526–3532. http://ieeexplore.ieee.org/stamp/stamp.jsp?tp=&arnumber=6094790&isnumber=6094399

95. Wang, X., Tieu, K., Grimson, E.: Learning semantic scene models by trajectory analysis. In: Proceedings of European Conference on Computer Vision, pp. 110–123 (2006)

96. Warren, C.W.: Multiple path coordination using artificial potential fields. In: Proceedings of IEEE Conference on Robotics and Automation, 13–18 May 1990, vol. 1, pp. 500–505. http://ieeexplore.ieee.org/stamp/stamp.jsp?tp=&arnumber=126028&isnumber=3534

97. Wu, B., Nevatia, R.: Tracking of multiple, partially occluded humans based on static body part detection. In: Proceedings of IEEE International Conference on Computer Vision and Pattern Recognition, pp. 951–958 (2006)

98. Wu, S., et al.: Chaotic invariants of Lagrangian particle trajectories for anomaly detection in crowded scenes. In: IEEE Conference on Computer Vision and Pattern Recognition (CVPR), 13–18 June 2010, pp. 2054–2060. http://ieeexplore.ieee.org/stamp/stamp.jsp?tp=&arnumber=5539882&isnumber=5539770

99. Yamaguchi, K., Berg, C., Ortiz, L.E., Berg, T.L.: Who are you with and where are you going? In: IEEE Conference on Computer Vision and Pattern Recognition (CVPR), 20–25 June 2011, pp.1345–1352. http://ieeexplore.ieee.org/stamp/stamp.jsp?tp=&arnumber=5995468&isnumber=5995307

100. Yersin, B., Maim, J., Ciechomski, P., Schertenleib, S., Thalmann, D.: Steering a virtual crowd based on a semantically augmented navigation graph. In: VCROWDS (2005)

101. Yu, Q., Terzopoulos, D.: A decision network framework for the behavioral animation of virtual humans. In: Proceedings of the 2007 ACM SIGGRAPH/Eurographics Symposium on Computer Animation (SCA '07), pp. 119–128. Eurographics Association, Aire-la-Ville (2007)

102. Zhan, B., Monekosso, D., Remagnino, P., Velastin, S., Xu, L.-Q.: Crowd analysis: a survey. Mach. Vis. Appl. **19**(5), 345–357 (2008)

103. Zhao, T., Nevatia, R., Wu, B.: Segmentation and tracking of multiple humans in crowded environments. IEEE Trans. Pattern Anal. Mach. Intell. **30**(7), 1198–1212 (2008)

104. Zipf, G.K.: Human Behavior and the Principle of Least Effort. Addison-Wesley Press, Cambridge (1949)

Part I
Crowd Simulation and Behavior Modeling

Chapter 2
On Force-Based Modeling of Pedestrian Dynamics

Mohcine Chraibi, Andreas Schadschneider, and Armin Seyfried

Abstract A brief overview of mathematical modeling of pedestrian dynamics is presented. Hereby, we focus on space-continuous models which include interactions between the pedestrian by forces. Conceptual problems of such models are addressed. Side-effects of spatially continuous force-based models, especially oscillations and overlapping which occur for erroneous choices of the forces, are analyzed in a quantitative manner. As a representative example of force-based models the Generalized Centrifugal Force Model (GCFM) is introduced. Key components of the model are presented and discussed. Finally, simulations with the GCFM in corridors and bottlenecks are shown and compared with experimental data.

2.1 Introduction

The study of pedestrian dynamics has gained special interest due to the increasing number of mass events, where several thousand people gather in restricted areas. In order to understand the laws that govern the dynamics of a crowd several experiments were performed and evaluated. A brief overview can be found in

M. Chraibi (✉)
Jülich Supercomputing Centre, Forschungszentrum Jülich GmbH, 52425 Jülich, Germany
e-mail: m.chraibi@fz-juelich.de

A. Schadschneider
Institute for Theoretical Physics, Universität zu Köln, 50937 Köln, Germany
e-mail: as@thp.uni-koeln.de

A. Seyfried
Jülich Supercomputing Centre, Forschungszentrum Jülich GmbH, 52425 Jülich, Germany

Computer Simulation for Fire Safety and Pedestrian Traffic, Bergische Universität Wuppertal, Pauluskirchstraße 7, 42285 Wuppertal, Germany
e-mail: a.seyfried@fz-juelich.de

S. Ali et al. (eds.), *Modeling, Simulation and Visual Analysis of Crowds*, The International Series in Video Computing 11, DOI 10.1007/978-1-4614-8483-7_2, © Springer Science+Business Media New York 2013

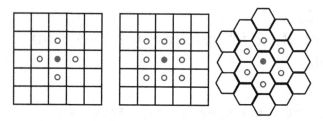

Fig. 2.1 *Left*: Von-Neumann neighborhood. *Middle*: Moore neighborhood. *Right*: Hexagonal neighborhood

[26]. Due to ethical and technical limitations, experimental studies with large numbers of pedestrians are often restricted to controlled labor experiments in specific geometries e.g., bottlenecks [3,10,12,14,15,28,29,35], T-junctions [37] and corridors [1,5,31,38,39]. Nevertheless, those experiments are beneficial to study quantitative and qualitative properties of pedestrian dynamics. Furthermore, they provide an empirical basis for model development and validation. In fact, validated models can be used to extrapolate the empirical knowledge to cover situations that are difficult to produce with experiments.

Several mathematical models have been developed. Based on their properties, existing models can be categorized into different classes [26]. An increasingly important type of model is based on individual description of pedestrians by means of intrinsic properties and spatial interactions between individuals. Those models state that phenomena which emerge at a macroscopic level arise as a result of interactions at a microscopic level.

Probably, the most investigated microscopic models are the Cellular Automata models (CA), which are "mathematical idealizations of physical systems in which space and time are discrete, and physical quantities take on finite set of discrete values." [34] In the simplest case, CA models decompose space into a rectangular or hexagonal lattice with a cell size of $40 \times 40\,\mathrm{cm}^2$. The state of each cell is described by a discrete variable; "1" for occupied and "0" for empty. It is updated in time according to a set of predefined (stochastic) rules depending on the states of the cells in a certain neighborhood. Depending on the system different neighborhoods can be defined. Figure 2.1 depicts schematically three of the most common neighborhoods used in CA applied to pedestrian dynamics. The full specification of the dynamics of a CA model requires to specify the order in which cells are updated. The most common update strategy is the parallel or synchronous update where all cells are updated at the same time.

CA models describe properties of pedestrian traffic fairly well. However, the discretization of space is not always possible in sensible way. For more details the reader is referred to [27].

Another type of microscopic models which, contrary to CA models, is defined in a continuous space, are force-based models. Force-based models describe the movement of individuals by means of non-linear second-order differential equations. In this chapter, we address properties of force-based models. The question of their realism and ability to describe pedestrian dynamics is discussed in the following.

2.2 Force-Based Models

As early as 1950s, several second-order models has been developed to study traffic dynamics [21–23]. By means of differential equations the *change* of the system with respect to time can be described microscopically by those models. Following Newtonian dynamics, *change* of state results from the existence of exterior forces. As such it can be concluded that the origin of force-based modeling can be traced back to the beginning of the 1950s. An explicit formulation of this forced-based principle in pedestrian dynamics was expressed in [11], who presented a CA-model that "hypothesizes the existence of repulsive forces between pedestrians so that as the subject approaches another pedestrian the 'potential energy' of his position rises and the 'kinetic energy' of his speed drops" [11]. However, the first space-continuous force-based model was introduced by Hirai et al. [8].

Further models for pedestrian dynamics that are based on this force-Ansatz followed [6, 7, 13, 18, 30].

2.2.1 Definition and Issues

Given a pedestrian i with coordinates $\vec{R_i}$ one defines the set of all pedestrians that influence pedestrian i at a certain moment as $\mathcal{N}_i$ and the set of walls or boundaries that act on i as $\mathcal{W}_i$. In general the forces defining the equation of motion are split into driving and repulsive forces. The repulsive forces model the collision-avoidance performed by pedestrians and should in principle guarantee a certain volume exclusion for each pedestrian. The driving force, on the other hand, models the intention of a pedestrian to move to a certain destination and walk with a desired speed.

Formally the movement of each pedestrian is defined by the equation of motion

$$m_i \frac{d}{dt^2} \vec{R_i} = \vec{F_i} = \vec{F_i}^{\text{drv}} + \sum_{j \in \mathcal{N}_i} \vec{F_{ij}}^{\text{rep}} + \sum_{w \in \mathcal{W}_i} \vec{F_{iw}}^{\text{rep}}, \qquad (2.1)$$

where $\vec{F_{ij}}^{\text{rep}}$ denotes the repulsive force from pedestrian j acting on pedestrian i, $\vec{F_{iw}}^{\text{rep}}$ is the repulsive force emerging from the obstacle w and $\vec{F_i}^{\text{drv}}$ is a driving force and m_i is the mass of pedestrian i. In [8] the equation of motion (2.1) contains a coefficient of viscosity. However, the influence of this coefficient was not investigated.

For a system of n pedestrians we define the state vector $\overrightarrow{X}(t)$ as

$$\overrightarrow{X}(t) := \begin{pmatrix} \overrightarrow{R_1}(t) \\ \vdots \\ \overrightarrow{R_n}(t) \\ \overrightarrow{v_1}(t) \\ \vdots \\ \overrightarrow{v_n}(t) \end{pmatrix}. \tag{2.2}$$

According to Eq. (2.1) the change of $\overrightarrow{X}(t)$ over time is described by:

$$\frac{d}{dt}\overrightarrow{X}(t) = \begin{pmatrix} \overrightarrow{v}(t) \\ \overrightarrow{F}(t)/m \end{pmatrix}, \tag{2.3}$$

with

$$\overrightarrow{F}(t) = \begin{pmatrix} \overrightarrow{F_1} \\ \vdots \\ \overrightarrow{F_n} \end{pmatrix}, \quad \overrightarrow{v}(t) = \begin{pmatrix} \overrightarrow{v_1} \\ \vdots \\ \overrightarrow{v_n} \end{pmatrix} \quad \text{and} \quad m_i = m \quad \forall i \in [1, n]. \tag{2.4}$$

The state vector at time $t + \Delta t$ is then obtained by integrating (2.3):

$$\overrightarrow{X}(t + \Delta t) = \int_t^{t+\Delta t} \begin{pmatrix} \overrightarrow{v}(\tilde{t}) \\ \overrightarrow{F}(\tilde{t})/m \end{pmatrix} d\tilde{t} + \overrightarrow{X}(t). \tag{2.5}$$

In general the integral in (2.5) may not be expressible in closed form and must be solved numerically.

Force-based models are able to describe qualitatively and quantitatively some aspects of pedestrian dynamics. Nevertheless, they have some conceptual problems. The first problem is Newton's third law. According to this principle two particles interact by forces of equal magnitudes and opposite directions. For pedestrians this law is unrealistic since e.g. normally a pedestrian does not react to pedestrians behind him/her. Even if the angle of vision is taken into account, the forces mutually exerted on each other are not of the same magnitude. In classical mechanics the acceleration of a particle is linear in the force acting on it. Consequently the acceleration resulting from several forces is summed up from accelerations computed from each force. The superposition-principle however, leads to some side-effects when modeling pedestrian dynamics, especially in dense situations where unrealistic backwards movement or high velocities can occur.

Further problems are related to the Newtonian equation of motion describing particles with inertia. This could lead to *overlapping* and *oscillations* of the modeled pedestrians.

On one hand, the particles representing pedestrians can excessively overlap and thus violate the principle of volume exclusion. On the other hand, pedestrians can be pushed backwards by repulsive forces and so perform an oscillating movement towards the exit. This leads to unrealistic behavior especially in evacuation scenarios where a forward movement is dominating. Depending on the strength of the repulsive forces, overlapping and oscillations of pedestrians can be mitigated. However, since both phenomena are related to the repulsive forces this can not be achieved *simultaneously* in a satisfactory way. Reducing the overlapping-issue by increasing the strength of the repulsive forces would lead to an increase of the oscillations in the system. On the other hand, reducing the strength of the repulsive forces may solve the problem of oscillations, but at the same time increase the tendency of overlapping.

In order to solve this overlapping-oscillations duality one can introduce extra rules. One possible solution may be avoiding oscillations by choosing adequate values of the repulsive forces and deal with overlapping among pedestrians with an "overlap-eliminating" algorithm [13]. In [36] a "collision detection technique" was introduced to modify the state variables of the system each time pedestrians overlap with each other. The other possible solution goes in the opposite direction, namely avoiding overlapping by strong repulsive forces and simply eliminate oscillations by setting the velocity to zero [7, 16].

Even if those extra rules may solve the problematic duality, it seems that they are redundant since interactions among pedestrians are no longer expressed only by repulsive forces. This redundancy adds an amount of complexity to the model and is clearly in contradiction to the minimum description length principle [24]. Besides, it is unclear how the modification of the state vector $X(t)$ (2.2) influences the stability of the Eq. (2.5). For those reasons, it is necessary to investigate solutions for the overlapping-oscillations duality without dispensing with the simplicity of the model as originally described with the equation of movement (2.3).

In order to understand the relation between overlapping and oscillations with the repulsive force and hence investigate solutions for the aforementioned problem, we first try to quantify those phenomena and study their behavior with respect to the strength of the repulsive force.

2.2.2   Overlapping

Overlapping is a simulation-specific phenomenon that arises in some models. Unlike CA-models, where volume exclusion is given with the discretization of the space, in poorly calibrated force-based models, unrealistic overlapping between pedestrians are not excluded (Fig. 2.2).

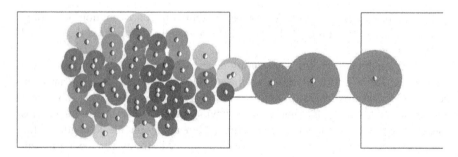

Fig. 2.2 Evacuation through a bottleneck. The simulation screen-shot highlights the problem of excessive overlapping

In order to measure the overlapping that arise during a simulation an "overlapping-proportion" is defined as

$$o^{(v)} = \frac{1}{n_{ov}} \sum_{t=0}^{t_{end}} \sum_{i=1}^{N} \sum_{j>i}^{N} o_{ij},$$

(2.6)

with

$$o_{ij} = \frac{A_{ij}}{\min(A_i, A_j)} \leq 1,$$

(2.7)

where N is the number of simulated pedestrians and t_{end} the duration of the simulation. A_{ij} is the overlapping area of the geometrical forms representing i and j with areas A_i and A_j, respectively. n_{ov} is the cardinality of the set

$$\mathscr{O} := \{o_{ij} : o_{ij} \neq 0\}.$$

(2.8)

For $n_{ov} = 0$, $o^{(v)}$ is set to zero.

2.2.3 Oscillations

Oscillations are backward movements fulfilled by pedestrians when moving under high repulsive forces. Figure 2.3 shows a simulation where pedestrians are force to move in the opposite direction of the exit.

For a pedestrian with velocity $\vec{v_i}$ and desired velocity $\vec{v_i^0}$ the "oscillation-proportion" is defined as

$$o^{(s)} = \frac{1}{n_{os}} \sum_{t=0}^{t_{end}} \sum_{i=1}^{N} S_i,$$

(2.9)

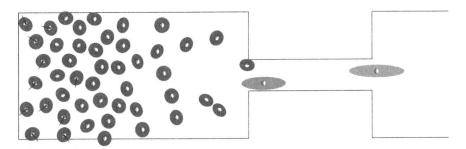

Fig. 2.3 Evacuation through a bottleneck. The simulation screen-shot highlights the problem of oscillations. Note the pedestrians near the walls have the wrong orientation

where S_i quantifies the oscillation-strength of pedestrian i and is defined as follows:

$$S_i = \frac{1}{2}(-s_i + |s_i|),\tag{2.10}$$

with

$$s_i = \frac{\vec{v_i} \cdot \vec{v_i}^0}{\|\vec{v_i^0}\|^2},\tag{2.11}$$

and n_{os} is the cardinality of the set

$$\mathscr{S} := \{s_i : s_i \neq 0\}.\tag{2.12}$$

Here again $o^{(s)}$ is set to zero if $n_{os} = 0$. Note that S_i in Eq. (2.10) is zero if the angle between the velocity and the desired velocity is less that $\pi/2$. This means a realistic deviation of the velocity from the desired direction is not considered as an "oscillation".

The proportions $o^{(v)}$ and $o^{(s)}$ are normalized to 1 and describe the evolution of the overlapping and oscillations during a simulation. The change of $o^{(v)}$ and $o^{(s)}$ is measured with respect to the strength of the repulsive force η. This dependence as well as the overlapping-oscillation duality is showcased in Fig. 2.4.

Increasing the strength of the repulsive force (η) to make pedestrians "impenetrable" leads to a decrease of the overlapping-proportion $o^{(v)}$. Meanwhile, the oscillation-proportion $o^{(s)}$ increases, thus the system tends to become unstable. Large values of the oscillation-proportion $o^{(s)}$ imply less stability. For $s_i = 1$ one has $\vec{v_i} = -\vec{v_i}^0$, i.e., a pedestrian moves backwards with desired velocity. Even values of s_i higher than 1 are not excluded and can occur during a simulation.

It should be mentioned that the proportions $o^{(v)}$ and $o^{(s)}$ introduced here are diagnostic tools that help calibrating the strength of the repulsive force in order to minimize overlapping as well as oscillations.

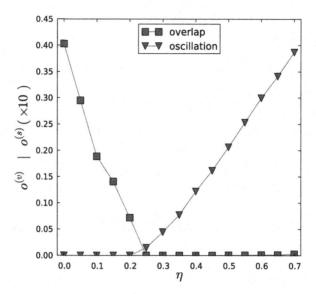

Fig. 2.4 The change of the overlapping-proportion (2.6) and the oscillation-proportion (2.9) in dependence of the repulsive force strength. For each η, 200 simulations were performed

2.3 The Generalized Centrifugal Force Model (GCFM)

The GCFM [2] describes the two-dimensional projection of the human body, by means of ellipses with velocity-dependent semi-axes. It takes into account the distance between the "edges" of the pedestrians as well as their relative velocities. An elliptical volume exclusion has several advantages over a circular one. Because a circle is symmetric with respect to its center, it is inconsistent with the asymmetric space requirement of pedestrians in their direction of motion and transverse to it. One possible remedy would be allowing the center of mass to be different from the geometrical center of the circle. Whether this leads to realistic compliance with the volume exclusion is not clear and should be studied in more detail.

As a force-based model, the GCFM describes the time evolution of pedestrians by a system of superposing short-range forces. Besides the geometrical shape of modeled pedestrians, it emphasizes the relevance of clear model definition without any hidden restrictions on the state variables. Furthermore, quantitative validation, with help of experimental data taken from different scenarios, plays a key role in the development of the model.

2.3.1 Volume Exclusion of Pedestrians

As mentioned earlier, one drawback of circles that impact negatively the dynamics is their rotational symmetry with respect to their centers. Therefore, they occupy the same amount of space in all directions. In single file movement this is irrelevant since the circles are projected to lines and only the required space in movement direction matters. However, for two-dimensional movement, a rotational symmetry has a negative impact on the dynamics of the system due to unrealistically large lateral space requirements.

In [4] Fruin introduced the "body ellipse" to describe the plane view of the average adult male human body. Pauls [19] presented ideas about an extension of Fruin's ellipse model to better understand and model pedestrian movement as density increases. Templer [32] noticed that the so called "sensory zone", which can be interpreted as a "safety" space between pedestrians and other objects in the environment to avoid physical conflicts and for "psychocultural reasons", varies in size and takes the shape of an ellipse. In fact, ellipses are closer to the projection of required space of the human body on the plane, including the extent of the legs during motion and the lateral swaying of the body. Introducing an elliptical volume exclusion for pedestrians has the advantage over circles (or points) to adjust independently the two semi-axes of the ellipse such that one- and two-dimensional space requirement is described with higher fidelity.

Given a pedestrian i, an ellipse with center (x_i, y_i), major semi-axis a and minor semi-axis b can be defined. a models the space requirement in the direction of movement,

$$a = a_{\min} + \tau_a v_i \qquad (2.13)$$

with two parameters $a_{\min}$ and τ_a.

Fruin [4] observed body swaying during both human locomotion and while standing. Pauls [20] remarks that swaying laterally should be considered while determining the required width of exit stairways. In [10], characteristics of lateral swaying are determined experimentally. Observations of experimental trajectories in [10] indicate that the amplitude of lateral swaying varies from a maximum $b_{\max}$ for slow movement and gradually decreases to a minimum $b_{\min}$ for free movement when pedestrians move with their free velocity. Thus with b the lateral swaying of pedestrians is defined as

$$b = b_{\max} - (b_{\max} - b_{\min})\frac{v_i}{v_i^0}. \qquad (2.14)$$

Since a and b are velocity-dependent, the inequality

$$b \leq a \qquad (2.15)$$

does not always hold for the ellipse i. In the rest of this work we denote the semi-axis in the movement direction by a and its orthogonal semi-axis by b.

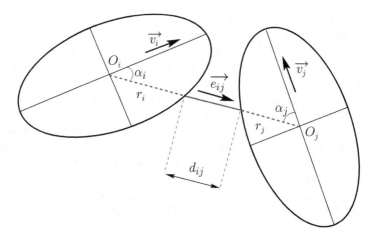

Fig. 2.5 d_{ij} is the distance between the borders of the ellipses i and j along a line connecting their centers

2.3.2 Repulsive Force

Assuming the direction connecting the positions of pedestrians i and j is given by

$$\vec{R_{ij}} = \vec{R_j} - \vec{R_i}, \qquad \vec{e_{ij}} = \frac{\vec{R_{ij}}}{\|\vec{R_{ij}}\|}, \tag{2.16}$$

the repulsive force reads

$$\vec{F_{ij}}^{\text{rep}} = -m_i k_{ij} \frac{(\eta \|\vec{v_i^0}\| + v_{ij})^2}{d_{ij}} \vec{e_{ij}}, \tag{2.17}$$

with the effective distance between pedestrians i and j

$$d_{ij} = \|\vec{R_{ij}}\| - r_i(v_i) - r_j(v_j). \tag{2.18}$$

r_i is the polar radius of pedestrian i (Fig. 2.5).

This definition of the repulsive force reflects several aspects. First, the force between two pedestrians decreases with increasing distance. In the GCFM it is inversely proportional to their distance (2.18). Furthermore, the repulsive force takes into account the relative velocity v_{ij} between pedestrians i and pedestrian j. The following special definition ensures that for constant d_{ij} slower pedestrians are less affected by the presence of faster pedestrians than by that of slower ones:

$$v_{ij} = \Theta\left((\vec{v_i} - \vec{v_j}) \cdot \vec{e_{ij}}\right) \cdot (\vec{v_i} - \vec{v_j}) \cdot \vec{e_{ij}}, \tag{2.19}$$

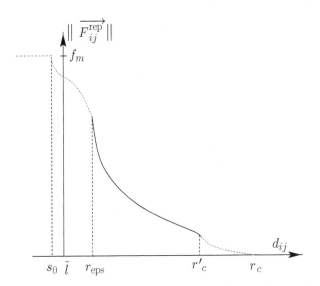

Fig. 2.6 The interpolation of the repulsive force between pedestrians i and j Eq. (2.17) depending on d_{ij} and the distance of closest approach $\tilde{l}$ [40]. As the repulsive force also depends on the relative velocity v_{ij}, this figure depicts the curve of the force for $v_{ij} = $ const. The *right* and *left dashed curves* are defined by a Hermite-interpolation at r_c and r'_{eps}. The wall-pedestrian interaction has an analogous form

with $\Theta()$ is the Heaviside function.

As in general pedestrians react only to obstacles and pedestrians that are within their perception, the reaction field of the repulsive force is reduced to the angle of vision (180°) of each pedestrian, by introducing the coefficient

$$k_{ij} = \Theta(\vec{v_i} \cdot \vec{e_{ij}}) \cdot (\vec{v_i} \cdot \vec{e_{ij}}) / \| \vec{v_i} \| . \tag{2.20}$$

The coefficient k_{ij} is maximal when pedestrian j is in the direction of movement of pedestrian i and minimal when the angle between j and i is bigger than 90°. Thus the strength of the repulsive force depends on the angle.

The interaction of pedestrians with walls is similar to Eq. (2.17). In GCFM walls are treated as three static pedestrians. The number of points is chosen to avoid "going through" walls for pedestrians that are walking almost parallel to walls.

To enhance the numerical behavior of the function (2.17) at small distances a Hermite-interpolation is performed. Furthermore, the force range is reduced to a certain distance r_c. This is especially necessary to avoid summing over distant pedestrians. Figure 2.6 depicts a possible curve of the repulsive force extended by the above mentioned right and left Hermite-interpolation (dashed curves).

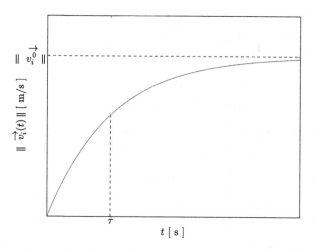

Fig. 2.7 Expected evolution of a pedestrian's velocity with respect to time

2.3.3 Driving Force

From a mathematical standpoint the acceleration of pedestrians may be of different nature e.g., Dirac-like, linear or exponential. According to [21], the later type is more realistic and can take the following expression:

$$\vec{v_i}(t) = \vec{v_i^0} \cdot \left(1 - \exp\left(-\frac{t}{\tau} \right) \right), \tag{2.21}$$

with τ a time constant. Figure 2.7 shows the evolution of the velocity in time. See Fig. 2.7.

Differentiation of Eq. (2.21) with respect to t yields

$$\frac{d}{dt}\vec{v_i}(t) = \frac{1}{\tau} \cdot \vec{v_i^0} \exp\left(-\frac{t}{\tau} \right). \tag{2.22}$$

From Eq. (2.21) one gets

$$\vec{v_i^0} \exp\left(-\frac{t}{\tau} \right) = \vec{v_i^0} - \vec{v_i}(t). \tag{2.23}$$

Combining (2.22) and (2.23) and considering Newton's second law, the force acting on i with mass m_i is

$$\vec{F_i}^{\,\mathrm{drv}} = m_i \frac{\vec{v_i^0} - \vec{v_i}}{\tau}. \tag{2.24}$$

This mathematical expression of the driving force, is systematically used in all known force-based models and describes well the free movement of pedestrians. In [33] is has been reported that evaluation of empirical data yields $\tau = 0.61\,\text{s}$. A different value of $0.54\,\text{s}$ was reported in [17].

2.4   Steering Mechanisms

In this section the effects of the desired direction on the dynamics by measuring the outflow from a bottleneck with different widths is studied. Two different methods for setting the direction of the desired velocity are discussed.

2.4.1   Directing Towards the Middle of the Exit

This is probably the most obvious mechanism. Herein, the desired direction $\overrightarrow{e_i^0}$ for pedestrian i is permanently directed towards a reference point that exactly lies on the middle of the exit. In some situations it happens that pedestrians can not get to the chosen reference point without colliding with walls. To avoid this and to make sure that all pedestrians can "see" the middle of the exit the reference point e_1 is shifted by half the minimal shoulder length $b_{\min} = 0.2\,\text{m}$ (Fig. 2.8).

Figure 2.9 shows a simulation with 180 pedestrians with this steering mechanism. Even if the entrance of the bottleneck is relatively wide, because of the steering the pedestrians do not make optimal use of the full width and stay oriented towards the middle of the bottleneck.

2.4.2   Mechanism with Directing Lines

In this section we introduce a mechanism that is, unlike the previous one, applicable to all geometries even if the exit point is not visible. Three different lines are

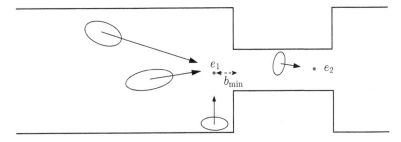

Fig. 2.8 All pedestrians are directed towards the reference points e_1 and e_2

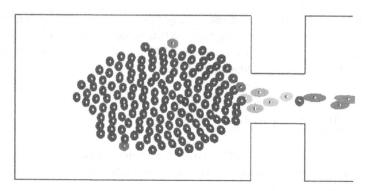

Fig. 2.9 Screen-shot of a simulation. Width of the bottleneck $w = 2.5\,\mathrm{m}$

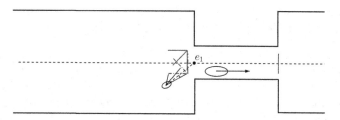

Fig. 2.10 Guiding line segments in front of the generated

defined (Fig. 2.10) which allow to "ease" the movement of pedestrians through the bottleneck. The nearest point from each pedestrian to those lines define its desired direction.

The blue line set (down the dashed line segment) is considered by pedestrians in the lower half and the red line set by pedestrians in the upper half of the bottleneck. For a pedestrian i at position p_i we define the angle

$$\theta_i = \arccos\left(\frac{\overrightarrow{p_i e_1} \cdot \overrightarrow{p_i l_{ij}}}{\| \overrightarrow{p_i e_1} \| \cdot \| \overrightarrow{p_i l_{ij}} \|}\right),\tag{2.25}$$

with l_{ij} the nearest point of the line j to the pedestrian i.

The next direction is then chosen as

$$\overrightarrow{e_i^0} = \frac{\overrightarrow{p_i l_{ij}}}{\| \overrightarrow{p_i l_{ij}} \|}\tag{2.26}$$

with j such that $\theta_j = \min\{\theta_1, \theta_2, \theta_3\}$. The direction lines are shifted in x- and y-direction by $b_{\min}$ to mitigate blocking in the corners.

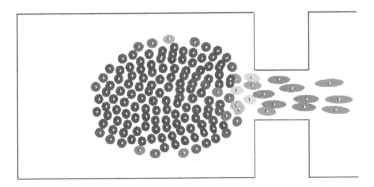

Fig. 2.11 Screenshot of a simulation with directing lines. Width of the bottleneck $w = 2.5$ m

Figure 2.11 shows the form of the jam in front of the bottleneck for $w = 2.5$ m. In comparison to the first steering mechanism, where pedestrians were direction towards the middle of the bottleneck, here pedestrians make better use of the whole width, which influences the qualitative behavior of pedestrians positively.

2.5 Simulation Results

The free parameters of the model are systematically calibrated by considering single file movement, two dimensional movement in corridors, bottlenecks and corners. In this chapter only simulations results in wide corridors and bottlenecks are presented.

The initial value problem in Eq. (2.1) was solved using an Euler scheme with fixed-step size $\Delta t = 0.01$ s. First the state variables of all pedestrians are determined. Then the update to the next step is performed. Thus, the parallelism of the update is ensured.

The desired speeds of pedestrians are Gaussian distributed with mean $\mu = 1.34$ m/s and standard deviation $\sigma = 0.26$ m/s. Since there is no uniquely accepted experimental value for the time constant τ in the driving force Eq. (2.24), we set for simplicity $\tau = 0.5$ s, i.e. $\tau \gg \Delta t$. The mass m_i is set to unity. In all following simulations the set of parameters is not changed.

To compare the presented steering mechanisms several simulations in a bottleneck are performed. For each mechanisms only the width of the bottleneck is varied from 1 to 2.4 m.

On the basis of a quantitative analysis, the importance of the steering of pedestrians for the observed behavior can be estimated. In the following, for each mechanism the flow through bottlenecks of varying width w is measured. The flow is measured directly at the entrance of the bottleneck according to

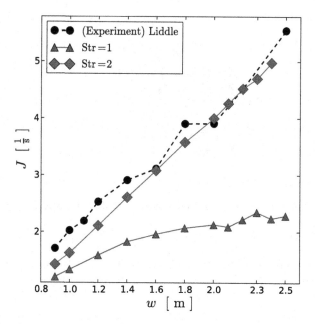

Fig. 2.12 Flow through a bottleneck with different widths

$$J = \frac{N_{\Delta t} - 1}{\Delta t}, \tag{2.27}$$

with $N_{\Delta t} = 180$ pedestrians and Δt the time necessary that all pedestrians pass the measurement line. In Fig. 2.12 the resulting flow in comparison with experimental data is presented.

Keeping the same values of model parameters, the fundamental diagram in a corridor with closed boundary conditions is measured. Here again, for the sake of comparison simulations with circles and ellipses are performed. Results are then validated against experimental data (Figs. 2.13 and 2.14).

2.6 Conclusion and Outlook

In this chapter a brief overview of force-based modeling of pedestrian dynamics is given. Force-based models continuously describe in space the movement of pedestrians by means of differential equations. One can track the origin of this Ansatz back to early 1950s, where first models were developed to describe lane-movement in traffic flow. Since then, force-based models have been successful in describing fairly well the dynamics of pedestrians. Nevertheless, several problems arise from the analogy to Newtonian dynamics. Therefore, principles like superposition, *actio et reactio* should be revised when applied to pedestrian dynamic.

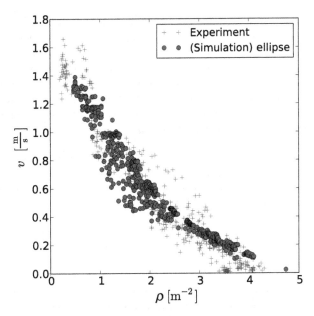

Fig. 2.13 Density-velocity relation with *ellipses* in a corridor of dimensions $25 \times 1\,\mathrm{m}^2$ in comparison with experimental data obtained in the HERMES-project [9, 25]

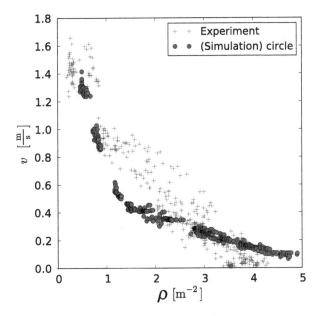

Fig. 2.14 Density-velocity relation with *circles* in a corridor of dimensions $25 \times 1\,\mathrm{m}^2$ in comparison with experimental data obtained in the HERMES-project [9, 25]. In these simulations b is set to be equal to a

By considering the GCFM as an example for force-based models, several important aspects of force-based models were addressed. First, the definition of the repulsive force is presented. By means of a Hermite-interpolation it was possible to overcome the instability of the force at small distances and restrict its range to a maximum distance. Second, several steering mechanism in the driving force are discussed. Finally, simulation results in corridors and bottlenecks are compared to experimental data. It was shown, that it is possible to describe quantitatively pedestrian dynamics in several geometries with *one* set of parameters.

Acknowledgements This work is within the framework of two projects. The authors are grateful to the Deutsche Forschungsgemeinschaft (DFG) for funding the project under Grant-No. SE 1789/1-1 as well as the Federal Ministry of Education and Research (BMBF) for funding the project under Grant-No. 13N9952 and 13N9960.

References

1. Chattaraj, U., Chakroborty, P., Seyfried, A.: Empirical studies on pedestrian motion through corridors of different geometries. In: Proceedings of Transportation Research Board 89th Annual Meeting, CD Rom, Washington, D.C (2010)
2. Chraibi, M., Seyfried, A., Schadschneider, A.: The generalized centrifugal force model for pedestrian dynamics. Phys. Rev. E **82**, 046111 (2010)
3. Daamen, W., Hoogendoorn, S.: Capacity of doors during evacuation conditions. Procedia Eng. **3**, 53–66 (2010)
4. Fruin, J.J.: Pedestrian Planning and Design. Elevator World, New York (1971)
5. Hankin, B.D., Wright, R.A.: Passenger flow in subways. Oper. Res. Soc. **9**(2), 81–88 (1958)
6. Helbing, D.: Collective phenomena and states in traffic and self-driven many-particle systems. Comput. Mater. Sci. **30**, 180–187 (2004)
7. Helbing, D., Molnár, P.: Social force model for pedestrian dynamics. Phys. Rev. E **51**, 4282–4286 (1995)
8. Hirai, K., Tirui, K.: A simulation of the behavior of a crowd in panic. Syst. Control **21**, 409–411 (1977)
9. Holl, S., Seyfried, A.: Hermes – an evacuation assistant for mass events. inSiDe **7**(1), 60–61 (2009)
10. Hoogendoorn, S., Daamen, W.: Pedestrian behavior at bottlenecks. Transp. Sci. **39**(2), 147–159 (2005)
11. Gipps, P.G., Marksjö, B.: A micro-simulation model for pedestrian flows. Math. Comput. Simul. **27**, 95–105 (1985)
12. Kretz, T., Grünebohm, A., Schreckenberg, M.: Experimental study of pedestrian flow through a bottleneck. J. Stat. Mech. Theory Exp. **2006**(10), P10014 (2006)
13. Lakoba, T.I., Kaup, D.J., Finkelstein, N.M.: Modifications of the Helbing-Molnár-Farkas-Vicsek social force model for pedestrian evolution. Simulation **81**(5), 339–352 (2005)
14. Liddle, J., Seyfried, A., Klingsch, W., Rupprecht, T., Schadschneider, A., Winkens, A.: An experimental study of pedestrian congestions: influence of bottleneck width and length. ArXiv e-prints (2009)
15. Liddle, J., Seyfried, A., Steffen, B., Klingsch, W., Rupprecht, T., Winkens, A., Boltes, M.: Microscopic insights into pedestrian motion through a bottleneck, resolving spatial and temporal variations. ArXiv e-prints (2011)
16. Löhner, R.: On the modelling of pedestrian motion. Appl. Math. Model. **34**(2), 366–382 (2010)

17. Moussaïd, M., Helbing, D., Garnier, S., Johansson, A., Combe, M., Theraulaz, G.: Experimental study of the behavioural mechanisms underlying self-organization in human crowds. Proc. R. Soc. B **276**(1668), 2755–2762 (2009)
18. Parisi, D.R., Dorso, C.O.: Morphological and dynamical aspects of the room evacuation process. Physica A **385**(1), 343–355 (2007)
19. Pauls J.L.: Suggestions on evacuation models and research questions. In: Shields, T.J. (ed.) Human Behaviour in Fire. Interscience Communications, London (2004)
20. Pauls J.L.: Stairways and Ergonomics. ASSE, Des Plaines (2006)
21. Pipes, L.A.: An operational analysis of traffic dynamics. J. Appl. Phys. **24**(3), 274–281 (1953)
22. Reuschel, A.: Fahrzeugbewegungen in der Kolonne. Z. Öster. Ing. Arch. **4**, 193–214 (1950)
23. Reuschel, A.: Fahrzeugbewegungen in der Kolonne bei gleichförmig beschleunigtem oder verzögertem Leitfahrzeug. Z. Öster. Ing. Arch. **59**, 73–77 (1950)
24. Rissanen, J.: Minimum-description-length principle. In: Encyclopedia of Statistical Sciences. John Wiley & Sons, Inc., Hoboken (2004)
25. Schadschneider, A.: I'm a football fan ··· get me out of here. Phys. World **21**, 21–25 (2010)
26. Schadschneider, A., Klingsch, W., Klüpfel, H., Kretz, T., Rogsch, C., Seyfried, A.: Evacuation Dynamics: Empirical Results, Modeling and Applications. In: Encyclopedia of Complexity and System Science, vol. 5, pp. 3142–3176. Springer, Berlin/Heidelberg (2009)
27. Schadschneider, A., Chowdhury, D., Nishinari, K.: Stochastic Transport in Complex Systems. From Molecules to Vehicles. Elsevier, Amsterdam (2010)
28. Seyfried, A., Passon, O., Steffen, B., Boltes, M., Rupprecht, T., Klingsch, W.: New insights into pedestrian flow through bottlenecks. Transp. Sci. **43**(3), 395–406 (2009)
29. Seyfried, A., Schadschneider, A.: Empirical results for pedestrian dynamics at bottlenecks. In: Wyrzykowski, R., Dongarra, J., Karczewski, K., Wasniewski, J. (eds.) Parallel Processing and Applied Mathematics. Volume 6068 of Lecture Notes in Computer Science, pp. 575–584. Springer, Berlin/Heidelberg (2010)
30. Seyfried, A., Steffen, B., Lippert, T.: Basics of modelling the pedestrian flow. Physica A **368**, 232–238 (2006)
31. Suma, Y., Yanagisawa, D., Nishinari, K.: Anticipation effect in pedestrian dynamics: modelling and experiments. Physica A **391**, 248–263 (2012)
32. Templer, J.A.: The Staircase: Studies of Hazards, Falls, and Safer Design. MIT, Cambridge (1992)
33. Werner, T., Helbing, D.: The social force pedestrian model applied to real life scenarios. In: Galea, E.R. (ed.) Pedestrian and Evacuation Dynamics, pp. 17–26. CMS, London (2003)
34. Wolfram, S.: Statistical mechanics of cellular automata. Rev. Mod. Phys. **55**, 601–644 (1983)
35. Yanagisawa, D., Kimura, A., Tomoeda, A., Ryosuke, A., Suma, Y., Ohtsuka, K., Nishinari, K.: Introduction of frictional and turning function for pedestrian outflow with an obstacle. Phys. Rev. E **80**(3), 036110 (2009)
36. Yu, W.J., Chen, L.Y., Dong, R., Dai, S.Q.: Centrifugal force model for pedestrian dynamics. Phys. Rev. E **72**(2), 026112 (2005)
37. Zhang, J., Klingsch, W., Schadschneider, A., Seyfried, A.: Transitions in pedestrian fundamental diagrams of straight corridors and t-junctions. J. Stat. Mech. (2011). doi:10.1088/1742-5468/2011/06/P06004
38. Zhang, J., Klingsch, W., Seyfried, A.: High precision analysis of unidirectional pedestrian flow within the Hermes project. In: The Fifth Performance-based Fire Protection and Fire Protection Engineering Seminars, Guangzhou, China (2010)
39. Zhang, J., Klingsch, W., Schadschneider, A., Seyfried, A.: Ordering in bidirectional pedestrian flows and its influence on the fundamental diagram. J. Stat. Mech. **2**, P02002 (2012)
40. Zheng, X., Palffy-Muhoray, P.: Distance of closest approach of two arbitrary hard ellipses in two dimensions. Phys. Rev. E **75**(6), 061709 (2007)

Chapter 3
Connection Between Microscopic and Macroscopic Models

Jan-Frederik Pietschmann

Abstract This chapter is devoted to the detailed study of the relation between a microscopic cellular automation and a macroscopic partial differential equation model for the movement of pedestrians. We describe the mathematical tools allowing to derive the macroscopic from the microscopic model. Such a connection between discrete, particle based and continuous, density based models can help to improve the understanding of basic properties of human crowds. We exemplify this by applying our results to typical cases. The first one is the formation of lanes in bi-directional flow. The second is the analysis of the fundamental diagram. Our analysis provides (at least qualitatively) a connection between these phenomena and model parameters. We conclude by pointing out a number of possible directions of future research.

3.1 Introduction

A good understanding of the collective behavior of large human crowds (crowd motion) is of importance for several reasons. First of all, a growing fraction of humanity is living in urban regions. These regions especially include facilities such as airports or shopping malls in which a large number of people is concentrated in a relatively small place. Appropriate mathematical models can help to optimize these buildings in order to avoid congestion and allow for faster operation. Even more important, they can be used to create and validate (using numerical simulations) evacuation plans which are of course of highest importance.

Therefore the ultimate goal in terms of mathematical modeling is to develop a model which is able to describe (at least qualitatively) the behavior of human

J.-F. Pietschmann (✉)
Numerical Analysis and Scientific Computing, Department of Mathematics, TU Darmstadt,
Dolivostr. 15, 64293 Darmstadt
e-mail: pietschmann@mathematik.tu-darmstadt.de

S. Ali et al. (eds.), *Modeling, Simulation and Visual Analysis of Crowds*, The International
Series in Video Computing 11, DOI 10.1007/978-1-4614-8483-7_3,
© Springer Science+Business Media New York 2013

crowds over a large range of situations. Such a model does not yet exist and in the following we shall outline the difficulties in creating it. A major issue here is obviously the complexity of the humans involved. Their behavior depends on the individual characteristics of each agent such as age, height, sex or even cultural heritage [8]. Furthermore the behavior of each individual may change drastically depending on the situation (e.g. normal walking versus panic). However, even if it is assumed that all people behave exactly the same way, the situation remains complicated. This becomes clearer by comparing a human crowd with a multi particle system from physics (e.g. an electron gas or a plasma). The usual strategy in physics to understand these complex systems is to start from a simple case, i.e. the interaction between only two particles. This process is governed by a simple physical law which then acts as a starting point for the understanding of the complete system using certain mathematical tools. In crowd motion, however, the interaction of a small number of people is already difficult to understand and therefore the principle "from simple to complex" is not applicable. As a result, most existing models are built upon simplified hypotheses and are mostly phenomenological.

3.2 Related Work

In crowd modelling, one can distinguish between two general approaches: micro-scopic and macroscopic models. In the microscopic framework, people are treated as individual entities (particles). The evolution of the particles in time is determined by physical and social laws which describe the interaction among the particles as well as their interactions with the physical surrounding. Examples for microscopic methods are social force models (see [17] and the references therein), cellular automata, [12, 28], queuing models, [40], or continuum dynamic approaches like [37]. Social force models are also popular in computer vision, see [24, 27, 30, 39]. For an extensive review of different microscopic approaches we refer to [14]. In contrast to microscopic models, macroscopic models treat the whole crowd as an entity without considering the movement of single individuals. The crowd is often represented by a density function depending on ('continuous') space and time. Classical approaches use well known concepts from fluid or gas dynamics, see [18]. More recent models are based on optimal transportation methods [26], mean field games, cf. [22] (see [23] for a general introduction) or non-linear conservation laws [11]. In [31], an approach based on time-evolving measures is presented. We finally note that crowd motion models share many features with traffic models, cf. [2].

We remark that, that due to the difficulties mentioned above, there is in many cases no connection between microscopic and macroscopic models. However, such an approach would be useful: While microscopic models are closest to observations and are a very intuitive approach, many interesting quantities (such as density or velocity) are macroscopic. Also collective self-organization phenomena (e.g. lane formation) appears on a macroscopic scale.

In this work we describe in detail how such a connection can be derived between a cellular automaton model introduced in [21] and a system of partial differential equations similar to the ones in [4, 6]. We will discuss lane formation and the analysis of the fundamental diagram as two possible applications.

3.2.1 Notation

Throughout this chapter we use the following notation:

$\mathbb{R}$ Set of real numbers,

$\mathbb{R}^+$ Set of positive real numbers,

$\mathbb{N}$ Set of positive integers,

∂_t partial derivative with respect to t,

∇ Vector containing all partial derivatives with respect to the space variable x,
        i.e. $\nabla = (\partial_{x_1}, \partial_{x_2}, \ldots, \partial_{x_n})^T$ for $x \in \mathbb{R}^n$,

$\nabla \cdot$ divergence operator, i.e. for $V = (V_1, \ldots, V_n)^T \in \mathbb{R}^n$ we have
        $\nabla \cdot V(x) = \sum_{i=1}^n \partial_{x_i} V_i(x)$, $x \in \mathbb{R}^n$

∇^2 The Laplace operator, i.e. $\nabla^2 u(x) = \sum_{i=1}^n \partial_{x_i x_i} u(x)$.

3.3 The Microscopic Model for Two Species

We start from a cellular automaton model for human crowd motion introduced in [21]. The model is based on an asymmetric simple exclusion process (ASEP) on a two-dimensional grid of size $m_x \cdot n_y$ (the size of one cell is typically about $40 \times 40 \, \text{cm}^2$, cf. [33], originating from a maximal density of 6.25 people per m^2, cf. [38]). Given a discrete time step, the model provides for each individual in a given cell the probability to jump into a neighboring cell. This probability is determined by several factors: First of all, individuals are not allowed to jump to an occupied cell (size exclusion, cf. [36]). Furthermore, there exist two driving forces, called "floor fields", cf. [7], a static field S and a dynamic field D on which the jump-probability depends exponentially. The static field S provides individuals with a sense of their environment, increasing towards locations they want to reach, such as doors. The dynamic field D is created by the particles themselves and accounts for herding effects. This is a key feature of the model and one goal of this chapter is to examine its impact on the formation of lanes. Being zero at the initial time, the value of D is increased whenever a particle leaves a cell, modeling the tendency of people to follow others. Note that the concept of floor fields is also used in computer vision, cf. [1]. It is straightforward to extend this model to multiple species, each of them coupled to its own dynamic and static field. In the following, we consider the case of two species (labeled red or r and blue or b). For simplicity, we shall only explain the model for particles of group r with corresponding fields D_r and S_r. The probability of a particle to jump into a neighboring cell i, j is given by

$$(P_r)_{i,j} = (N_r)_{i,j} \exp\left(k_D (D_r)_{i,j}\right) \exp\left(k_S (S_r)_{i,j}\right)(1 - r_{i,j} - b_{i,j}). \qquad (3.1)$$

The term $(1 - r_{i,j} - b_{i,j})$ accounts for the size exclusion effect rendering the probability zero if a cell is occupied. The positive constants k_D and k_S regulate the relative influence of the two floor fields. Finally, $(N_r)_{i,j}$ is a normalization factor given by

$$(N_r)_{i,j}^{-1} = \sum_{k=\{i-1,i+1\}} \sum_{l=\{j-1,j+1\}} e^{k_D (D_r)_{k,l}} e^{k_S (S_r)_{k,l}}. \qquad (3.2)$$

The dynamic field D_r is zero at the beginning of a simulation. In every step, it is updated using the following rules

- It is increased by one whenever a particle left a cell, i.e.

$$(D_r)_{i,j}^{k+1} = \begin{cases} (D_r)_{i,j}^k + 1 \text{ if } (r_{i,j}^k - r_{i,j}^{k+1}) = 1 \\ (D_r)_{i,j}^k \qquad \text{otherwise} \end{cases} \qquad (3.3)$$

- If $D_r \geq 1$, it decreases by a given probability $\delta > 0$, i.e. given a random number p

$$(D_r)_{i,j}^{k+1} = \begin{cases} (D_r)_{i,j}^k - 1 \text{ if } p < \delta \\ (D_r)_{i,j}^k \qquad \text{otherwise} \end{cases} \qquad (3.4)$$

- The diffusion is implemented in the following way: With a probability of $\kappa/4$, $\kappa \in \mathbb{R}^+$ a particle jumps to one of its neighboring fields. With probability $(1 - \kappa)$, it stays at its place.

Note that these rules imply that the value of D_r is always a non-negative integer.

3.4 Derivation of the Macroscopic Model

In this section we describe the derivation of a (system of) partial differential equations from the modified ASEP introduced above, cf. [4,36]. The resulting model is macroscopic is the following sense: In the ASEP, pedestrians are represented by particles and at each time step, the position of each particle is known. On the contrary, the PDE model describes the evolution of densities. These densities can be understood as the average number of pedestrians at a certain point in time and space. Therefore, this model is not able to produce trajectories of individual pedestrian. On the other hand, due to the intrinsic averaging, global quantities (which are for example necessary to compute the fundamental diagram) can be extracted easily.

For the sake of clarity, we perform the derivation in one space dimension only. It is analogous in higher dimensions.

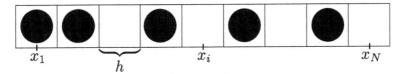

Fig. 3.1 The microscopic setting in one space dimension

3.4.1 Setup

In one space dimension, the model reduces to a row of N cells of width h, as shown in Fig. 3.1. We scale the width of the cells such that the total length of the row is one, i.e. chose h such that $hN = 1$. Furthermore, we denote by x_i the midpoint of cell i, $i = 1,\ldots,N$. In the microscopic model, all quantities are discrete: r_i and b_i denote the number of red and blue particles in cell i, respectively. The values of the floor fields are assumed to be constant in each cell resulting in a finite number of values S_i and D_i. On the other hand, the densities appearing in the macroscopic model are functions of continuous space and time variables ($x \in [0,1]$, $t \in \mathbb{R}^+$). Therefore, we introduce the functions

$$
\begin{aligned}
\tilde{S}_r &: [0,1] && \to \mathbb{R}^+ & \tilde{S}_b &: [0,1] && \to \mathbb{R}^+ \\
\tilde{D}_r &: [0,1] \times \mathbb{R}^+ \to \mathbb{R}^+ & \tilde{D}_b &: [0,1] \times \mathbb{R}^+ \to \mathbb{R}^+ \\
\tilde{r} &: [0,1] \times \mathbb{R}^+ \to [0,1] & \tilde{b} &: [0,1] \times \mathbb{R}^+ \to [0,1].
\end{aligned}
\tag{3.5}
$$

To connect these functions with their discrete analogues (i.e. $\tilde{S}$ with S_i, $\tilde{D}$ with D_i, etc.), we define them as constant on each cell. For example

$$
\tilde{S}_r(x) = \begin{cases} S_1, & \text{if } 0 \le x < h, \\ \ldots, \\ S_i, & \text{if } hi \le x < h(i+1), \\ \ldots \end{cases}
\tag{3.6}
$$

In other words, the functions are piecewise constant and in each cell attain the value of their discrete analogue. With these definitions, the probability of a particle to jump into the cell with midpoint x_i is given by

$$
\tilde{P}_c(x_i,t_k) = N e^{k_{D_c}\tilde{D}_c(x_i,t_k)} e^{k_{S_c}\tilde{S}_c(x_i)}\left(1 - \tilde{r}_i(x_i,t_k) - \tilde{b}_i(x_i,t_k)\right), \quad c = r, b,
\tag{3.7}
$$

with

$$
\tilde{N}_c(x_i,t_k) = \frac{1}{\sum_{k=i-1}^{i+1} e^{k_{D_c}\tilde{D}_c(x_i,t_k)} e^{k_{S_c}\tilde{S}_c(x_i)}}, \quad c = r, b.
$$

Note that even though these expressions look very similar to (3.1) and (3.2) they are different as they are valid for arbitrary $x \ne x_i$, $i = 1,\ldots,N$ and $t \ne t_k$, $k \in \mathbb{N}_+$.

Closure Assumption A priori, only $n_{i,j}$, i.e. the information whether cell (i,j) is occupied or not is known. However, in (3.7) we replaced $n_{i,j}$ by $\tilde{r}_i(x_i,t_k) + \tilde{b}_i(x_i,t_k)$, i.e. the **probability** of the cell being occupied or not. This is needed to obtain a closed equation and therefore called a "closure assumption". In cases in which the macroscopic limit can be justified rigorously, this closure assumption turns out to be the right one, cf. [13], which motivates our choice.

3.4.2   Master Equation

We will now formulate the dynamics of the macroscopic model by using a so-called master-equation, see for example [3]. This equation describes how the probabilities to find a particle in a certain cell evolve in time. To increase readability, we will neglect the tildes from the previous section and write r, b, S, D instead of $\tilde{r}, \tilde{b}, \tilde{S}, \tilde{D}$. For red and blue particles, the corresponding master equations are given by

$$r(x_i,t_{k+1}) = P_r(x_i,t_k)(r(x_{i-1},t_k) + r(x_{i+1},t_k)) \tag{3.8}$$
$$+ r(x_i,t_k)(1 - P_r(x_{i-1},t_k) - P_r(x_{i+1},t_k)),$$
$$b(x_i,t_{k+1}) = P_b(x_i,t_k)(b(x_{i-1},t_k) + b(x_{i+1},t_k)) \tag{3.9}$$
$$+ b(x_i,t_k)(1 - P(x_{i-1},t_k) - P(x_{i+1},t_k)),$$

These equations states the probability of finding a particle at position x_i at time t_{k+1} given the state of the system at time t_k. Since the ASEP allows particles to jump only one cell per time step, we only need to take the cells i, $i-1$ and $i+1$ into account. Roughly speaking, the master equation consists of two terms: The probability of a particle at cell $i \pm 1$ to jump into cell i (first term) and probability for a particle to stay in cell i (i.e. one minus the probability to jump out of the cell, second term). Note that since the probabilities are real numbers in $[0,1]$, the values of r and b do not need to be discrete integers anymore. For corresponding equations for the dynamic fields read as

$$D_r(x_i,t_{k+1}) = D_r(x_i,t_k) + (\Delta t)r(x_i,t_k)(P_r(x_{i-1},t_k) + P_r(x_{i+1},t_k)) \tag{3.10}$$
$$- \delta D_r(x_i,t_k),$$
$$D_b(x_i,t_{k+1}) = D_b(x_i,t_k) + (\Delta t)b(x_i,t_k)(P_b(x_{i-1},t_k) + P_b(x_{i+1},t_k)) \tag{3.11}$$
$$- \delta D_b(x_i,t_k).$$

i.e. value of D increases, whenever a particle leaves an occupied field and decreases with rate $\delta > 0$.

We are now ready to perform the limiting procedure which transforms the ASEP into a system of partial differential equations. We will only consider red particles from now on as the strategy is exactly the same for the blue ones. The mathematical strategy to obtain the microscopic model is to perform the limit $\Delta t \to 0$, $h \to 0$. In other words: We let the width of the cells and the length of the discrete time steps tend to zero. Since the total length is fixed by $hN = 1$ this means that the number of cells tend to infinity. If we multiply the master equation for r, (3.8), by $\frac{1}{\Delta t}$ we obtain

$$
\frac{1}{\Delta t}\left(r(x_i,t_{k+1}) - r(x_i,t_k)\right) = \frac{1}{\Delta t}(P_r(x_i,t_k)(r(x_{i-1},t_k) + r(x_{i+1},t_k))
$$
$$
+ r(x_i,t_k)(1 - P_r(x_{i-1},t_k) - P_r(x_{i+1},t_k))),
$$

we see that in the limit $\Delta t \to 0$, the left hand side will converge to $\partial_t r$. However, the resulting equation is still discrete in space and it is not clear what happens to the right hand side in the limit. We therefore perform a Taylor expansion of the right hand side. Remember that we are planing to let h tend to zero, therefore it is reasonable to assume it to be small which justifies the Taylor expansion. We first take a closer look at Eq. (3.12). Taylor expansion of the right hand side around the point $x = x_i$ yields

$$
r(x_i, t + \Delta t) - r(x_i, t)
$$
$$
= h^2 P(x_i,t)\partial_x r(x_{i+1},t) - h^2 r(x_i,t)\partial_x P(x_{i+1},t)
$$
$$
= h^2 P(x_i,t)\partial_x r(x_i,t) - h^2 r(x_i,t)
$$
$$
\partial_x\left(Ne^{k_{S_r}S_r}e^{k_{D_r}D_r}\left[(1 - r(x_i,t))(\partial_x D + \partial_x S) - \partial_x r(x_i,t)\right]\right)
$$
$$
= h^2\partial_x\left(Ne^{k_{S_r}S_r}e^{k_{D_r}D_r}\partial_x r(x_i,t)\right)
$$
$$
- h^2\partial_x\left(Ne^{k_{S_r}S_r}e^{k_{D_r}D_r}r(x_i,t)(1 - r(x_i,t))(\partial_x D + \partial_x S)\right).
$$

The final step before we actually pass to the limit is to chose the relation between Δt and h, the so-called scaling, [13]. If we divide again by Δt, we see that the quantity $\frac{h^2}{\Delta t}$ appears on the right hand side. Thus we chose $\frac{h^2}{\Delta t} =: P = \text{const} > 0$. The constant P acts as a diffusion coefficient. Then, the left hand side will converge to the time derivative $\partial_t r$, while the right hand side does not depend on h or Δt anymore (up to higher order terms which vanish as $h \to 0$). Thus passing to the limit $h, \Delta t \to 0$ we obtain

$$
\partial_t r + \frac{P}{3}\partial_x\left(r(1 - r)\left(k_{S_r}\partial_x S_r + k_D\partial_x D_r\right)\right) = \frac{P}{3}\partial_{xx}r. \tag{3.12}
$$

We made use of the fact that

$$F_i(h^2) := e^{k_{D_r}D_r(x_i)}e^{k_{S_r}S_r(x_i,t)}N_i = \frac{e^{k_{D_r}D_r(x_i,t)}e^{k_{S_r}S_r(x_i)}}{3e^{k_{D_r}D_r(x_i,t)}e^{k_{S_r}S_r(x_i)} + \mathcal{O}(h^2)} \xrightarrow{h\to 0} \frac{1}{3}. \qquad (3.13)$$

For (3.10), we apply the same procedure yielding

$$\frac{D_r(x_i, t+\Delta t) - D_r(x_i, t)}{\Delta t}$$

$$= r_i(t)(P(x_{i+1}, t) + P(x_{i-1}, t)) - \delta D_r(x_i, t_k)$$

$$= r(x_i, t)(F_{i+1}(h^2)(1 - r(x_{i+1}, t)) + F_{i-1}(h^2)(1 - r(x_{i-1}, t))) - \delta D_r(x_i, t_k)$$

$$= r(x_i, t)\left(F_{i+1}(h^2)(1 - r(x_i, t)) + F_{i-1}(h^2)(1 - r(x_i, t))\right) - \delta D_r(x_i, t_k)$$

$$+ r(x_i, t)\left(h\left(F_{i+1}(h^2)\frac{\partial r(x_i, t)}{\partial x} - F_{i-1}(h^2)\frac{\partial r(x_i, t)}{\partial x}\right)\right) - \delta D_r(x_i, t_k)$$

$$+ \mathcal{O}(h^2)$$

In the limit Δt, $h^2 \to 0$, the last term on the right hand side vanishes and we obtain

$$\partial_t D_r = -\delta D_r + \frac{2}{3}r(1 - \rho).$$

As it is well known, cf. [9, 10], that the diffusion algorithm described in Sect. 3.3 yields, in the continuum limit a term $\kappa \partial_{xx} D_r$, we arrive at

$$\partial_t D_r = \kappa \partial_{xx} D_r - \delta D_r + \frac{2}{3}r(1 - \rho). \qquad (3.14)$$

3.4.3   The Macroscopic PDE Limit

As the limiting procedure can be performed in arbitrary spatial dimensions, we finally obtain the following non-linear Nernst-Planck type equations for the densities r and b

$$\partial_t r = \nabla \cdot P((1 - \rho)\nabla r + r\nabla\rho + r(1 - \rho)\nabla(k_{S_r}S_r + k_{D_r}D_r)), \qquad (3.15)$$

$$\partial_t b = \nabla \cdot P((1 - \rho)\nabla b + b\nabla\rho + b(1 - \rho)\nabla(k_{S_r}S_b + k_{D_r}D_b)), \qquad (3.16)$$

coupled to

$$\partial_t D_r = \kappa \nabla^2 D_r - \delta_1 D_r + r(1 - \rho), \qquad (3.17)$$

$$\partial_t D_b = \kappa \nabla^2 D_b - \delta_1 D_b + b(1 - \rho), \qquad (3.18)$$

with $x \in \Omega \subset \mathbb{R}^n, t > 0$. Appropriate initial and boundary conditions depend on the application and will be discussed below. We remark that neglecting the dynamic fields, this model has been analyzed extensively in [4,5].

3.5 Application I: Linear Stability Analysis and the Formation of Lanes

Video recordings of real crowds show that they exhibit a wide range of what is called *collective phenomena*. One common example among them is *lane formation*, see for example [15] and [17]. Pedestrians with the same desired walking direction prefer to walk in lanes. Typically, the number of lanes depends on the width of the street and on the density of pedestrians. One possible explanation for lane formation is as follows: Pedestrians walking against the stream have a high relative velocity. As a consequence, these pedestrians change their walking direction sideways to avoid collisions, which finally leads to separation [16]. The formation of lanes for two species (but without static field) has already been briefly demonstrated numerically in [32]. In this section, we provide simulation results of the microscopic model demonstrating the occurrence of lanes. This motivates the use linear stability analysis of our macroscopic model (3.15)–(3.18) to obtain insight into the role of the dynamic fields in this process.

3.5.1 Monte Carlo Simulations

We performed simulations of the above model on a 20×100 cell grid. We used a Mersenne twister, cf. [25], to create the pseudo-random numbers needed. The main issue here is to deal with so-called "conflicts", i.e. the case when two particles want to jump into the same cell. In our implementation, we followed the strategy described in [20]. The basic idea is the following: A new parameter $\lambda \in [0,1]$ in introduced. If two or more particles want to jump to the same cell, this new parameter determines their behavior: With probability λ, none of the particles jumps and the cell remains empty. With probability $(1 - \lambda)$, one particle is chosen randomly and jumps into the target cell. In our set-up, red particles enter the domain from the left and blue particles from the right. Both species are supplemented with a static field transporting them through the channel. For this simulation, we chose the following parameters: $\delta = 0.05, k_D = 1.0, k_S = 7, \kappa = 0.5$. The diffusion coefficients of r and b are chosen as 0.0005 in x- and 0.1 in y-direction. The boundary conditions are implemented as follows: In each step, for each cell on the left boundary, a random number is generated. If this number is below a given value bc_l, a virtual particle is created. This particle evolves due to the usual probabilities given by the model and can either jump into the domain or vanishes. On

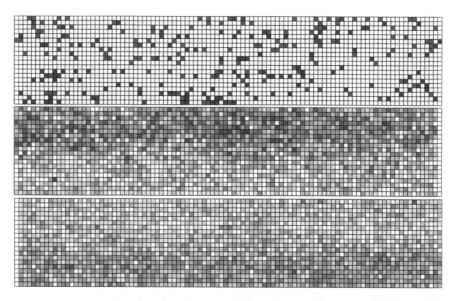

Fig. 3.2 (Color online) Results of the Monte Carlo simulations: Snapshot of a single simulation after 1,000 steps (*top*); density of *red* particles after 2,000 steps, averages over 35 runs (*bottom*); density of *blue* particles after 2,000 steps, averages over 35 runs (*middle*)

the right side, the boundary conditions are implemented in the same way with a corresponding boundary value bc_r. In our experiment, we added small perturbations in y-direction, i.e.

$$bc_l = \tilde{bc}_l + 0.04 \sin\left(\frac{2k\pi}{n_x}i\right), i = 1, \ldots, n_x, \tag{3.19}$$

$$bc_r = \tilde{bc}_r + 0.04 \sin\left(\frac{2k\pi}{n_x}i + \pi\right), i = 1, \ldots, n_x. \tag{3.20}$$

Here, we chose $\tilde{bc}_l = \tilde{bc}_r = 0.06$. In Fig. 3.2 (top), we show a snapshot of one simulation demonstrating the formation of two lanes. Figure 3.2 (bottom), we show the average density of red particles at step 2,000 averages over 35 simulations, in Fig. 3.2 (middle), we same is shown for the blue species.

3.5.2 Linear Stability Analysis

The idea is as follows: We consider an equilibrium state of the system in which both red and blue particles are uniformly distributed within the whole domain. Then, we add a small, asymmetric perturbation and observe the system's reaction as time evolves. The question is under which conditions the perturbations do not smooth

Fig. 3.3 Geometry of the domain on which the linear stability analysis is performed

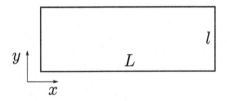

out, but grow in time. The procedure is summarized in Fig. 3.4. We remark we do not assume that individuals have a tendency to prefer a special walking site.

To be able to obtain an explicit condition, we make the following assumptions:

- The static floor fields are acting only in x-direction, we assume the special case

$$\nabla S_r = (1,0), \quad \nabla S_b = (-1,0), \tag{3.21}$$

  meaning that the red and blue individuals have opposite preferred walking directions.
- The diffusion of particles in x-directions vanishes, which is reasonable in case of pedestrians walking along a corridor, as it is unlikely for them to go randomly forward or backward.
- The diffusion of the dynamic floor fields D_c, $c = r, b$ vanishes in y-direction, corresponding to small movements orthogonal to the walking direction.
- The evolution of the dynamic floor fields is slow compared to that of the densities r and b. Therefore, we only consider the stationary version of (3.17) and (3.18).
- The coupling constants to static and dynamic field are equal for both species, i.e.

$$k_{S_r} = k_{S_b} = k_S, \quad k_{D_r} = k_{D_b} = k_D.$$

- All calculations are performed on the two-dimensional domain $\Omega = [0,L] \times [0,l] \subset \mathbb{R}^2$ depicted in Fig. 3.3.

Under these assumptions, Eqs. (3.15)–(3.18) reduce to

$$\partial_t r = P\partial_y \left((1-b)\partial_y r + r\partial_y b\right) + Pk_S\partial_x(r(1-\rho)) + Pk_D\nabla(r(1-\rho)\nabla D_r) \tag{3.22}$$

$$= \nabla \cdot (-J_r)$$

$$\partial_t b = P\partial_y \left((1-r)\partial_y b + b\partial_y r\right) - Pk_S\partial_x(b(1-\rho)) + Pk_D\nabla(b(1-\rho)\nabla D_b) \tag{3.23}$$

$$= \nabla \cdot (-J_b),$$

$$0 = \kappa\partial_{xx}D_r - \delta_1 D_r + r(1-\rho), \tag{3.24}$$

$$0 = \kappa\partial_{xx}D_b - \delta_1 D_b + b(1-\rho) \tag{3.25}$$

Hence, the total flux of red and blue particles, respectively, is given by

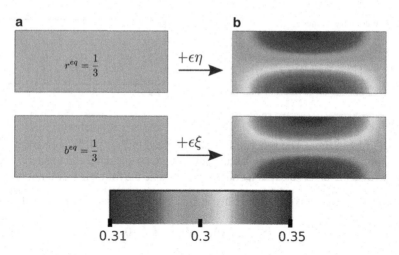

Fig. 3.4 (**a**) Constant densities r^{eq}, b^{eq}, (**b**) small perturbations $\epsilon\xi$ and $\epsilon\eta$ added

$$J_r = -P \begin{pmatrix} k_S r(1-b) + k_D (r(1-\rho)\partial_x D_r \\ (1-b)\partial_y r - r\partial_y \rho + k_D (r(1-\rho)\partial_y D_r \end{pmatrix},$$

and

$$J_b = -P \begin{pmatrix} k_S b(1-r) + k_D (b(1-\rho)\partial_x D_b \\ (1-r)\partial_y b - b\partial_y \rho + k_D (b(1-\rho)\partial_y D_b \end{pmatrix}.$$

We supplement this system with the following boundary conditions: We assume constant influxes of r and b at the left or right end of the corridor, respectively. Furthermore, we say that the particles are leaving the domain at with a constant velocity v_0 and with a rate proportional to their density. This leads to

$$J_r \cdot \begin{pmatrix} -1 \\ 0 \end{pmatrix} = J_r^{in} = \text{const and } J_b \cdot \begin{pmatrix} -1 \\ 0 \end{pmatrix} = b v_0.$$

At $x = L$, we obtain

$$J_r \cdot \begin{pmatrix} 1 \\ 0 \end{pmatrix} = r v_0 \text{ and } J_b \cdot \begin{pmatrix} 1 \\ 0 \end{pmatrix} = J_b^{in} = \text{const.}$$

We assume no-flux boundary conditions in y-direction:

$$J_r \cdot \begin{pmatrix} 0 \\ 1 \end{pmatrix} = J_b \cdot \begin{pmatrix} 0 \\ 1 \end{pmatrix} = 0 \qquad \text{for } y = 0, l,$$

meaning that particles can only enter or leave the corridor though the left or right entrance. Now we are able to perform the linear stability analysis explained above: We call the constant equilibrium states r^{eq} and b^{eq} as well as D_r^{eq} and D_b^{eq}. Then we add the following perturbations:

$$r^{eq} \to r^{eq} + \varepsilon \xi, \qquad\qquad\qquad b^{eq} \to b^{eq} + \varepsilon \eta,$$

$$D_r^{eq} \to D_r^{eq} + \varepsilon \Psi_r, \qquad\qquad\qquad D_b^{eq} \to D_b^{eq} + \varepsilon \Psi_b,$$

where $\xi(x,t)$, $\eta(x,t)$, $\Psi_r(x,t)$ and $\Psi_b(x,t)$ denote the time and space dependent perturbations and $\varepsilon \ll 1$. If we insert this ansatz into the system of equations (3.22)–(3.25) (with given external potentials) and neglect all terms of order ε^2 and higher (first order approximation), we obtain the following linear system describing the evolution of the perturbations:

$$\partial_t \xi = P((1 - b^{eq})\partial_{yy}\xi - r^{eq}\partial_{yy}\eta) + Pk_S\partial_x((1 - \rho^{eq})\xi - r^{eq}(\xi - \eta))$$
$$\qquad + Pk_D r^{eq}(1 - \rho^{eq})(\partial_{xx} + \partial_{yy})\Psi_r,$$

$$\partial_t \eta = P((1 - r^{eq})\partial_{yy}\eta - b^{eq}\partial_{yy}\xi) - Pk_S\partial_x((1 - \rho^{eq})\eta - b^{eq}(-\xi + \eta))$$
$$\qquad + Pk_D b^{eq}(1 - \rho^{eq})(\partial_{xx} + \partial_{yy})\Psi_b,$$

$$0 = \kappa\partial_{xx}\Psi_r - \delta\Psi_r + (1 - \rho^{eq} - r^{eq})\xi + r^{eq}\eta,$$

$$0 = \kappa\partial_{xx}\Psi_b - \delta\Psi_b + (1 - \rho^{eq} - b^{eq})\eta + b^{eq}\xi.$$

The first two equations of this system can be written in the form

$$\partial_t \begin{pmatrix} \xi \\ \eta \end{pmatrix} = A \begin{pmatrix} \xi \\ \eta \end{pmatrix},$$

with a 2×2 matrix A that depends on P, r^{eq}, b^{eq}, k_{S_r}, k_{S_b}, k_{D_r}, k_{D_b}, κ and δ. From this representation it is clear that the evolution of ξ and η is determined by the largest eigenvalue of A: If it is negative, ξ and η will converge to zero as $t \to \infty$ meaning that the system will return to its constant stationary state. If the eigenvalue is positive, ξ and η will stay positive which corresponds to the formation of lanes. In order to calculate an explicit condition for the sign of the eigenvalues, we consider cosinusoidal shaped perturbation in the y direction (as depicted in Fig. 3.4), hence the number of (possible) lanes is given by the mode of the cosine. We are able to predict for several densities and geometries of the domain how many lanes are formed.

The detailed derivation of this condition and the condition itself is stated in the appendix. In the following, we will use it to analyze to interesting special cases.

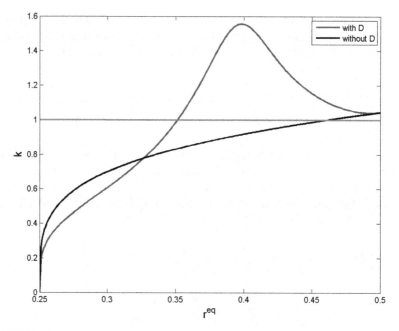

Fig. 3.5 Number of lanes versus density

3.5.3 Number of Lanes for Varying Density

We are able to predict instabilities, thus forming of lanes and the number of lanes, depending on the equilibrium density r^{eq}. In Fig. 3.5, the number of lanes k is plotted versus the density. In this setup, we choose as length l in y-direction to be 7 m. The length L in Fig. 3.5 is 100 m. The decay parameter δ is given by 0.05, and $k_D = 1$. We set $k_S = 7$. Without the dynamic fields D_c, $c = r, b$, (i.e. $k_D = 0$) the first lane in each direction arises at densities of approximately 0.45 in each direction. If we include the dynamic fields, the first lanes are formed at densities of approximately 0.35. Hence, the inclusion of the dynamic fields leads increases the tendencies to follow others.

3.5.4 Number of Lanes for Varying Length

Figure 3.6 shows the number of lanes plotted versus channel length L. We chose the same parameters as before, the density is set to $r^{eq} = 0.33$. It is obvious that the herding behavior does not lead to an increase in the number of lanes. As expected, the tendency to follow others is more pronounced, resulting in less lanes.

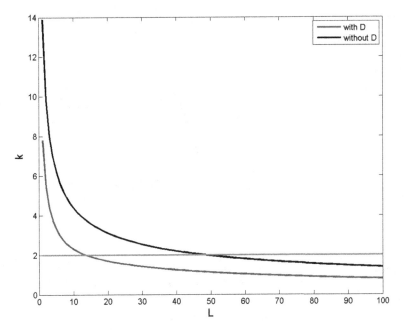

Fig. 3.6 Number of lanes versus length L

3.6 Application II: The Fundamental Diagram

The relation between density ρ and velocity v or total flux (or flow) j, also known as *fundamental diagram*, is a key property of pedestrian movement, [34, 35]. In particular, it can be used to estimate the capacity of facilities such as corridors or streets. There is a lot of experimental data available for different situations, starting from the classical review of Weidmann [38] up to many more recent studies [8, 19, 35]. Thus, the fundamental diagram can also be used as a test to validate crowd motion models. One advantage of our macroscopic model is that the density-velocity relationship in encoded in the structure of the partial differential equation. If we consider the model for one species, e.g. red particles, only and neglect the dynamic field Eqs. (3.15)–(3.18) reduce to

$$\partial_t r = \nabla^2 r - \mathrm{div}(r(1-r)(k_{S_r}\nabla S_r)). \tag{3.26}$$

This corresponds to a uni-directional flow of one species. In this model, the total flux j is given by

$$j := r(1-r)(k_S\nabla S) - \nabla r.$$

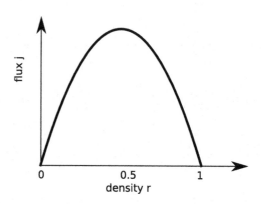

Fig. 3.7 Fundamental diagram for uni-directional flow as predicted by the macroscopic model without dynamic field

Using the relation $r\nabla \ln(r) = \nabla r$, the flux can be rewritten in the form rv, i.e.

$$j = rv = r((1-r)(k_S\nabla S) - \nabla ln(r)).$$

If we assume a constant density r and a constant preferred walking direction ∇S, we recover a linear fundamental diagram $v \sim (1-r)$ and the density-flow relationship is given by $r(1-r)$ as shown in Fig. 3.7. Even though this is far away from detailed experimental results, the key properties are still present: The flow increases with the density until it reaches a maximum. When the density increases further, i.e. the situation becomes more crowded, the individuals are not able to move at their preferred walking speed anymore and the flux decreases until the critical density, at which no movement is possible anymore, is reached. This is of particular interest since situations involving high densities are hard to realize experimentally.

Recent experimental results for bi-directional flow (i.e. two species of people walking in opposite directions), have shown to exhibit a significantly different fundamental diagram compared to uni-directional flow, cf. [41, Fig. 6]. In particular, the maximum is lower but shifted to higher densities. In fact, there seems to be a flat region. One possible explanation is that individuals that are walking towards each other have a stronger dependency to avoid the other. This can easily be incorporated into the microscopic ASEP model by allowing red and blue particles to "switch" their places, as sketched in Fig. 3.8. Such a mechanism is also known from models for intracellular transport, cf. [29]. Thus a rewarding direction of future research is to derive and analyze the corresponding macroscopic formulation for this modified model and compare the resulting fundamental diagram with the experimental results. Another extension is to include the dynamical floor fields and produce fundamental diagrams using numerical simulations.

3.7 Conclusion and Future Work

We established a link between an extended floor field model and a system of non-linear partial differential equations. Using this connection, we were able to give conditions on the formation of lanes depending on the density of particles

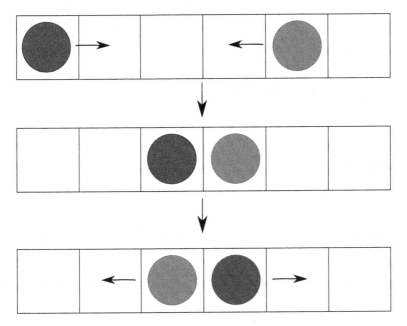

Fig. 3.8 Two particles switching there places in the cellular automata model

as well as the geometry on the domain. The effect of the floor field presenting herding behavior is that lanes are formed at lower densities. A logical next step would be to systematically verify these conditions using the Monte Carlo simulations described above. Also, it would be worthwhile to perform detailed simulations on the macroscopic model (using, e.g. a finite difference scheme) and to compare these results with the microscopic simulations. This would lead to a unified understanding of lane formation in these kind of models. Furthermore, we used the macroscopic description to extract fundamental properties of pedestrian movement, namely the fundamental diagram. We explained how the density-flow relationship can easily be obtained from the macroscopic and briefly mentioned possible extensions to cover recent experimental results regarding bi-directional flow.

Acknowledgements The author acknowledges helpful discussions with M. Burger and B. Schlake (both WWU Münster) as well as A. Seyfried (Jülich/Wuppertal).

Appendix

The linear stability approach described in Sect. 3.5 yields the following linear system for the perturbations

$$\partial_t \xi = P((1-b^{eq})\partial_{yy}\xi - r^{eq}\partial_{yy}\eta) + Pk_S\partial_x((1-\rho^{eq})\xi - r^{eq}(\xi-\eta)) \qquad (3.27)$$
$$+ Pk_D r^{eq}(1-\rho^{eq})(\partial_{xx}+\partial_{yy})\Psi_r,$$

$$\partial_t \eta = P((1-r^{eq})\partial_{yy}\eta - b^{eq}\partial_{yy}\xi) - Pk_S\partial_x((1-\rho^{eq})\eta - b^{eq}(-\xi+\eta)) \qquad (3.28)$$
$$+ Pk_D b^{eq}(1-\rho^{eq})(\partial_{xx}+\partial_{yy})\Psi_b,$$

$$0 = \kappa\partial_{xx}\Psi_r - \delta\Psi_r + (1-\rho^{eq}-r^{eq})\xi + r^{eq}\eta, \qquad (3.29)$$

$$0 = \kappa\partial_{xx}\Psi_b - \delta\Psi_b + (1-\rho^{eq}-b^{eq})\eta + b^{eq}\xi. \qquad (3.30)$$

We denote length of the domain in y-direction by l, the length in x-direction is denoted by L, see Fig. 3.3. The perturbations are assumed as

$$\xi = U(x)\cos\left(\frac{k\pi}{l}y\right)\exp(\lambda t), \qquad (3.31)$$

$$\eta = V(x)\cos\left(\frac{k\pi}{l}y\right)\exp(\lambda t), \qquad (3.32)$$

$$\Psi_r = Y_r(x)\cos\left(\frac{k\pi}{l}y\right)\exp(\lambda t), \qquad (3.33)$$

$$\Psi_b = Y_b(x)\cos\left(\frac{k\pi}{l}y\right)\exp(\lambda t). \qquad (3.34)$$

where $U(x)$, $V(x)$, $Y_r(x)$ and $Y_b(x)$ denote perturbations in the x-direction, and k denotes the mode of the perturbation in y-direction. From now on, we assume $r^{eq} = b^{eq}$. Inserting this ansatz into Eqs. (3.27) and (3.28) we obtain

$$\lambda/PU = -(1-r^{eq})\gamma U + r^{eq}\gamma V + k_S(1-3r^{eq})U' + k_S r^{eq}V'$$
$$- k_D r^{eq}(1-2r^{eq})(\gamma+\Gamma)Y_r \qquad (3.35)$$

$$\lambda/PV = -(1-r^{eq})\gamma V + r^{eq}\gamma U - k_S(1-3r^{eq})V' - k_S r^{eq}U'$$
$$- k_D r^{eq}(1-2r^{eq})(\gamma+\Gamma)Y_b, \qquad (3.36)$$

where we used $\gamma = \frac{k^2\pi^2}{l^2}$, $'$ denotes the derivative with respect to x, $\Gamma = \frac{\pi^2}{L^2}$ and we assume perturbations Y_i of a sinusoidal or cosinusoidal type $Y_i'' = -\Gamma Y_i$. The equations for U and V finally read, using (3.29) and (3.30):

$$[\lambda/P + (1-r^{eq})\gamma]U - r^{eq}\gamma V + k_D r^{eq}(1-2r^{eq})\frac{\gamma+\Gamma}{\kappa\Gamma+\delta}[(1-3r^{eq})U + r^{eq}V]$$
$$= k_S(1-3r^{eq})U' + \mu r^{eq}V', \qquad (3.37)$$

$$[\lambda/P + (1-r^{eq})\gamma]V - r^{eq}\gamma U + k_D r^{eq}(1-2r^{eq})\frac{\gamma+\Gamma}{\kappa\Gamma+\delta}[(1-3r^{eq})V + r^{eq}U]$$

$$= -k_S(1-3r^{eq})V' - \mu r^{eq}U'. \tag{3.38}$$

We denote $\Theta = \frac{\gamma+\Gamma}{\kappa\Gamma+\delta}$. The summation of (3.37) and (3.38) is given by

$$[\lambda/P + (1-2r^{eq})\gamma + k_D r^{eq}(1-2r^{eq})^2\Theta](U+V) \quad = k_S(1-4r^{eq})(U'-V'). \tag{3.39}$$

The derivatives of (3.37) and (3.38) are given by

$$[\lambda/P + (1-r^{eq})\gamma]U' - r^{eq}\gamma V' + k_D r^{eq}(1-2r^{eq})\Theta\left[(1-3r^{eq})U' + r^{eq}V'\right]$$

$$= k_S(1-3r^{eq})U'' + \mu r^{eq}V'', \tag{3.40}$$

$$[\lambda/P + (1-r^{eq})\gamma]V' - r^{eq}\gamma U' + k_D r^{eq}(1-2r^{eq})\Theta\left[(1-3r^{eq})V' + r^{eq}U'\right]$$

$$= -k_S(1-3r^{eq})V'' - \mu r^{eq}U''. \tag{3.41}$$

Subtracting these equation yields

$$[\lambda/P + \gamma + k_D r^{eq}(1-2r^{eq})(1-4r^{eq})\Theta](U'-V')$$

$$= k_S(1-2r^{eq})(U''+V''). \tag{3.42}$$

Combining (3.39) and (3.42) leads to

$$[\lambda/P + (1-2r^{eq})\gamma + k_D r^{eq}(1-2r^{eq})^2\Theta](U+V) = \tag{3.43}$$

$$k_S^2 \frac{(1-4r^{eq})(1-2r^{eq})}{\lambda/P + \gamma + k_D r^{eq}(1-2r^{eq})(1-4r^{eq})\Theta}(U''+V'').$$

In the following, we assume perturbations U and V in x-direction of a sinusoidal type, due to the homogeneous boundary conditions. This leads to

$$U'' = -\frac{m^2\pi^2}{L^2}U, \qquad V'' = -\frac{m^2\pi^2}{L^2}V,$$

where L denotes the length of the domain in x direction. In the following, we take $m = 1$, as we are only interested in lanes forming along the x-direction. We finally arrive at

$$0 = [\lambda^2/P^2 + 2\lambda/P[\gamma(1-r^{eq}) + k_D r^{eq}(1-2r^{eq})(1-3r^{eq})\Theta]$$

$$+ \gamma^2(1-2r^{eq}) + 2\gamma k_D r^{eq}(1-2r^{eq})^3\Theta$$

$$+ k_D^2(r^{eq})^2(1-2r^{eq})^3(1-4r^{eq})\Theta^2$$

$$+k_S^2\Gamma(1-4r^{eq})(1-2r^{eq})](U+V). \tag{3.44}$$

Accordingly, the equation for λ is given by

$$\lambda_{1/2} = -P[(1-r^{eq})\gamma + k_D r^{eq}(1-2r^{eq})(1-3r^{eq})\Theta]$$
$$\pm P\sqrt{(r^{eq})^2[\gamma - k_D r^{eq}(1-2r^{eq})\Theta]^2 - k_S^2\Gamma(1-4r^{eq})(1-2r^{eq})}. \tag{3.45}$$

The parameter λ is supposed to be real-valued for all k, particularly for $k=1$. From that we conclude

$$(r^{eq})^2[\gamma - k_D r^{eq}(1-2r^{eq})\Theta]^2 \geq k_S^2\Gamma(1-4r^{eq})(1-2r^{eq}). \tag{3.46}$$

As $r^{eq} \leq 1/2$, (3.46) is always fulfilled in case that $r^{eq} \geq 1/4$. This means that instabilities arise only in case $r^{eq} \geq 1/4$. To obtain instabilities increasing in time, $\lambda > 0$ has to be satisfied. This means

$$[(1-r^{eq})\gamma + k_D r^{eq}(1-2r^{eq})(1-3r^{eq})\Theta]^2$$
$$< (r^{eq})^2\gamma^2 - 2\gamma k_D(r^{eq})^3(1-2r^{eq})\Theta$$
$$+k_D^2(r^{eq})^4(1-2r^{eq})^2\Theta^2 - k_S^2\Gamma(1-4r^{eq})(1-2r^{eq}) \tag{3.47}$$

Assuming $(1-2r^{eq}) > 0$, which means that the overall density is below maximum, we obtain

$$\gamma^2 + 2\gamma k_D r^{eq}(1-2r^{eq})^2\Theta$$
$$+k_D^2(r^{eq})^2(1-2r^{eq})^2(1-4r^{eq})\Theta^2 + k_S^2\Gamma(1-4r^{eq}) < 0. \tag{3.48}$$

The mode of the cosinusoidal perturbation in y-direction is given by k, hence it gives the number of lanes of particles moving in opposite direction which are amplified during time. If $k=1$, we obtain one lane in each direction. Accordingly, we obtain as inequality for $\gamma = \frac{k^2\pi^2}{l^2}$

$$\gamma^2\left[1 + 2k_D r^{eq}(1-2r^{eq})^2\frac{1}{\kappa\Gamma+\delta}\right.$$
$$\left. + k_D^2 r^{eq2}(1-2r^{eq})^2(1-4r^{eq})\frac{1}{(\kappa\Gamma+\delta)^2}\right]$$
$$+\gamma\left[2k_D r^{eq}(1-2r^{eq})^2\frac{\Gamma}{\kappa\Gamma+\delta}\right.$$
$$\left. + 2k_D^2 r^{eq2}(1-2r^{eq})^2(1-4r^{eq})\frac{\Gamma}{(\kappa\Gamma+\delta)^2}\right]$$

$$+ k_D^2 r^{eq\,2} (1 - 2r^{eq})^2 (1 - 4r^{eq}) \frac{\Gamma^2}{(\kappa\Gamma + \delta)^2} + k_S^2 \Gamma (1 - 4r^{eq})$$

$$< 0. \tag{3.49}$$

The evaluation of (3.49) leads to a condition on k which determines under which conditions instabilities, which lead to lane formation, appear.

References

1. Ali, S., Shah, M.: Floor fields for tracking in high density crowd scenes. In: ECCV (2), pp. 1–14 (2008)
2. Aw, A., Rascle, M.: Resurrection of "second order" models of traffic flow. SIAM J. Appl. Math. **60**(3), 916–938 (electronic) (2000)
3. Bedeaux, D., Lakatos-Lindenberg, K., Shuler, K.E.: On the relation between master equations and random walks and their solutions. J. Math. Phys. **12**(10), 2116–2123 (1971)
4. Burger, M., Di Francesco, M., Pietschmann, J.-F., Schlake, B.: Nonlinear cross-diffusion with size exclusion. SIAM J. Math. Anal. **42**(6), 2842–2871 (2010)
5. Burger, M., Schlake, B., Wolfram, M.-T. Nonlinear Poisson-Nernst Planck equations for ion flux through confined geometries (2010, preprint)
6. Burger, M., Markowich, P., Pietschmann, J.-F.: Continuous limit of a crowd motion and herding model: analysis and numerical simulations. Kinet. Relat. Models **4**, 1025–1047 (2011)
7. Burstedde, C., Klauck, K., Schadschneider, A., Zittartz, J.: Simulation of pedestrian dynamics using a two-dimensional cellular automaton. Phys. A Stat. Mech. Appl. **295**(3–4), 507–525 (2001)
8. Chattaraj, U., Seyfried, A., Chakroborty, P.: Comparison of pedestrian fundamental diagram across cultures. Adv. Complex Syst. **12**(03), 393–405 (2009)
9. Chopard, B., Droz, M.: Cellular automata model for the diffusion equation. J. Stat. Phys. **64**, 859–892 (1991)
10. Chopard, B., Droz, M.: Cellular Automata Modeling of Physical Systems. Cambridge University Press, Cambridge/New York (2005)
11. Colombo, R.M., Rosini, M.D.: Pedestrian flows and non-classical shocks. Math. Methods Appl. Sci. **28**(13), 1553–1567 (2005)
12. Fukui, M., Ishibashi, Y.: Self-organized phase transitions in CA-models for pedestrians. J. Phys. Soc. Jpn. **8**, 2861–2863 (1999)
13. Giacomin, G., Lebowitz, J.L.: Phase segregation dynamics in particle systems with long range interactions. I. Macroscopic limits. J. Stat. Phys. **87**(1–2), 37–61 (1997)
14. Helbing, D.: Traffic and related self-driven many-particle systems. Rev. Mod. Phys. **73**(4), 1067–1141 (2001)
15. Helbing, D., Molnár, P.: Social force model for pedestrian dynamics. Phys. Rev. E **51**(5), 4282–4286 (1995)
16. Helbing, D., Farkas, I., Vicsek, T.: Freezing by heating in a driven mesoscopic system. Phys. Rev. Lett. **84**, 1240–1243 (2000)
17. Helbing, D., Farkas, I.J., Molnar, P., Vicsek, T.: Simulation of pedestrian crowds in normal and evacuation situations. In: Schreckenberg, M., Sharma, S.D. (eds.) Pedestrian and Evacuation Dynamics, pp. 21–58. Springer, Berlin (2002)

18. Henderson, L.F.: The statistics of crowd fluids. Nature **229**, 381–383 (1971)
19. Hoskin, K.J., Spearpoint, M.: Crowd characteristics and egress at stadia. In: Shields, T.J. (ed.) Human Behaviour in Fire. Intersience, London (2004)
20. Kirchner, A.: Modellierung und statistische Physik biologischer und sozialer Systeme. PhD thesis (in german) (2002)
21. Kirchner, A., Schadschneider, A.: Simulation of evacuation processes using a bionics-inspired cellular automaton model for pedestrian dynamics. Phys. A Stat. Mech. Appl. **312**(1–2), 260–276 (2002)
22. Lachapelle, A., Wolfram, M.-T.: On a mean field game approach modeling congestion and aversion in pedestrian crowds. Transp. Res. B Methodol. **45**, 1572–1589 (2011)
23. Lasry, J.-M., Lions, P.-L.: Mean field games. Jpn. J. Math. **2**(1), 229–260 (2007)
24. Leal-Taixé, L., Pons-Moll, G., Rosenhahn, B.: Everybody needs somebody: modeling social and grouping behavior on a linear programming multiple people tracker. In: IEEE International Conference on Computer Vision Workshops (ICCVW). 1st Workshop on Modeling, Simulation and Visual Analysis of Large Crowds (2011)
25. Matsumoto, M., Nishimura, T.: Mersenne twister: a 623-dimensionally equidistributed uniform pseudo-random number generator. ACM Trans. Model. Comput. Simul. **8**(1), 3–30 (1998)
26. Maury, B., Roudneff-Chupin, A., Santambrogio, F.: A macroscopic crowd motion model of the gradient-flow type. M3AS **20**(10), 1787–1821 (2010)
27. Mehran, R., Oyama, A., Shah, M.: Abnormal crowd behavior detection using social force model. In: CVPR, pp. 935–942 (2009)
28. Muramatsu, M., Nagatani, T.: Jamming transition in two-dimensional pedestrian traffic. Physica A **275**, 281–291 (2000)
29. Painter, K.: Continuous models for cell migration in tissues andï£¡applications to cell sorting via differential chemotaxis. Bull. Math. Biol. **71**, 1117–1147 (2009)
30. Pellegrini, S., Ess, A., Schindler, K., van Gool, L.: You'll never walk alone: modeling social behavior for multi-target tracking. In: International Conference on Computer Vision (2009)
31. Piccoli, B., Tosin, A.: Pedestrian flows in bounded domains with obstacles. Contin. Mech. Thermodyn. **21**(2), 85–107 (2009)
32. Schadschneider, A.: Simulation of evacuation processes using a bionics-inspired cellular automaton model for pedestrian dynamics. In: Fukui, M., Sugiyama, Y., Schreckenberg, M., Wolf, D. (eds.) Traffic and Granular Flow '01, pp. 1160–1168. Springer, Berlin/Heidelberg/Nagoya (2002)
33. Schadschneider, A., Klingsch, W., Kluepfel, H., Kretz, T., Rogsch, C., Seyfried, A.: Evacuation dynamics: empirical results, modeling and applications. In: Meyers, R.A. (ed.) Encyclopedia of Complexity and System Science, vol. 3, p. 3142. Springer, New York/London (2009)
34. Seyfried, A., Schadschneider, A.: Fundamental diagram and validation of crowd models. In: Umeo, H., Morishita, S., Nishinari, K., Komatsuzaki, T., Bandini, S. (eds.) Cellular Automata. Volume 5191 of Lecture Notes in Computer Science, pp. 563–566. Springer, Berlin/Heidelberg (2008)
35. Seyfried, A., Steffen, B., Klingsch, W., Lippert, T., Boltes, M.: The fundamental diagram of pedestrian movement revisited – empirical results and modelling. In: Schadschneider, A., Pöschel, T., Kühne, R., Schreckenberg, M., Wolf, D.E. (eds.) Traffic and Granular Flow'05, pp. 305–314. Springer, Berlin/Heidelberg (2007)
36. Simpson, M., Landman, K., Hughes, B.: Diffusing populations: ghosts or folks? Australas. J. Eng. Educ. **15**, 59–67 (2009)
37. Treuille, A., Cooper, S., Popovic, Z.: Continuum crowds. ACM Trans. Graph. **25**, 1160–1168 (2006). Proceedings of SCM SIGGRAPH 2006
38. Weidmann, U.: Transporttechnik der Fussgänger - Transporttechnische Eigenschaften des Fussgängerverkehrs (Literaturstudie). Literature Research 90, Institut füer Verkehrsplanung, Transporttechnik, Strassen- und Eisenbahnbau IVT an der ETH Zürich, March 1993. in German

39. Yamaguchi, K., Berg, A., Ortiz, L., Berg, T.: Who are you with and where are you going? In: IEEE Conference on Computer Vision and Pattern Recognition (CVPR), 2011, pp. 1345–1352, June 2011
40. Yuhaski, S.J., Jr., Smith, J.M.: Modeling circulation systems in buildings using state dependent queueing models. Queueing Syst. Theory Appl. **4**(4), 319–338 (1989)
41. Zhang, J., Klingsch, W., Schadschneider, A., Seyfried, A.: Ordering in bidirectional pedestrian flows and its influence on the fundamental diagram. J. Stat. Mech. Theory Exp. **2012**(02), P02002 (2012)

Chapter 4
Analysis of Crowd Dynamics with Laboratory Experiments

Maik Boltes, Jun Zhang, and Armin Seyfried

Abstract For the proper understanding and modelling of crowd dynamics, reliable empirical data is necessary for analysis and verification. Laboratory experiments give us the opportunity to selectively analyze parameters independently of undesired influences and adjust them to high densities seldom seen in field studies. The setup of the experiments, the extraction of the trajectories of the pedestrians and the analysis of the resulting data are discussed.

Two strategies for the time-efficient automatic collection of accurate pedestrian trajectories from stereo recordings are presented. One strategy uses markers for detection and the other one is based on a perspective depth field. Measurement methods for quantities like density, velocity and specific flow are compared. The fundamental diagrams from trajectories for different experiments are analyzed.

4.1 Introduction

Design of egress routes for buildings and large scale events is one application for models of pedestrian streams [1–5]. Typical questions regarding the capacity of facilities for pedestrians are: Is the width of a door or a corridor large enough to evacuate a certain amount of people in a given time? How long does it take to clear a building? The methods and tools available to evaluate or to dimension these

M. Boltes (✉) • J. Zhang
Jülich Supercomputing Centre, Forschungszentrum Jülich GmbH, 52425 Jülich, Germany
e-mail: m.boltes@fz-juelich.de; ju.zhang@fz-juelich.de

A. Seyfried
Jülich Supercomputing Centre, Forschungszentrum Jülich GmbH, 52425 Jülich, Germany

Computer Simulation for Fire Safety and Pedestrian Traffic, Bergische Universität Wuppertal, Pauluskirchstraße 7, 42285 Wuppertal, Germany
e-mail: a.seyfried@fz-juelich.de

S. Ali et al. (eds.), *Modeling, Simulation and Visual Analysis of Crowds*, The International
Series in Video Computing 11, DOI 10.1007/978-1-4614-8483-7_4,
© Springer Science+Business Media New York 2013

facilities can be categorized in legal regulations, handbooks [6–8] and computer simulations [9–12]. Legal regulations are based on prescriptive methods using static rules which depend on the occupancy of the building. Two examples of rules are the minimal width of doors in dependence of the number of people in the room and the maximal length of an escape route. But static rules cannot consider the dynamics of an evacuation process and methods with a higher fidelity are necessary to resolve the development over time. Handbooks for example, use macroscopic models of streams and provide a description of the evacuation process in time and space. These models forecast when and where congestions will occur. However, macroscopic models give only a coarse description of pedestrian flow by modeling the streams as an entity with quasi-constant size and density. Microscopic models instead describe the individual movement of all pedestrians and are thus able to resolve the dynamics on an individual scale. Regardless of the fidelity of the models the basic aim should be to describe quantitatively the transport properties of pedestrian streams.

Quantities describing the performance of facilities and the transport properties of pedestrian streams are borrowed from the physics of fluids. The flow J gives the throughput at a certain cross section and is defined as the number of persons N passing the cross section in the time interval Δt. To express the degree of congestion the density $\rho = N/A$, which is the number of persons in an area A, is used. The velocity v specifies how long it takes to reach the exit of the building. These quantities depend on each other and the empirical relation between them ($J(\rho)$ or $v(\rho)$) is commonly called the fundamental diagram. This relation is a basis for quantifying the transport properties of driven systems in stationary states. One major problem is that the empirical data base is rudimentary and inconsistent. Even basic questions and relations are queried and discussed contradictory in the literature [13]. E.g., how the maximal possible flow through a bottleneck depends on the width of the opening [14, 15], at which density jams occur [16] or whether the fundamental diagrams for uni- or bidirectional streams differ or not [17].

Several research groups started in the last decade performing experiments to address these as well as other questions. Before introducing our contribution we give an overview of the activities by countries and by citing the recent articles without any claim to comprehensiveness. These are in China [18–25], in France the Pedigree project [26–29], in the Netherlands [30, 31], in Japan [32–35] and in Germany [36, 37]. We want to note that these experiments and field studies cover in their differentness the complexity and diversity of pedestrian traffic.

Since the year 2005 we performed in cooperation with the universities of Cologne and Wuppertal more than 300 experiments to improve the database for model validation. We focused on experiments under well controlled laboratory conditions due to several reasons. Pedestrians are subject to a lot of influences which cannot be controlled in field studies. To study the influence of one single parameter it is helpful to control external influences (light, sound, ground, etc.), boundaries and initial conditions. Even under laboratory conditions this is difficult to achieve, see e.g. [31]. Moreover, the variability allows a survey of a parameter range e.g. for the bottleneck width or length, or the density inside a corridor. Another reason for performing controlled experiments is the interest in high densities, which are seldom observable in field studies. With increasing number of participants we improved the methods to extract the individual walking paths automatically from video footage.

For the design of the experiments we started with simple geometries (corridors, bottlenecks, etc.) and flow types (unidirectional, bidirectional). Then we extended the experiments to consider more complex scenarios with bends, stairs and merging streams. The experiments are designed to ensure that the influence of one parameter on the quantity of interest can be studied. The data can then be used to develop and to systematically validate mathematical models.

4.2 Experiments and Data Capturing

4.2.1 Experiment Overview

Performing experiments under laboratory conditions gives the opportunity to analyze parameters of interest under well defined constant conditions. For self-initiated experiments the location and the structure of the test persons (e.g. culture, fitness, age, gender, size) can be determined. For this reason, series of pedestrian experiments have been designed and carried out since 2005.

The first experiments were designed to study the relation between density and velocity of single file movement [38]. With the same experimental setup the influence of motivation and culture [39] was analyzed.

After the one dimensional experiments, we have made experiments on plane ground focusing on bottlenecks and corridors. 99 runs with up to 250 test persons have been performed [40].

For the development of an evacuation assistant in the project Hermes [41] the experimental database was expanded by experiments on different types of stairs, straight corridors, corners and T-junctions. 170 runs with up to 350 people have been made in artificial environments, and additionally inside the facilities of the stadium for which the evacuation system has been developed.

In Sect. 4.3 some details of these experiments will be described.

For capturing the experiments by video recordings the cameras can be chosen appropriate to the coverage area and ceiling height. Overhead recordings perpendicular to the floor allow a view without occlusion for a range of body heights, so that an individual detection and tracking without estimation of the persons' route can be performed. To get constant lighting conditions the experiments have primarily been made indoor with uniform artificial light. The extraction process of the route of every person is outlined in the following sections.

4.2.2 Trajectory Extraction

The goal of the extraction process are trajectories, $p_i(t)$, $i \in [1,N]$, as exact as possible for all N persons at any time t in particular in crowded scenes. For this reason markers are used to improve the robustness of the automatic extraction wherever applicable.

For the same reason of exactness all automatic results are inspected by humans, who are able to correct the trajectories directly within our software. Under laboratory conditions almost no error occurs, but in real facilities like stairways in a stadium the number of incorrect detections increases. Problems faced include the varying lighting conditions and the distance of up to 13 m to the head of the persons. The manual visual inspection of the automatically extracted trajectories is one reason for the off-line detection from video recordings. The more important reason for the downstream extraction from recordings is the possible later visual analysis of effects evaluated from the trajectory data. For these advantages we acknowledge the problems of huge recording space requirements and privacy protection.

Before extracting metric information the video has to be calibrated. For the correction of the lens distortion a model of a pinhole camera with distortion is adopted (considering radial and tangential distortion up to fourth order). The perpendicular view and cameras with quadratic pixel allow an easy specification of a pixel to meter ratio considering the perspective view. For more information we refer to [42].

4.2.2.1 Detection with Markers

For the description of the detection process we restrict to one experiment performed in the project Hermes [41]. Details of the artificial setup of this experiment, a merging flow through a T-junction with a corridor width of 2.4 m and 303 test persons, can be found in Sect. 4.3.3.

The experiments have been recorded with two synchronized stereo cameras of type Bumblebee XB3 (manufactured by Point Grey). For the T-junction experiment they were mounted $a = 784$ cm above the floor with the viewing direction perpendicular to the floor.

The overlapping field of view of the stereo system is $\alpha = 64°$ at the average head distance of about 6 m from the cameras. Thus all pedestrians with the discovered height range can be seen without occlusion at any time. The cameras have a resolution of $1,280 \times 960$ pixels and a frame rate of 16 frames per second, $\Delta t = 1/16$.

The marker has a simple structure to detect it from distances up to 13 m. All pedestrians wear a white bandana with a centered black dot of 4 cm diameter (Fig. 4.1).

Fig. 4.1 (Color online) *Left*: Rectified image of one stereo camera of a T-junction experiment. *Right*: Color coded disparity map restricted to the distance of the upper body part (570–735 cm). The background is greyed out

The recognition of the marker is done by detecting directed isolines of the same brightness and subsequent analysis of the size, shape, arrangement and orientation of approximating ellipses.

Perspective Depth Field

For the detection with markers the perspective depth field, which can be obtained from a stereo camera, is only used for background subtraction and the measurement of the head distance to the camera.

This depth field h contains the distance to every pixel of the camera and is inversely proportional to the disparity map, $d \propto 1/h$, which describes the pixel offset of both camera views of the stereo camera for every pixel. The disparity map is calculated with the semi-global block matching algorithm [43] implemented in the computer vision library OpenCV [44]. The mask size for the matched blocks has been set to 11 to get a smooth depth field. The drawback of this is a blurry depth field where the shape of objects is less sharp. Figure 4.1 shows on the right an overlay of the disparity map on the left picture. The disparity map is restricted to the distance of the upper body part color coded from red to blue according to the distance of 570–735 cm to the camera. The greyed out part indicates the background, which determination is described below.

Background Subtraction

A prior background subtraction reduces the number of false positive detections. For the background subtraction the camera distance h is used directly without generating a rectified depth field. No laborious plan-view statistics is needed because of the perpendicular view of the stereo recordings. Pixels with the coordinate $u = (u_x, u_y) \in \mathbb{R}^2$ at frame f are part of the background and thus are ignored in the detection process, if

$$h_{bg}(u) - h(u, f) < 40 \text{cm}. \tag{4.1}$$

The perspective depth field of the background h_{bg} is captured once with the scene deserted or is set to a cautiously adapted maximum distance during all frames

$$h_{bg}(u) \approx \max_{f}(h(u, f)). \tag{4.2}$$

The distance threshold of 40 cm cannot be increased for robustness, since people near walls would be eliminated because of the omitted plan-view statistic. This effect can already be seen in Fig. 4.1 at the walls forming the junction. Small regions of missing values inside h_{bg} are interpolated linearly within the row. Small regions inside the segmented foreground are added to the background to erase noise and regions that cannot be occupied by a person.

3D Position

To calculate the position in 3D real world a coordinate transformation from pixel positions to real positions and an inverse perspective transformation have to be performed. Therefor the distance to the camera is needed. For planar experiments, where we used monocular cameras, the color of a part of the marker corresponds to a height range [42]. But because we also made experiments at stairs within the Hermes project this approach cannot be used anymore. The distance or the height of the pedestrian for planar experiments, $h'(u_i) = a - h(u_i)$, respectively is now set according to the disparity of the center pixel of the black dot, $u_i \in \mathbb{R}^2, i \in [1, N]$. Only with the pedestrians' height the correct position on the plane ground can be calculated. Because of the perspective distortion the maximum error without considering the height would be [42]

$$\frac{\max_i(h'(u_i)) - \min_i(h'(u_i))}{2} \tan \frac{\alpha^\circ}{2} \approx 16 \text{cm}. \tag{4.3}$$

The height is oscillating according to the step frequency. The height is minimal at the position where body shifting moves from one to the other leg [45].

The distribution of the maximum height of each pedestrian matches exactly the distribution of the height obtained by the evaluation of questionnaires handed out to them.

4.2.2.2   Detection Without Markers

We are also developing a markerless detection, which facilitates field studies and the easier realization of moderated experiments in real environments. Studying the influence of group structures (e.g. football fan or visitor of a classical concert, sober or drunken persons) realistically can only be accomplished there. With moderated experiments in real environments, where we use independent gatherings which happen anyway (e.g. works meetings) we can increase the amount of trajectory data with less time and effort. Besides this, the markerless detection can further improve the robustness of the marker based detection described below.

Related Work

Techniques for the detection without markers for single pedestrians in crowds using monocular cameras are not as robust as techniques using stereo cameras. Publications like [46–51] all report a false detection rate of more than 10 %. Typically the decrease of the false detection rate induces the increase of false positive detections. For our purpose this means a lot of manual work, because we need nearly no error to get reliable data for further analysis.

Detection techniques, such as [52] or [53] for stereo cameras, depend on accurate segmentation of foreground objects from the background. For dense crowds such as in our experiments these methods would not be applicable or would only detect groups of people. Other techniques use motion patterns of human beings [54, 55] like periodic leg movement or additionally take skin colour [56] into account or use a face detector [57], which is only applicable from side view because of visibility. The side view is also needed by Hou and Pang [58], because they assign a region to one person, if the region has the same distance to the camera. In our experiments the video recordings were done overhead to avoid occlusions, because we want to know the detailed position of every person at any time also in crowded scenes, so that often no extremity or skin is visible. This perpendicular view also disengages us from a decelerating plan-view statistic like in [59].

Algorithms using motion like [60] cannot be adopted, because in our experiments dense situations and thus stagnant flow often occur.

In [61] a method for people tracking in dense situations with multiple cameras is suggested. The combined data from several views is used to calculate the height and thus the position of peoples' head.

A robust detection and tracking algorithm also for crowded scenes is described in [62]. The detection process is based on a clustering procedure using bio-metrically inspired constraints.

In [63] the detection is done by searching for clusters inside the point cloud of a depth map, which are arranged in a sphere that is proportional in size to human heads. For people walking close together van Oosterhout et al. obtain a precision of 0.97 taking tracking into account, which is nearly as high as our precision.

Detection Process

For the following example we use the same recordings as for the detection with markers, but ignore the markers.

To detect markerless pedestrians in dense crowds we utilize the depth field described in Sect. 4.2.2.1. The identification of the people is done only using the shape of the top part of their body especially the head and shoulders. If we want to identify people only by their shape, the background subtraction has to be performed previously as described before, because of the possible occurrence of similar formed objects.

Directed Isolines and Approximating Ellipses

To extract features identifying pedestrians inside the depth field, directed isolines of the same distance to the camera are used. The step size of the iso-value scanning the depth field is 5 cm. Beforehand the depth field is adapted by replacing values covered by the background mask with the furthest value which belongs to the foreground.

In Fig. 4.2 red isolines surround regions further away. They can be ignored. Green isolines encircle regions which are nearer to the camera. To improve the visibility of the isolines the color coding of the disparity map is replaced by a grey scale one.

The remaining isolines enclosing a minimum and maximum of pixels and with a small ratio between the length of the isoline and the enclosed area (to eliminate isolines with big dents) are approximated by ellipses. The ellipses allow an easier access to the global shape. Small ellipses with a large eccentricity are discarded. Large ellipses can have a bigger eccentricity to enclose multiple people. The used values for the selection of isolines and ellipses have been chosen heuristically considering the number of pixel covered by one person.

Ellipses Pyramid

By scanning the depth field downwards in steps of 5 cm a pyramid of ellipses for every person is build up (Fig. 4.3). Thus for every new depth level an ellipse is assigned to that pyramid, whose center is inside the new ellipse. If no pyramid fits, the new ellipse starts a new pyramid. For multiple ellipses on the same level covering the identical pyramid that one is chosen, whose center is the closest to the nearest pyramid center. New ellipses can cover multiple pyramids, if the pyramids have already a substantial number of ellipses from previous depth levels. Otherwise the

Fig. 4.2 (Color online) *Left*: Isolines of the same distance to the camera at intervals of 5 cm (colored according to the orientation) drawn on the grey scaled perspective depth field. *Right*: Pyramidal grouped ellipses identifying pedestrians. *Green ellipses* correspond to the isoline nearest to the camera, following by *red* and *blue*

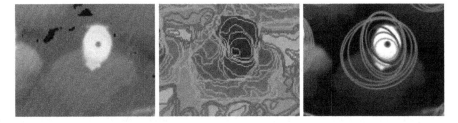

Fig. 4.3 Zoomed view of a part of Figs. 4.1 and 4.2: disparity, isolines and pyramid of approximating ellipses

small pyramids are rejected or, if there are only small ones, the pyramid with the closest center is chosen.

At the end we neglect pyramids with a

- Small number of ellipses,
- Large second ellipse (corresponding to the head),
- Small third ellipse (to reject e.g. lifted arms),
- Small last ellipse (corresponding to the body).

We prefer a strict deletion to avoid false detections since it is not necessary to detect a person every frame for tracking. The values again are chosen heuristically taking peoples' shape into account. The first ellipse is not analyzed, because the location varies too much, especially because of the different depth the heads are detected the

Fig. 4.4 *Left*: Tracked people by markers and their smooth path during the last second. *Right*: Tracked people without markers and their more unsettled path during the last second

first time. This is also the reason why the center of the second ellipse represents a pedestrian and thus is tracked. The resulting pyramids are shown on the right of Fig. 4.2. The ellipses are colored according to their level in the pyramid. The topmost ellipse has a green, the second a red and the third a blue color. The latter ellipses can cover more than one person.

After analyzing the complete video recording trajectories are rejected, which do not cross the whole test area, or have only few frames where the supposed pedestrian is identified. The right of Fig. 4.4 shows the tracked people and their unsettled path during the last second in comparison to the left picture showing the path of the people detected by marker.

4.2.2.3 Tracking

For tracking of detected pedestrians the robust pyramidal iterative Lucas Kanade feature tracker [64] is used. This tracker extends the Lucas Kanade method for calculating the optical flow by introducing successive Gaussian pyramids of successive images B and propagating tracking results from a low resolution level to the next higher level as an initial guess.

The tracker searches with sub-pixel accuracy in regions of same size in recursive Gaussian pyramids

$$B^L(u_x, u_y) = \sum_{i=-1}^{1} \sum_{j=-1}^{1} 2^{-(2+|i|+|j|)} B^{L-1}(2u_x - i, 2u_y - j) \quad (4.4)$$

with $B^0 = B$ and $u^L = 2^{-L}u$, starting in B^{L-1} with $u^{L-1} = 2u^L$.

The size of the tracked region is adapted to the head size, which can be deduced from the persons' height or the distance to the camera. The number of pyramidal level L is set to four. The size of the last level B^{L-1} is 50 % bigger than the head length of around 21 cm, so that the region of the first level B^0 has the size m of the marker used: $m = 1.5 \cdot 21\,\text{cm}/2^3 \approx 4\,\text{cm}$.

If the result of a tracked head is not feasible we extrapolate the next position and for the detection with marker adjust the position to the center of the marker considering the pixel brightness.

Merging Trajectories

The precision of the trajectories considering markers is sufficiently high, so that overlapping camera views allow a combination of trajectory sets. Since we use stereo recordings we also synchronize all cameras. Thus we do not need a temporal adjustment and have only to minimize the distance of the pedestrians' positions for every frame. By using the method of least squares we find the associated trajectory pairs for each overlapping view and minimize the average distance error between the two point clouds by adapting the extrinsic parameter set for one view. The method for finding the least square solution and searching for the optimal translation vector and rotation matrix is based on the single value decomposition [65]. After applying the transformation the average error describing the distance between a corresponding trajectory pair is $1 \pm 0.4\,\text{cm}$ and the maximum error $5 \pm 2.6\,\text{cm}$ for all planar experiments measured in 2D. For experiments in 3D space, we get an average error of $3 \pm 0.5\,\text{cm}$ and a maximum error of $8 \pm 2.6\,\text{cm}$. The maximum error appears towards the boundary of the camera's view. To reduce the influence of this error we interpolate linearly between two matching trajectories, so that the trajectory to the respective boundary has less influence on the resulting behavior of the combined trajectory. All trajectories in the Sect. 4.3 are combined results of two camera views.

4.2.3 Results of Trajectory Extraction

Quantitative results of the detection with and without markers are shown in Table 4.1. The one misleading match of the detection using markers traces back to an area similar to the easy structured marker. The strict heuristic of the markerless detection deletes three correct trajectories.

The combination of both methods has no false detection. It uses the detection with markers and accepts the detection only, if the center of the dot is inside the second highest ellipse of the pyramid of that pedestrian.

Figure 4.4 shows the way of each detected pedestrian during the last second (left for the detection with and right without using markers). Even if the detection result is good for both methods, the quality of the method using markers results

Table 4.1 Comparative results of the detection methods

Method	Detected	False positive	False negative	avg. acc. [m/s^2]
With markers	304	1	0	1.2 ± 0.7
Markerless	300	0	3	6.5 ± 6.8
Combination	303	0	0	1.2 ± 0.7

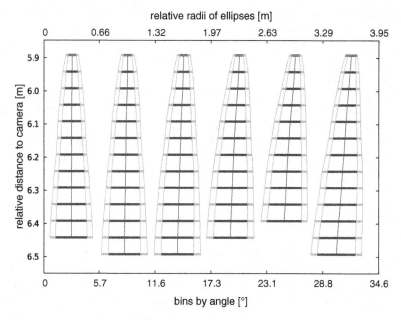

Fig. 4.5 Side view of mean pyramids binning by the angle to the optical axis. The *horizontal lines representing the ellipses* are surrounded by *two lines* at the major and minor radius. The *middle vertical lines show the pyramid axis* concatenating the center of the ellipses stack

in smoother trajectories. The smoothness is important for the analysis, e.g. for the microscopic velocity calculation. To quantify the smoothness the average microscopic acceleration can be used (see Table 4.1). For successive detected points, X_j, along a trajectory, p_i, the microscopic acceleration at position X_j is

$$\|(X_{j+1} - X_j) - (X_j - X_{j-1})\| / (\Delta t)^2. \tag{4.5}$$

Figure 4.5 shows six bins of all pyramids detected during the experiment according to their angle to the optical axis from side view. The lowest ellipses are neglected, if they cover more than one person. All pyramids have been adjusted to one camera distance before a mean pyramid for each bin was calculated. The middle lines concatenate the center of the ellipses of successive height level and depict the pyramid axes. The outer two lines surround the ellipses stack at the major and minor radius. One can see that the center line of the mean pyramids is tilted according to the angle, but less than one would expect from the perspective view, because the based isolines of the latter ellipses have to cover the higher ones due to the not performed plan view statistic. The radii increase slightly for larger angle. The size of the radii can be read from the top diagram axis.

4.3   Experimental Setup of the Studied Experiments

Four pedestrian experiments will be introduced in this section. Experiments of uni-, bidirectional and merging flow were performed in hall 2 of the fairground Düsseldorf (Germany) in 2009 with up to 400 students (age: 25 ± 5.7 years old, height: 1.76 ± 0.09 m, free velocity: 1.55 ± 0.18 m/s), whereas the experiment of bottleneck flow were performed in 2006 in the wardroom of the "Bergische Kaserne Düsseldorf" with a test group that was comprised of soldiers. For the Hermes project an announcement was put in universities to recruit the people who would like to participate in the experiments with 50 Euro per day. Consequently, most of the participants were students. No selection of participants was undertaken. During experiments, they were asked to move normally and purposely but without pushing. They were free to speak and the sound was also recorded.

4.3.1   Unidirectional Flow

Figure 4.6 shows the sketch and a snapshot of the experiment to study unidirectional flow in a straight corridor with open boundaries. Three corridor widths (1.8, 2.4 and 3.0 m) were chosen and 28 runs were carried out in all. To regulate the pedestrian density in the corridor, the widths of the entrance b_{entrance} and the exit b_{exit} were changed in each run. Figure 4.7 shows the trajectories extracted from two runs of the experiment. For more details of the setup and data capturing we refer to [42,66].

4.3.2   Bidirectional Flow

Figure 4.8 shows the sketch and a snapshot of the experimental setup to study bidirectional flow in a straight corridor. 22 runs were performed with corridor width of 3.0 *m* and 3.6 m respectively. The width of the left entrance b_l and the right entrance b_r were changed in each run to regulate the density inside the corridor and the ratio of the opposing streams. To vary the degree of disorder, the participants get different instructions on which exit to choose. Three different types of setting were adopted among these experiments:

$b_l = b_r$, *choose exits freely*: In this type of experiment, the widths of entrance b_l and b_r were set as the same. The test persons were not given any instruction about

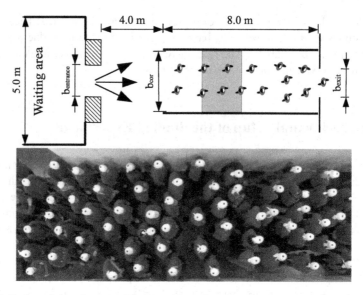

Fig. 4.6 Setup and snapshot of unidirectional flow experiment. Note that the gray area in the sketch shows the location of the measurement area in the analysis

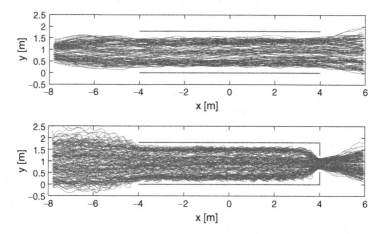

Fig. 4.7 Trajectories of pedestrians in two runs of unidirectional flow experiment. The distances between the edge of trajectories and the boundary are not the same in various density situations

exit chosen and they can choose the exit freely. Five runs of experiment were carried out with this conditions in a corridor with width of 3.6 m.

$b_l = b_r$, *specify exits in advance*: Again the same width b_l and b_r were chosen in the experiments. But the instruction to the test persons at the beginning of the experiments were changed. The participants were asked to choose an exit at the end of the corridor according to a number given to them in advance. The persons with

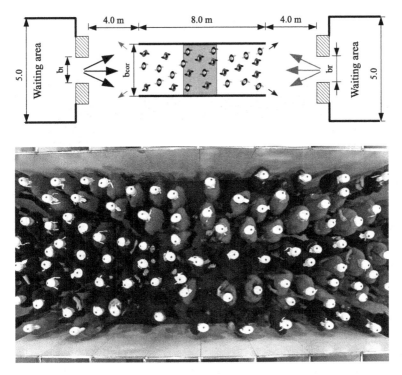

Fig. 4.8 Setup and snapshot of bidirectional flow experiment. Note that the gray area in the sketch shows the locations of the measurement areas in the analysis. Lane formation can be observed from the snapshot

odd numbers should choose the left exit in the end, while ones with even numbers were asked to choose the exit in the right side.

$b_l \neq b_r$, *specify exits in advance*: In this case the widths of entrances b_l and b_r were different and the participants were instructed to choose an exit at the end of the corridor according to a number as the last experiment.

Figure 4.9 shows the pedestrian paths for two runs of the experiment with and without instruction. More details of the experiment setup can be found in [17].

4.3.3   Merging Flow

Figure 4.10 shows the sketch of the experiment to study merging flow in a T-junction and a snapshot. Two pedestrian streams from the opposite sides of T-shaped corridor join together and form a single stream. In these experiments, all three parts of the corridor were set to the same width b_{cor}. 12 runs were carried out with b_{cor} of 2.4 and 3.0 m respectively. To regulate the pedestrian density, the width of the entrance was changed in each run. The left and right entrances were always set as the same

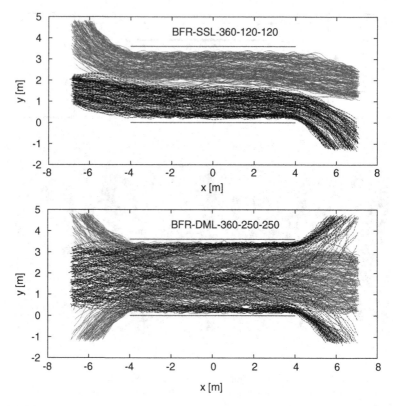

Fig. 4.9 Trajectories of pedestrians in two runs of bidirectional flow experiment. Different types of lane formation, stable separated lanes and dynamical multi-lanes, can be observed

width $b_{entrance}$. In this way, we guarantee the symmetry of the two branches of the stream. The number of pedestrians in the left and right branch of the T-junction was approximately equal. The number was set to a value that the overall duration of all experiments is similar and is long enough to assure a stationary state. Figure 4.11 shows the paths of the pedestrians in T-junction at low and high density conditions.

4.3.4 Bottleneck Flow

Figure 4.12 shows a still and a sketch of the setup. The experimental setup allows to analyze the influence of the bottleneck width and length (Fig. 4.13). In one experiment the width b was varied (from 0.9 to 2.5 m) at fixed corridor length. In

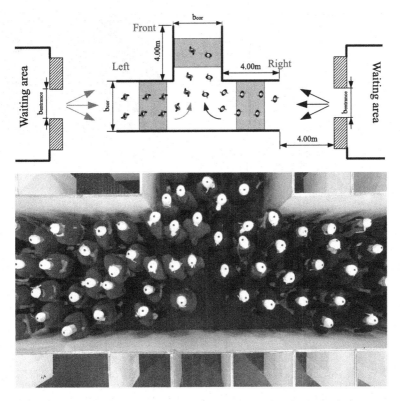

Fig. 4.10 Setup and a snapshot of merging flow experiment in a T-junction. Note that the gray areas in the sketch shows the locations of the measurement areas in the analysis

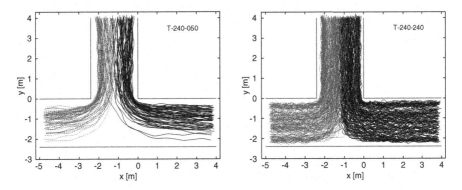

Fig. 4.11 Trajectories of pedestrians in two runs of merging flow experiment in a T-junction. The utilizations of the space near the merging area are different under various densities

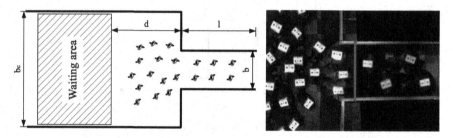

Fig. 4.12 Setup and snapshot of bottleneck flow experiment. For this experiment the length l and width b of the bottleneck were changed in each run to analyze the flow through it

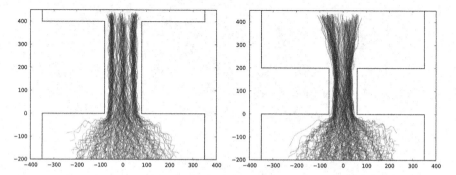

Fig. 4.13 Trajectories of pedestrians in two runs of bottleneck flow experiment. The influence of bottleneck length and width on pedestrian movement as well as lane formation can be observed

the other experiment the corridor length l was changed (0.06, 2.0, 4.0 m) while the width was fixed at $b = 1.2$ m. For more details of the experimental setup and data capturing we refer to [40, 67].

4.4 Measurement Methods

The trajectories gained by the methods described in Sect. 4.2.2 are the basis to measure the fundamental diagram $J(\rho)$ or $v(\rho)$. In this section, we study how the way of analyzing the trajectories influence the relation between v, ρ and J. To determine these variables one can choose time-averaged density, velocity or flow. However, various definition of methods may arouse different measurement errors. Here we use four different definitions including macroscopic and microscopic methods to measure observable like flow, velocity and density.

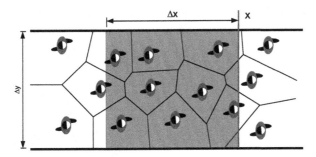

Fig. 4.14 Illustration of different measurement methods. *Method A* is a kind of local measurement at cross-section with position x averaged over a time interval Δt, while *Methods B–D* measure at a certain time and average the results over space Δx. Note that for *Method D*, the Voronoi diagrams are generated according to the spatial distributions of pedestrians at a certain time

- Method A

  For *MethodA*, a reference location x in the corridor is taken (as shown in Fig. 4.14) and mean values of flow and velocity are calculated over time Δt. We refer to this average by $\langle\rangle_{\Delta t}$. The time t_i and the velocity v_i of each pedestrian passing x can be determined directly. Thus, the flow over time $\langle J\rangle_{\Delta t}$ and the time mean velocity $\langle v\rangle_{\Delta t}$ can be calculated as

$$\langle J\rangle_{\Delta t} = \frac{N_{\Delta t}}{t_{N_{\Delta t}} - t_{1_{\Delta t}}} \quad \text{and} \quad \langle v\rangle_{\Delta t} = \frac{1}{N_{\Delta t}} \sum_{i=1}^{N_{\Delta t}} v_i(t) \tag{4.6}$$

  where $N_{\Delta t}$ is the number of persons passing the location x during the time interval Δt. $t_{1_{\Delta t}}$ and $t_{N_{\Delta t}}$ are the times when the first and last pedestrians pass the location in Δt. They could be different from Δt. The time mean velocity $\langle v\rangle_{\Delta t}$ is defined as the mean value of the instantaneous velocities $v_i(t)$ of the $N_{\Delta t}$ persons according to equation (4.7). We calculate $v_i(t)$ by use of the displacement of pedestrian i in a small time interval $\Delta t'$ (Note that $\Delta t \gg \Delta t'$)around t:

$$v_i(t) = \frac{\|\mathbf{x_i}(t + \Delta t'/2) - \mathbf{x_i}(t - \Delta t'/2)\|}{\Delta t'} \tag{4.7}$$

- Method B

  The second method measures the mean value of velocity and density over space and time. The spatial mean velocity and density are calculated by taking a segment with length Δx in the corridor as the measurement area. The velocity $\langle v\rangle_i$ of each person is defined as the length Δx of the measurement area divided by the time he or she needs to cross the area (see Eq. (4.8)),

$$\langle v\rangle_i = \frac{\Delta x}{t_{i,\text{out}} - t_{i,\text{in}}} \tag{4.8}$$

where $t_{i,\text{in}}$ and $t_{i,\text{out}}$ are the times a person i enters and exits the measurement area, respectively. The density ρ_i for each person is calculated with equation (4.9):

$$\langle\rho\rangle_i = \frac{1}{t_{i,\text{out}} - t_{i,\text{in}}} \cdot \int_{t_{i,\text{in}}}^{t_{i,\text{out}}} \frac{N'(t)}{\Delta x \cdot \Delta y} dt \qquad (4.9)$$

b_{cor} is the width of the measurement area while $N'(t)$ is the number of person in this area at a time t.

- Method C

    With the third measurement method, let's call it classical method, the density $\langle\rho\rangle_{\Delta x}$ is defined as the number of pedestrians divided by the area of the measurement section:

$$\langle\rho\rangle_{\Delta x} = \frac{N}{\Delta x \cdot \Delta y} \qquad (4.10)$$

The spatial mean velocity is the average of the instantaneous velocities $v_i(t)$ for all pedestrians in the measurement area at time t:

$$\langle v\rangle_{\Delta x} = \frac{1}{N} \sum_{i=1}^{N} v_i(t) \qquad (4.11)$$

- Method D

    This method is based on Voronoi diagrams [68] which are a special kind of decomposition of a metric space determined by distances to a specified set of objects in the space. To each such object one associates a corresponding Voronoi cell. The distance from the set of all points in the Voronoi cell to the given object is not greater than their distance to the other objects. At any time the positions of the pedestrians can be represented as a set of objects, from which the Voronoi diagrams (see Fig. 4.14) are generated. The cell area, A_i, can be thought as the personal space belonging to each pedestrian i. Then, the density and velocity distribution of the space ρ_{xy} and v_{xy} are defined as

$$\rho_{xy} = 1/A_i \quad and \quad v_{xy} = v_i(t) \qquad \text{if } (x,y) \in A_i \qquad (4.12)$$

where $v_i(t)$ is the instantaneous velocity of each person, see Eq. (4.7). The Voronoi density and velocity for the measurement area is then defined as [69]

$$\langle\rho\rangle_v = \frac{\iint \rho_{xy} dx dy}{\Delta x \cdot \Delta y} \qquad (4.13)$$

$$\langle v\rangle_v = \frac{\iint v_{xy} dx dy}{\Delta x \cdot \Delta y} \qquad (4.14)$$

4.5   Results of Analysis

4.5.1   Effects of Measurement Methods

To analyze the effect of measurement methods, we calculate the fundamental diagram from unidirectional experiments with corridor width $b_{cor} = 1.8$ m. For *Method A* we choose the time interval $\Delta t = 10$ s, $\Delta t' = 0.625$ s (corresponding to ten frames) and the measurement position at $x = 0$ (see Fig. 4.7). For the other three methods a rectangle with a length of 2 m from $x = -2$ m to $x = 0$ and a width of the corridor is chosen as the measurement area. We calculate the densities and velocities each frame with a frame rate of 16 fps. All data presented below are obtained from some set of trajectories. To determine the fundamental diagram only data at the stationary state, which were selected manually by analyzing the time series of density (see Fig. 4.15), were considered. For *Method D* we use one frame per second to decrease the number of data points and to represent the data more clearly.

Figure 4.16 shows the relationship between the density and flow obtained from different methods. Using *Method A* the flow and mean velocity can be obtained directly. To get the relationship between density and flow, the equation

$$\rho = \langle J \rangle_{\Delta t} / (\langle v \rangle_{\Delta t} \cdot b_{cor})$$
(4.15)

was adopted to calculate the density. For the *Method B, C* and *D* the mean density and velocity can be obtained directly since they are mean values over space. There

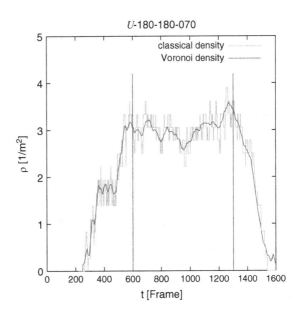

Fig. 4.15 Selection of stationary state from time series of density. The begin and the end of the stationary state are selected manually and represented by two *vertical lines*

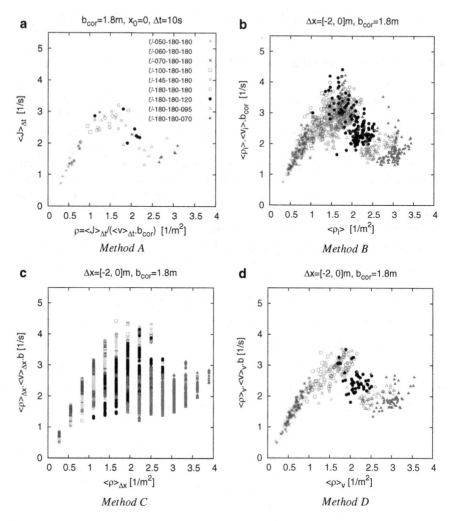

Fig. 4.16 The relationships between density and flow measured at the same set of trajectories but with different methods. The density in (**a**) is calculated indirectly using $\rho = J/(b \cdot \Delta x)$, while the flows in (**b**), (**c**) and (**d**) are obtained by adopting the equation $J = \rho vb$. The legends in (**b**), (**c**) and (**d**) are the same as in (**a**)

exists a similar trend of the fundamental diagram obtained using the different methods. The pedestrian flow shows small fluctuations at low densities and high fluctuations at high densities. The fluctuations for *Method A* and *D* are smaller than that for other methods. However, there is a major difference between the results. The fundamental diagrams obtained using *Method A* and *C* are smooth, while that obtained with *Method B* and *D* show a clear discontinuity at a density of about $2\,\mathrm{m}^{-2}$. The average over a time interval of *Method A* and the large scatter of *Method C* blur this discontinuity. *Method D* can reduce the density and velocity scatter [69].

Fig. 4.17 Comparison of the fundamental diagrams of unidirectional flow in corridors for different widths

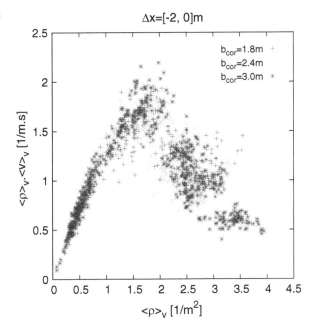

The reduced fluctuation of *Method D* is combined with a good resolution in time and space, which reveal a phenomenon that is not observable with *Method A* and *C*. Consequently, we mainly use the Voronoi method to analyze these experiments in the following part.

4.5.2 Comparison of Fundamental Diagrams for Various Flows

Figure 4.17 shows the relationship between density specific flow obtained from Voronoi method. The fundamental diagrams of unidirectional pedestrian flow in the same type of corridor but with three different widths are compared. It can be seen that they agree well with each other. The specific flow in the corridors is independent on the corridor width. At about $\rho = 2.0\,\mathrm{m}^{-2}$, the specific flow reaches the maximum value which is named the capacity of a facility. This result is in conformance with Hankin's findings [70]. He found that above a certain minimum of about 4 ft (about 1.22 m) the maximum flow in subways is directly proportional to the width of the corridor. In the range of densities reached in the experiment, our results seem to support the specific flow concept that the specific flow $J_s = J/b$ is independent on the width of the facility.

Further, we compare the fundamental diagram of uni- and bidirectional flows in Fig. 4.18. At densities of $\rho < 1.0\,\mathrm{m}^{-2}$, no significant difference exists. For $\rho >$

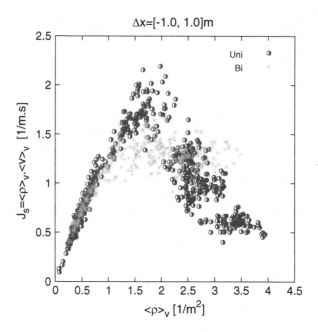

$\Delta x=[-1.0, 1.0]m$

Fig. 4.18 Comparison of the fundamental diagrams of unidirectional flow and bidirectional flow

$1.0\,m^{-2}$, however, the difference between the uni- and bidirectional flow becomes more pronounced and a qualitative difference can be observed. In the bidirectional case a plateau is formed starting at a density $\rho \approx 1.0\,m^{-2}$ where the flow becomes almost independent of the density. Such plateaus are typical for systems which contain 'defects' which limit the flow and have been observed e.g. on bidirectional ant trails [71] where they are a consequence of the interaction of the ants. In our experiments the defects are conflicts of persons moving in the opposite direction. These conflicts only happen between two persons but the reduction of the velocity influences the following people. One of the remarkable things is that the data of the unidirectional flow for $\rho > 2.0\,m^{-2}$ are obtained by slide change of the experiment setup. To reach densities $\rho > 2.0\,m^{-2}$ for unidirectional experiment, a bottleneck at the end of the corridor is builded. This may limit the comparability of fundamental diagrams for $\rho > 2.0\,m^{-2}$.

With the Voronoi method the measurement area could be chosen smaller than the pedestrians. We calculate the Voronoi density, velocity and specific flow over small regions ($\Delta x \times \Delta y = 10 \times 10\,cm$) for each frame and average them over the stationary state separately. Then the profile of density, velocity and specific flow over the experimental area are obtained (see Fig. 4.19). These profiles provide new insights into the spatial characteristics and sensitivity of the quantities to other potential factors. The density profile shows conspicuous high densities at the corner of the T-junction, indicating critical spots under crowded conditions. The region with the highest density is located at a small triangle area, where the left and right branches

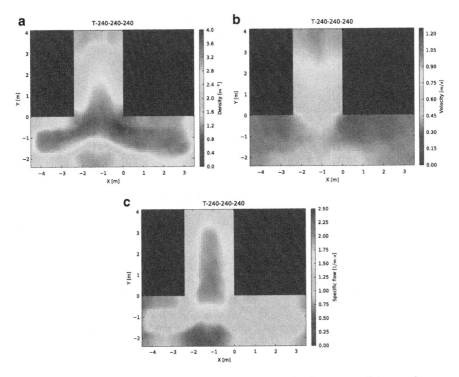

Fig. 4.19 The profiles of density, velocity and flow in T-junction for one run of the experiments. (**a**) Density profile. (**b**) Velocity profile. (**c**) Specific flow profile

merge. Moreover the density profile shows obvious boundary effects. Except for the merging area at the corner, the densities in the middle of the corridors are significantly higher than near the boundaries. The spatial variation of the velocity is different. Boundary effect does not occur for the velocity and the profile is independent from the corridor width especially in the exit corridor. But the velocity becomes larger after the merging of the streams and increases persistently along the movement direction. The specific flow profile shows that the highest flow occurs at the center of the exit corridor. The region of highest flow protrudes from the exit corridor into the area where the two branches start to merge. This indicates that the merging process in front of the exit corridor leads to a flow restriction. Causes for the restriction of the flow must be located outside the region of highest flow. These profiles demonstrate that density and velocity measurements are sensitive to the size and location of the measurement area. For the comparison of measurements (e.g. for model validation or calibration) it is necessary to specify precisely the size and position of the measurement area.

In Fig. 4.20, we compare the fundamental diagrams of merging flow in T-junction with corridor width $b_{cor} = 2.4$ m. The data assigned with 'T-left' and 'T-right' are measured in the areas where the streams prepare to merge, while the data assigned

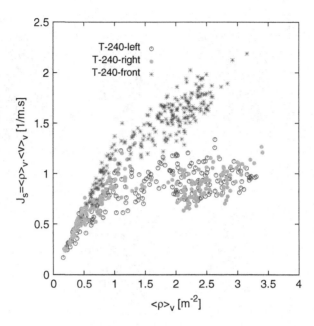

Fig. 4.20 Comparison of the fundamental diagrams of merging flow in different parts of a T-junction

with 'T-front' are measured in the region where the streams have already merged. The locations of the measurement areas are illustrated in Fig. 4.10. For ease of comparison, we choose these measurement areas with the same size ($4.8\,\mathrm{m}^{-2}$). One finds that the fundamental diagrams of the left and right branches match well. That means, the right or left turning of the stream dose not have influence on the fundamental diagram. However, for densities $\rho > 0.5\,\mathrm{m}^{-2}$ the velocities in the 'right' and 'left' part of the T-junction (T-left and T-right) are significantly lower than the velocities measured after the merging of the streams (T-front). This discrepancy becomes more distinct in the relation between density and specific flow. In the main stream (T-front), the specific flow increases with the density ρ till $2.5\,\mathrm{m}^{-2}$. While in the branches, the specific flow nearly remains constant for density ρ between 1.5 and $3.5\,\mathrm{m}^{-2}$. Thus, there seems no unique fundamental diagram which describes the relation between velocity and density for the complete system. For this difference, we can only offer assumptions regarding the causes. One is based on behavior of pedestrians. Congestion occurs at the end of the branches, where the region of maximum density appears. Pedestrians stand in a jam in front of the merging and could not perceive where the congestion disperse or whether the jam lasts after the merging. In such situation, it is questionable whether an urge or a push will lead to a benefit. Thus an optimal usage of the available space becomes unimportant. Otherwise, the situation totally changes if the location of dissolution becomes apparent. Then a certain urge or an optimal usage of the available space

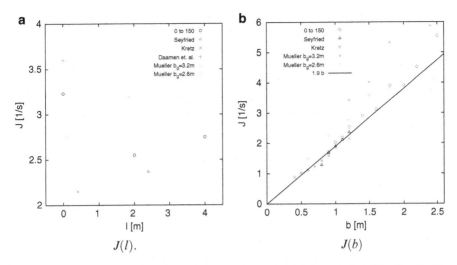

Fig. 4.21 (a) Variation of the flow J with bottleneck length l and (b) variation of the flow J with bottleneck width b

makes sense and could lead to a benefit. They will move in a relatively active way. That's maybe the reason why the velocities after merging are higher than that in front of merging at the same density. Whether this explanation is plausible could be answered by a comparison of these data with experimental data at a corner without the merging.

In Fig. 4.21, the flow from bottleneck experiment is compared with previous measurements using *Method A*. The black line in Fig. 4.21b represents a constant specific flow of $1.9\,(ms)^{-1}$. The difference between the flow at $l = 0.06$ and $l = 2.0, 4.0\,m$ is $\Delta J \simeq 0.5\,s^{-1}$. The data points of Müller's experiments [72] lie significantly above the black line. The Müller experimental setup features a large initial density of around $6\,m^{-2}$ and an extremely short corridor. The discrepancy between the Müller data and the empirical $J = 1.9b$ line is roughly $\Delta J \simeq 0.5\,s^{-1}$. This difference can be accounted to the short corridor length, but may also be due to the higher initial density in the Müller experiment.

4.6   Conclusion

We have performed a large series of experiments to selectively analyze parameters independent of undesired influences.

Different strategies with and without markers for collecting precise trajectories out of overhead video recordings of these experiments and for future field studies have been discussed. Using the perspective depth field directly from stereo recordings is a fast method for the perpendicular view. The automatic extraction of all trajectories which we need especially to verify microscopic models and to analyze the movement microscopically has small error.

To obtain smoother trajectories without markers a subsequent smoothing could be performed or the axes of the pyramidal ellipses stack could be taken into account to get more stable points along the pedestrians' routes.

Experiments of uni- and bidirectional flow in a straight corridor, merging flow in a T-junction and pedestrian flow through bottlenecks have been presented in more detail. Four different measurement methods for obtaining quantities from pedestrian trajectories are adopted and their influences on the fundamental diagram have been investigated. It is found that the results obtained from different methods agree well in the density ranges observed in the experiment. The main differences are the range of the fluctuations and the resolution in time and space. However, the Voronoi method permits the deepest and most precise insight into the temporal progress and spatial distribution of the velocity, density and flow.

Some selected analysis results are presented. It is shown that fundamental diagrams for the same type of facility but different corridor widths agree well and can be unified in one diagram for the specific flow. From the comparison of the fundamental diagrams between straight corridor and T-junction, it is indicated that the fundamental diagrams for different facilities are not comparable. Besides, the measurement of density and velocity strongly depends on the size and location of the measurement area, which can be observed form the profiles of the density, velocity and specific flow measured with the Voronoi method. The influence of the length and width of a bottleneck on the flow is shown and compared with previous studies.

Acknowledgements This study was performed within the project funded by the German Research Foundation (DFG) KL 1873/1-1 and SE 1789/1-1 and the project Hermes funded by the Federal Ministry of Education and Research (BMBF) Program on "Research for Civil Security – Protecting and Saving Human Life".

References

1. Schreckenberg, M., Scharma, S.D. (eds.): Pedestrian and Evacuation Dynamics. Springer, Berlin/Heidelberg (2002)
2. Galea, E.R. (ed.): Pedestrian and Evacuation Dynamics. CMS, London (2003)
3. Klingsch, W.W.F., Rogsch, C., Schadschneider, A., Schreckenberg, M. (eds.): Pedestrian and Evacuation Dynamics. Springer Berlin/Heidelberg (2010). http://www.springer.com/math/applications/book/978-3-642-04503-5
4. Peacock, R.D., Kuligowski, E.D., Averill, J.D. (eds.): Pedestrian and Evacuation Dynamics. Springer, Berlin/Heidelberg (2011). doi:10.1007/978-1-4419-9725-8
5. Grayson, S., Inter Science Communications Limited, Babrauskas, V., Building Research Establishment, UK, National Fire Protection Association, USA, National Institute for Standards & Technology, Building, Fire Research Laboratory, USA, Society of Fire Protection Engineers, USA, SP Technical Institute of Sweden: Interflam 2007: Proceedings of the Eleventh International Conference, no. Bd. 1. Interscience Communications (2007). http://books.google.de/books?id=0vszPwAACAAJ

6. Predtechenskii, V.M., Milinskii, A.I.: Planning for Foot Traffic Flow in Buildings. Amerind Publishing, New Dehli (1978). Translation of Proekttirovanie Zhdanii s Uchetom Organizatsii Dvizheniya Lyuddskikh Potokov, Stroiizdat Publishers, Moscow, 1969
7. Nelson, H.E., Mowrer, F.W.: In: DiNenno, P.J. (ed.) SFPE Handbook of Fire Protection Engineering, 3rd edn., chap. 14, pp. 367–380. National Fire Protection Association, Quincy (2002)
8. Weidmann, U.: Transporttechnik der Fussgänger. Tech. Rep. Schriftenreihe des IVT Nr. 90, Institut für Verkehrsplanung, Transporttechnik, Strassen- und Eisenbahnbau, ETH Zürich, ETH Zürich (1993). Zweite, ergänzte Auflage
9. Thompson, P.A., Marchant, E.W.: Fire Saf. J. 24(2), 131 (1995). doi:10.1016/0379-7112(95)00019-P, http://www.sciencedirect.com/science/article/pii/037971129500019P
10. TraffGo HT GmbH: Handbuch PedGo 2, PedGo Editor 2 (2005). http://www.evacuation-simulation.com
11. Kretz, T., Hengst, S., Vortisch, P.: In: Sarvi, M. (ed.) International Symposium of Transport Simulation (ISTS08). Monash University, Melbourne (2008). http://arxiv.org/abs/0805.1788
12. Hostikka, S., Korhonen, T., Paloposki, T., Rinne, T., Matikainen, K., Heliövaara, S.: Development and validation of fds+evac for evacuation simulations. Tech. Rep. VTT Technical Research Centre of Finland (2008)
13. Schadschneider, A., Klingsch, W., Kluepfel, H., Kretz, T., Rogsch, C., Seyfried, A.: Encyclopedia of Complexity and System Science, vol. 5, chap. Evacuation Dynamics: Empirical Results, Modeling and Applications, pp. 3142–3176. Springer, Berlin/Heidelberg (2009)
14. Seyfried, A., Passon, O., Steffen, B., Boltes, M., Rupprecht, T., Klingsch, W.: Transp. Sci. 43(3), 395 (2009). doi:10.1287/trsc.1090.0263
15. Liddle, J., Seyfried, A., Steffen, B., Klingsch, W., Rupprecht, T., Winkens, A., Boltes, M.: Microscopic insights into pedestrian motion through a bottleneck, resolving spatial and temporal variations. (2011). http://arxiv.org/abs/1105.1532
16. Seyfried, A., Portz, A., Schadschneider, A.: In: Bandini, S., Manzoni, S., Umeo, H., Vizzari, G. (eds.) Cellular Automata, 9th International Conference on Cellular Automata for Reseach and Industry, ACRI, Ascoli Piceno, September 2010. Lecture Notes in Computer Science, vol. 6350, pp. 496–505. Springer, Berlin/Heidelberg (2010). doi:10.1007/978-3-642-15979-4_53
17. Zhang, J., Klingsch, W., Schadschneider, A., Seyfried, A.: J. Stat. Mech. Theory Exp. 2012(2), P02002 (2012). http://stacks.iop.org/1742-5468/2012/i=02/a=P02002
18. Wong, S.C., Leung, W.L., Chan, S.H., Lam, W.H.K., Yung, N.H., Liu, C.Y., Zhang, P.: J. Transp. Eng. 136(3), 234 (2010). doi:10.1061/(ASCE)TE.1943-5436.0000086
19. Liu, X., Song, W., Zhang, J.: Phys. A Stat. Mech. Appl. 388(13), 2717 (2009). doi:10.1016/j.physa.2009.03.017
20. Fang, Z., Song, W., Zhang, J., Wu, H.: Phys. A Stat. Mech. Appl. 389, 815 (2010). doi:10.1016/j.physa.2009.10.019
21. Ma, J., Song, W.G., Liao, G.X.: Chin. Phys. B 19(12), 128901 (2010). doi:10.1088/1674-1056/19/12/128901
22. Lam, W.H.K., Lee, J.Y.S., Cheung, C.Y.: Transportation 29, 169 (2002). doi:10.1023/A:1014226416702
23. Lee, J.Y.S., Lam, W.H.K.: Transp. Res. Rec. 1982, 122 (2006). doi:10.3141/1982-17
24. Ren-Yong, G., Wong, S.C., Yin-Hua, X., Hai-Jun, H., Lam, W.H.K., Keechoo, C.: Chin. Phys. Lett. 29(6), 068901 (2012). http://stacks.iop.org/0256-307X/29/i=6/a=068901
25. LI Xiang, D.L.Y.: Chin. Phys. Lett. 29(9), 98902 (2012). doi:10.1088/0256-307X/29/9/098902, http://cpl.iphy.ac.cn/EN/abstract/article_50369.shtml
26. Pedigree project: http://www.math.univ-toulouse.fr/pedigree/
27. Moussaïd, M., Helbing, D., Garnier, S., Johansson, A., Combe, M., Theraulaz, G.: Proc. R. Soc. B 276(1668), 2755 (2009). doi:10.1098/rspb.2009.0405
28. Jelić, A., Appert-Rolland, C., Lemercier, S., Pettré, J.: Properties of pedestrians walking in line: Stepping behavior Phys. Rev. E 86, 046111 (2012)

29. Moussaid, M., Guillot, E.G., Moreau, M., Fehrenbach, J., Chabiron, O., Lemercier, S., Pettre, J., Appert-Rolland, C., Degond, P., Theraulaz, G.: PLoS Computat. Biol. **8**, 1002442 (2012). http://hal.archives-ouvertes.fr/hal-00716032. Article published in PLoS Computational biology. Freely available here: http://www.ploscompbiol.org/article/info %3Adoi%2F10.1371%2Fjournal.pcbi.1002442 LPT-ORSAY 12–75 LPT-ORSAY 12–75
30. Hoogendoorn, S.P., Daamen, W.: Transp. Sci. **39**(2), 147 (2005). doi:10.1287/trsc.1040.0102
31. Daamen, W., Hoogendoorn, S.: Procedia Eng. **3**, 53 (2010). doi:10.1016/j.proeng.2010.07.007, http://www.sciencedirect.com/science/article/pii/S1877705810004765. First International Conference on Evacuation Modeling and Management
32. Yanagisawa, D., Kimura, A., Tomoeda, A., Ryosuke, N., Suma, Y., Ohtsuka, K., Nishinari, K.: Phys. Rev. E **80**, 036110 (2009)
33. Yanagisawa, D., Tomoeda, A., Nishinari, K.: Physical Review E **85**, 016111+ (2012). doi:10.1103/PhysRevE.85.016111, http://dx.doi.org/10.1103/PhysRevE.85.016111
34. Nagai, R., Fukamachi, M., Nagatani, T.: Physica A **367**, 449 (2006). doi:10.1016/j.physa.2005.11.031, http://dx.doi.org/10.1016/j.physa.2005.11.031
35. Isobe, M., Adachi, T., Nagatani, T.: Physica A **336**, 638 (2004). doi:10.1016/j.physa.2004.01.043
36. Kretz, T., Grünebohm, A., Schreckenberg, M.: J. Stat. Mech. **10**, P10014 (2006). doi:10.1088/1742-5468/2006/10/P10014
37. Plaue, M., Chen, M., Bärwolff, G., Schwandt, H.: In: Stilla, U., Rottensteiner, F., Mayer, H., Jutzi, B., Butenuth, M. (eds.) Photogrammetric Image Analysis. Lecture Notes in Computer Science, vol. 6952, pp. 285–296. Springer, Berlin/Heidelberg (2011). doi:10.1007/978-3-642-24393-6_24, http://dx.doi.org/10.1007/978-3-642-24393-6_24
38. Seyfried, A., Steffen, B., Klingsch, W., Boltes, M.: J. Stat. Mech. Theory Exp. P10002 (2005). doi:10.1088/1742-5468/2005/10/P10002
39. Chattaraj, U., Seyfried, A., Chakroborty, P.: Adv. Complex Syst. **12**(3), 393 (2009). doi:10.1142/S0219525909002209
40. Seyfried, A., Boltes, M., Kähler, J., Klingsch, W., Portz, A., Rupprecht, T., Schadschneider, A., Steffen, B., Winkens, A.: In: Klingsch, W.W.F., et al. (eds.) Pedestrian and Evacuation Dynamics, pp. 145–156. Springer Berlin/Heidelberg (2010). doi:10.1007/978-3-642-04504-2_11, http://arxiv.org/abs/0810.1945
41. Holl, S., Seyfried, A.: inSiDe **7**(1), 60 (2009). http://inside.hlrs.de/pdfs/inSiDE_spring2009.pdf
42. Boltes, M., Seyfried, A., Steffen, B., Schadschneider, A.: In: Klingsch, W.W.F., et al. (eds.) Pedestrian and Evacuation Dynamics, pp. 43–54. doi:10.1007/978-3-642-04504-2-3, http://www.springer.com/math/applications/book/978-3-642-04503-5
43. Hirschmüller, H.: IEEE Trans. Pattern Anal. Mach. Intell. **30**, 328 (2008). doi:http://doi.ieeecomputersociety.org/10.1109/TPAMI.2007.1166
44. Bradski, G.R., Pisarevsky, V.: IEEE Computer Society Conference on Computer Vision and Pattern Recognition, vol. 2, pp. 2796+ (2000). doi:http://doi.ieeecomputersociety.org/10.1109/CVPR.2000.854964, http://dx.doi.org/10.1109/CVPR.2000.854964
45. Boltes, M., Seyfried, A., Steffen, B., Schadschneider, A.: In: Peacock, R.D., et al. (eds.) Pedestrian and Evacuation Dynamics, pp. 751–754. Springer, Berlin/Heidelberg (2011). doi:10.1007/978-1-4419-9725-8
46. Leibe, B., Seemann, E., Schiele, B.: In: Computer Vision and Pattern Recognition, San Diego, pp. 878–885 (2005)
47. Brostow, G.J., Cipolla, R.: In: IEEE Computer Society Conference on Computer Vision and Pattern Recognition, New York, vol. 1, pp. 594–601. IEEE Computer Society, Washington, DC (2006). doi:10.1109/CVPR.2006.320, http://dl.acm.org/citation.cfm?id=1153170.1153531
48. Tu, P., Sebastian, T., Doretto, G., Krahnstoever, N., Rittscher, J., Yu, T.: In: European Conference on Computer Vision, Marseille, pp. 691–704 (2008)
49. Cheriyadat, A.M., Bhaduri, B.L., Radke, R.J.: In: Computer Vision and Pattern Recognition Workshops, Anchorage, vol. 0, pp. 1–8. IEEE Computer Society, Los Alamitos (2008). doi:http://doi.ieeecomputersociety.org/10.1109/CVPRW.2008.4562983

50. Saadat, S., Teknomo, K., Fernandez, P.: Fire Technol. 1–18 (2010). doi:10.1007/s10694-010-0174-9, http://dx.doi.org/10.1007/s10694-010-0174-9
51. Johansson, A., Helbing, D., Al-Abideen, H.Z., Al-Bosta, S.: Adv. Complex Syst. **11**, 4 (2008). http://www.citebase.org/abstract?id=oai:arXiv.org:0810.4590
52. Rittscher, J., Tu, P., Krahnstoever, N.: In: Schmid, C., Soatto, S., Tomasi, C. (eds.) IEEE Computer Society Conference on Computer Vision and Pattern Recognition, San Diego, vol. 2, pp. 486–493. IEEE Computer Society (2005)
53. Hu, W., Zhou, X., Tan, T., Lou, J., Maybank, S.: IEEE Trans. Pattern Anal. Mach. Intell. **28**(4), 663 (2006)
54. Cutler, R., Davis, L.: IEEE Trans. Pattern Anal. Mach. Intell. **22**(8), 781 (2000)
55. Pai, C., Tyan, H., Liang, Y., Liao, H., Chen, S.: Pattern Recognit. **37**(5), 1025 (2004). doi:10.1016/j.patcog.2003.10.005
56. Darrell, T., Gordon, G., Harville, M., Woodfill, J.: Int. J. Comput. Vis. **37**(2), 175 (2000)
57. Muñoz Salinas, R., Aguirre, E., García-Silvente, M.: Image Vis. Comput. **25**, 995 (2007). doi:10.1016/j.imavis.2006.07.012, http://portal.acm.org/citation.cfm?id=1235891.1236069
58. Hou, Y.L., Pang, G.K.H.: In: Real, P., Díaz-Pernil, D., Molina-Abril, H., Berciano, A., Kropatsch, W.G. (eds.) International Conference on Computer Analysis of Images and Patterns, Seville. Lecture Notes in Computer Science, vol. 6854, pp. 93–101. Springer (2011)
59. Harville, M.: Image Vis. Comput. **22**(2), 127 (2004). doi:10.1016/j.imavis.2003.07.009, http://www.sciencedirect.com/science/article/B6V09-49VCBKN-1/2/1663779d9256ba7d3a54634abe9e23c3
60. García-Martín, Á., Hauptmann, A., Martinez, J.M.: In: 8th IEEE International Conference on Advanced Video and Signal-Based Surveillance (AVSS), Klagenfurt, p. 5 (2011)
61. Eshel, R., Moses, Y.: IEEE Computer Society Conference on Computer Vision and Pattern Recognition, Anchorage, vol. 1 (2008). doi:http://doi.ieeecomputersociety.org/10.1109/CVPR.2008.4587539
62. Kelly, P., O'Connor, N.E., Smeaton, A.F.: Image Vis. Comput. **27**(10), 1445 (2009). doi:10.1016/j.imavis.2008.04.006
63. van Oosterhout, T., Bakkes, S., Kröse, B.: In: International Conference on Computer Vision Theory and Applications (VISAPP), Vilamoura, pp. 620–625 (2011)
64. Bouguet, J.Y.: OpenCV Documents (1999)
65. Arun, K.S., Huang, T.S., Blostein, S.D.: IEEE Trans. Pattern Anal. Mach. Intell. **9**, 698 (1987). doi:10.1109/TPAMI.1987.4767965, http://portal.acm.org/citation.cfm?id=28809.28821
66. Zhang, J., Klingsch, W., Schadschneider, A., Seyfried, A.: J. Stat. Mech. Theory Exp. (2011). http://arxiv.org/abs/1102.4766. ArXiv:1102.4766
67. Liddle, J., Seyfried, A., Klingsch, W., Rupprecht, T., Schadschneider, A., Winkens, A.: In: Traffic and Granular Flow (2009). http://arxiv.org/abs/0911.4350. ArXiv:0911.4350
68. Voronoi, G.M.: Journal für die reine und angewandte Mathematik **133**, 198 (1908)
69. Steffen, B., Seyfried, A.: Physica A **389**(9), 1902 (2010). doi:10.1016/j.physa.2009.12.015
70. Hankin, B.D., Wright, R.A.: Oper. Res. Q. **9**, 81 (1958)
71. John, A., Schadschneider, A., Chowdhury, D., Nishinari, K.: J. Theor. Biol. **231**, 279 (2004). http://arxiv.org/abs/cond-mat/0409458
72. Müller, K.: Zur Gestaltung und Bemessung von Fluchtwegen für die Evakuierung von Personen aus Bauwerken auf der Grundlage von Modellversuchen. Dissertation, Technische Hochschule Magdeburg (1981)

Chapter 5
Modeling a Crowd of Groups: Multidisciplinary and Methodological Challenges

Stefania Bandini and Giuseppe Vizzari

Abstract The main aim of the chapter is to introduce a recent and current trend of research in the modeling, simulation and visual analysis of crowds: the study of the impact of groups on the overall crowd dynamics, and its implications of the aforementioned research activities as well as their outcomes. In most situations, in fact, a crowd of pedestrians is more than a simple set of individuals, each interpreting the presence of the others in a uniform way, trying to preserve a certain distance from the nearest person. A crowd is rather a composite assembly of individuals, some of which are bound by different types of ties, not only representing the presence of other pedestrians as a repulsive force, influencing their attitude towards the movement in the environment. Current models for the simulation of crowds of pedestrians have just started to analyze this phenomenon, and we still lack a complete understanding of the implications of not considering it, either in a real simulation project supporting decision making activities of designers or planners, or in the analysis and automatic extraction of information, for instance from video footage of events or crowded environments.

5.1 Introduction

The modeling and simulation of pedestrians and crowds is a consolidated and successful application of research results in the more general area of computer simulation of complex systems. Results of different approaches from researchers in

S. Bandini (✉) • G. Vizzari
Department of Computer Science, Systems and Communication, Complex Systems and Artificial Intelligence (CSAI) Research Center, University of Milan – Bicocca,

Viale Sarca 336/14, 20126 Milano, Italy
e-mail: bandini@disco.unimib.it; vizzari@disco.unimib.it

S. Ali et al. (eds.), *Modeling, Simulation and Visual Analysis of Crowds*, The International Series in Video Computing 11, DOI 10.1007/978-1-4614-8483-7__5,

different disciplines, from physics and applied mathematics, to computer science, often influenced by (and sometimes in collaboration with) anthropological, psychological, sociological studies and the humanities in general, can be found in the literature. The level of maturity of these approaches was in some cases sufficient to lead to the design and development of commercial software packages, often offering interesting and advanced functionalities for the end user (e.g. CAD integration, CAD-like functionalities, advanced visualization and analysis tools) in addition to a simulation engine.[1] Nonetheless, as testified by a recent survey of the field [46] and by a report commissioned by the Cabinet Office [11], there is still room for innovations in models improving their performances both in terms of *effectiveness* in modeling pedestrians and crowd phenomena, in terms of *expressiveness* of the models (i.e. simplifying the modeling activity or introducing the possibility of representing phenomena that were still not modelled by existing approaches), in terms of *efficiency* of the simulation tools.

The unit of analysis of most the above mentioned approaches is represented by the single pedestrian, and this is also testified by the fact that most approaches claim to be agent–based (even though the different approaches do not necessarily employ agent models and/or technologies [4]): most pedestrians and crowd simulation approaches can be legitimately and safely classified in the category of micro-simulation. The analyses on simulation results are generally focused on aggregated data and emerging macro phenomena, such as average total travel times for specific classes of pedestrians, average or peak pedestrian densities in various points of the simulated environment. Generally, models do not include any meso-level [15] concept besides the aforementioned idea of class of pedestrians, i.e. a set of agents sharing behavioral rules and goals but otherwise completely unrelated.

The main aim of the chapter is to highlight a recent and current trend of research in modeling, simulation and visual analysis of crowds: the study of the impact of groups on the overall crowd dynamics, and its implications of the aforementioned research activities as well as their outcomes. In most situations, in fact, a crowd of pedestrians is more than a simple set of individuals, each interpreting the presence of the others in a uniform way, that is, trying to preserve a certain distance from the nearest person. A crowd is rather a composite assembly of individuals, some of which are bound by different types of ties, not only representing the presence of other pedestrians as a repulsive force, influencing their attitude towards the movement in the environment. Current models for the simulation of crowds of pedestrians have just started to analyze this phenomenon, and we still lack a complete understanding of the implications of not considering it, either in a real simulation project supporting decision making activities of designers or planners, or in the analysis and automatic extraction of information, for instance from video footage of events or crowded environments. The chapter, in addition to describing

---

[1] See http://www.evacmod.net/?q=node/5 for a large although not necessarily complete list of pedestrian simulation models and tools.

the current state of the art on this topic and discussing some recent results and open challenges, will also attempt to clarify that this research enterprize requires a coordinated and multidisciplinary effort along both the lines, that is, the synthesis and the analysis of crowds comprising groups of pedestrians. The chapter aims first of all at suggesting a relevant selection of literature in area of cultural studies and anthropology that represents a useful framework suggesting approaches both the modeling and to the analysis of the relevant phenomena. In particular, the work on proxemics by Edward T. Hall encompasses both a justification of the tendency of individuals to keep a certain distance from the others, unless they belong to a specific set of special persons (e.g. friends, relatives, loved ones).

The chapter then provides a thorough review of the current landscape in models for the simulation of crowd pedestrians that, to a different extent and in a more or less comprehensive way, extend the basic pedestrian models to provide an account for this sort of meso-level concept that is the notion of *group*.

A detailed example of one of these models, explicitly considering groups as first class abstractions and modelled entities that, on one hand, influence individuals in their decisions and, on the other, can represent an observed entity per se whose status depends on the individuals it is composed of. The model will be described in its principles and mechanisms, and it will be exemplified in an experimental situation and in a real world scenario. Some traditional approaches will be employed to evaluate, measure and describe the results of the simulated pedestrians' behaviors and some hypotheses will be done on the possibility to observe, characterize and possibly validate phenomena that are specifically related to groups and therefore not yet considered in the previous researches: new observations and metrics, in fact, must be defined and analyzed to root the results of the new models on actual data.

Finally, and as a consequence of this last consideration, this chapter presents a reflection on the current landscape in the area of crowd analysis, proposing a multidisciplinary research direction in which the efforts on crowd analysis and synthesis can benefit from the mutual challenges, methods and results.

5.2 Influential Contributions on Pedestrians and Crowd Modeling

In this section of the chapter we want to briefly introduce some selected contributions from disciplines in the humanities, and especially anthropology and sociology, that to a certain extent influenced previous pedestrian and crowd modeling approaches or that represented a useful resource in the development of innovative models considering groups as a first class abstraction influencing the overall system dynamics. In addition, we also report here some works describing reports on relevant observations that represent useful evidences and potentially also data to support innovative modeling and simulation efforts.

5.2.1 Proxemics

The term *proxemics* was first introduced by Edward T. Hall with respect to the study
of a set of measurable distances between people as they interact [23]. In his studies,
Hall carried out analysis of different situations in order to recognize behavioral
patterns. These patterns are based on people's culture as they appear at different
levels of awareness. In [22] Hall proposed a system for the notation of proxemic
behavior in order to collect data and information on people sharing a common space.
Hall defined proxemic behavior and four types of perceived distances: *intimate
distance* for embracing, touching or whispering; *personal distance* for interactions
among good friends or family members; *social distance* for interactions among
acquaintances; *public distance* used for public speaking. Perceived distances depend
on some additional elements which characterize relationships and interactions
between people: posture and sex identifiers, sociofugal-sociopetal (SFP) axis,
kinesthetic factor, touching code, visual code, thermal code, olfactory code and
voice loudness.

Proxemic behavior includes different aspects which could be useful and interesting to integrate in crowd and pedestrian dynamics simulation. In particular, the
most significant of these aspects being the existence of two kinds of distance:
physical distance and *perceived* distance. While the first depends on physical
position associated to each person, the latter depends on proxemic behavior based
on culture and social rules.

It must be noted that some recent research effort was aimed at evaluating the
impact of proxemics and cultural differences on the fundamental diagram [12],
a typical way of evaluating both real crowding situations and simulation results.
Moreover, first attempts to explicitly include proxemic considerations not only as
a background element in the motivations a behavioral model is based upon, but
rather as a concrete element of the model itself are present in the most recent
literature [33, 51].

5.2.2 Groups: Contributions from Anthropology

The term *group* appears in very different and varied contexts of the anthropological
literature, both ethnographic and theoretical [16]. The term, per se, is not endowed
with specific characteristics and it is generally accompanied by additional specifications such as "domestic group," "ethnic group," and so on. With reference to
the term in general and common sense usage, anthropology borrows sociological
considerations, defining a group as a set of individuals related by a common project,
a common identity, that can be perceived by the members of the group and by
external observers.

The common element between the different strains of research related to groups
is the topic of social cohesion that, to a certain extent, pervades the works of

researchers since the end of the Nineteenth century. The existence of a group considered as a set of (at least two) individuals does not necessarily imply the presence of a formal organization, even though this characteristic could represent a group classification criterion; other classification criteria are related to the degree of homogeneity/heterogeneity in the group, the mechanisms of recruitment of new members, the presence or absence of common interests towards goods (e.g., territory, domestic herds) or ritual knowledge. Generally groups are aimed at the execution of a plan for the achievement of some final goal. Therefore, groups exist since they carry out specific 'functions'; the latter can be classified into three types: executive, control and expressive functions. The executive aspect deals with the need of a group to successfully adapt to the natural and social environment in which it is set in order to achieve the goals of the group (e.g., the management of resources, the performance of some ritual). The control aspect deals with the enactment of mechanisms (e.g., behavioral norms, recruitment practices, rituals) for the preservation of group characteristics, namely structure and goals. The expressive aspect consists in the ability of the group to gratify on a psychological and emotional level of its own members.

5.2.3   Canetti's Crowd Theory

Elias Canetti's work [10] proposes a classification and an ontological description of the crowd phenomenon; this description represents the result of 40 years of empirical observations and studies from psychological and anthropological viewpoints. Elias Canetti can be considered as belonging to the tradition of social studies that consider the crowd as an entity dominated by uniform moods and feelings. This uniformity, the loss of individuality, however, are not the normal state of a set of pedestrians in an environment, although maybe densely populated.

The normal pedestrian behavior, according to Canetti, is based upon what can be called the *fear to be touched* principle:

> There is nothing man fears more than the touch of the unknown. He wants to see what is reaching towards him, and to be able to recognize or at least classify it.

> All the distance which men place around themselves are dictated by this fear.

The normal situation can however be interrupted by a *discharge*, a particular event, a situation, a specific context in which this principle is not valid anymore, since pedestrians are willing to accept being very close, within touch distance. Canetti provided an extensive categorization of the conditions, situations in which this happens and he also described the features of these situations and of the resulting types of crowds. Finally, Canetti also provides the concept of *crowd crystal*, a particular set of pedestrians that are part of a group willing to preserve its unity, despite crowd dynamics. Canetti's theory (and precisely the fear to be touched principle) is apparently compatible with Hall's proxemics, but it also provides additional concepts that are useful to describe phenomena that take place in several relevant crowding phenomena, especially from the Hajj perspective.

Recent developments aimed at formalizing, embedding and employing Canetti's crowd theory into computer systems (for instance, supporting crowd profiling and modeling) can be found in the literature [2, 3] and they represent a useful contribution to the present work.

5.2.4 Direct Observations

Direct observations, when carried out in a systematic way, represent a fundamental instrument aimed at, on one hand, at highlighting phenomena to be modelled, behavioral tendencies to be included in model mechanisms and, on the other, they also represent a way to acquire quantitative information on some pedestrian and crowd related phenomenon. Data and information deriving from the observation can directly support some specific form of decision by an expert designer or planner, or they can represent a useful element for the calibration of a simulation model.

Two relevant examples of direct observations that report relevant information from the perspective of a modeler trying to capture elements of the behavior of groups of pedestrians are represented by two video-based observational studies [48, 52]: the first paper, presents an analysis of three mixed-use (residential/retail) uncluttered urban environments close to the city centers of Edinburgh and York (essentially in free flow conditions), while the second analyzes an area between the check-in facilities and the security control at the Dresden International Airport. Both studies analyze the effect of the presence of groups, as well as other variables like gender, age and even travel purpose (only for the second observation). Both the observations conclude that members of groups tend to assume a lower speed than individuals, very likely due to the tendency of each group member of adapting his/her own movement to stay close to the other members. Similar considerations are also discussed in [17], where a situation in which authors were expecting groups to be less frequently identified and relevant, that is, an admission test to a programmed number university course. Another recent study in the vein of Hall's proxemics [13] supports the above interpretation and also adds an analysis of the spatial patterns and formations assumed by the group members.

While these works are extremely important in pointing out the fact that the presence of groups can have a noticeable influence on walking behavior of pedestrians, and therefore not considering this aspect can present a problem when making predictions on pedestrians' behaviors, they are not sufficient to actually characterize this impact in general, since they essentially analyze situations in which pedestrians have an almost unconstrained possibility to choose their walking direction and speed due to the low level of density. Moreover, the analyzed groups are in most cases of relatively small size: the analysis carried out in the context of the airport terminal considered that larger groups split into smaller ones and focused on the latter, not considering the potential influence of the larger group on the smaller ones. The real world scenario that will be analyzed in Sect. 5.6 will instead consider the presence of potentially large groups (i.e. 250 members), although it does not consider the presence of comprised smaller groups.

5.3 Pedestrians and Crowd Modeling Approaches

The aim of this section is to provide a compact but as comprehensive as possible overview of the different approaches to the representation and simulation of crowd dynamics: entire workshops and conferences attracting researchers from different disciplines are focused on this topic (see, e.g., the proceedings of the first edition of the International Conference on Pedestrian and Evacuation Dynamics [47] and consider that this event will reach the sixth edition in 2012), therefore we are not pretending to even mention the most significant approaches and model. We will try, instead, to present broad classes identified according to the way pedestrians are represented and managed, and in particular: (i) pedestrians as *particles subject to forces* of attraction/repulsion, (ii) pedestrians as particular *states of cells in a CA*, (iii) pedestrians as *autonomous agents*, situated in an environment.

5.3.1 Particle-Based Approach

A significant number of models and experiences of simulation of pedestrian dynamics are based on an analytical approach, considering pedestrian as particles subject to forces, and representing in this was the various forms of interaction between pedestrian and the environment (and also among pedestrians themselves, in the case of *active walker* models [26]). Forces of attraction lead the pedestrians/particles towards their destinations, whereas forces of repulsion are used to represent the tendency to stay at a distance from other points of the environment. This kind of effect was introduced by a relevant and successful example of this modeling approach, the *social force* model [25]; this approach introduces the notion of social force, representing the tendency of pedestrians to stay at a certain distance one from another; other relevant approaches take inspiration from fluid-dynamic [24] and magnetic forces [39] for the representation of mechanisms governing flows of pedestrians.

While this approach is based on a precise methodology and has provided relevant results, it represents pedestrian as mere particles, whose goals, characteristics and interactions must be represented by means of equations, and it is not simple thus to incorporate heterogeneity and complex pedestrian behaviors in this kind of model. Nonetheless, recent extensions of the basic social force model introduce a contribution to the general laws of motion representing a form of cohesion between members of a group [34,53]: the authors of these works focus on small unstructured groups and they analyze the impact of this modification to the starting model in low to moderate density scenarios.

5.3.2 Cellular Automata Approach

A different approach to crowd modeling is characterized by the adoption of Cellular Automata (CA), with a discrete spatial representation and discrete time-steps, to represent the simulated environment and the entities it comprises. The cellular space includes thus both a representation of the environment and an indication of its state, in terms of occupancy of the sites it is divided into, by static obstacles as well as human beings. Transition rules must be defined in order to specify the evolution of every cell's state; they are based on the concept of neighborhood of a cell, a specific set of cells whose state will be considered in the computation of its transition rule. The transition rule, in this kind of model, generates the illusion of movement, that is mapped to a coordinated change of cells state. To make a simple example, an atomic step of a pedestrian is realized through the change of state of two cells, the first characterized by an *"occupied"* state that becomes *"vacant"*, and an adjacent one that was previously *"vacant"* and that becomes *"occupied"*. This kind of application of CA-based models is essentially based on previous works adopting the same approach for traffic simulation [36].

Local cell interactions are thus the uniform (and only) way to represent the motion of an individual in the space (and the choice of the destination of every movement step). The sequential application of this rule to the whole cell space may bring to emergent effects and collective behaviors. Relevant examples of crowd collective behaviors that were modelled through CAs are the formation of lanes in bidirectional pedestrian flows [7], the resolution of conflicts in multidirectional crossing pedestrian flows [8]. In this kind of example, different states of the cells represent pedestrians moving towards different exits; this particular state activates a particular branch of the transition rule causing the transition of the related pedestrian to the direction associated to that particular state. Additional branches of the transition rule manage conflicts in the movement of pedestrians, for instance through changes of lanes in case of pedestrians that would occupy the same cell coming from opposite directions.

It must be noted, however, that the potential need to represent goal driven behaviors (i.e. the desire to reach a certain position in space) has often led to extend the basic CA model to include features and mechanisms breaking the strictly locality principle. A relevant example of this kind of development is represented by a CA based approach to pedestrian dynamics in evacuation configurations [45]. In this case, the cellular structure of the environment is also characterized by a predefined desirability level, associated to each cell, that, combined with more dynamic effects generated by the passage of other pedestrians, guide the transition of states associated to pedestrians. Recent developments of this approach introduce even more sophisticated behavioral elements for pedestrians, considering the anticipation of the movements of other pedestrians, especially in counter flows scenarios [38].

As for the particle–based approaches, also in CA pedestrians and crowd models the impact of the presence of groups has been recently investigated [44]: once again, group members have a tendency to stay close to each others, but this model also includes the possibility to represent leader and followers roles. The paper, however, does not present a validation against real data or an application in a real–world scenario.

5.3.3 Autonomous Agents Approach

Recent developments in this line of research (e.g. [14, 27]), introduce modifications to the basic CA approach that are so deep that the resulting models can be considered much more similar to agent–based and Multi Agent Systems (MAS) models exploiting a cellular space representing spatial aspects of agents' environment. A MAS is a system made up of a set of autonomous components which interact, for instance according to collaboration or competition schemes, in order to contribute in realizing an overall behavior that could not be generated by single entities by themselves. As previously introduced, MAS models have been successfully applied to the modeling and simulation of several situations characterized by the presence of autonomous entities whose action and interaction determines the evolution of the system, and they are growingly adopted also to model crowds of pedestrians [1, 6, 20, 50]. All these approaches are characterized by the fact that the agents encapsulate some form of behavior inspired by the above described approaches, that is, forms of attractions/repulsion generated by points of interest or reference in the environment but also by other pedestrians.

Some of the agent based approaches to the modeling of pedestrians and crowds were developed with the primary goal of providing a realistic 3D visualization of the simulated dynamics: in this case, the notion of realism includes elements that are considered irrelevant by some of the previous approaches, and it does not necessarily require the models to be validated against data observed in real or experimental situations. The approach described in [35] and in [49] is characterized by a very composite model of pedestrian behavior, including basic reactive behaviors as well as a cognitive control layer; moreover, actions available to agents are not strictly related to their movement, but they also allow forms of direct interaction among pedestrians and interaction with objects situated in the environment. Other approaches in this area (see, e.g., [40]) also define layered architectures including cognitive models for the coordination of composite actions for the manipulation of objects present in the environment. Another relevant agent–based effort described in [41], although adopting the social force model for some internal mechanisms (i.e. local collision avoidance), employs guidance fields to achieve a goal directed agent movement.

A recent effort [42] represents instead an attempt to define a model able to reproduce composite forms of groups related dynamics: this modeling effort, although it represents an interesting investigation of how expressive an agent–based approach to the modeling of pedestrian groups can be, was not validated in a real world scenario.

5.4 GA-PED Model

We will now briefly introduce a model based on simple reactive situated agents based on some fundamental features of CA approaches to pedestrian and crowd modeling and simulation, with specific reference to the representation and management of the simulated environment and pedestrians; in particular, the adopted approach is discrete both in space and in time. The present description of the model is simplified and reduced for sake of space, reporting only a basic description of the elements required to understand its basic mechanisms; an extended version of the model description can be found in [5].

5.4.1 Environment

The environment in which the simulation takes place is a lattice of cells, each representing a portion of the simulated environment and comprising information about its current state, both in terms of physical occupation by an obstacle or by a pedestrian, and in terms of additional information, for instance describing its distance from a reference point or point of interest in the environment and/or its desirability for pedestrians following a certain path in the environment.

The scale of discretization is determined according to the principle of achieving cells in which at most one pedestrian can be present; traditionally the side of a cell is fixed at 40 or 50 cm, respectively determining a maximum density of 4 and 6.5 pedestrian per square meter. The choice of the scale of discretization also influences the length of the simulation turn: the average speed of a pedestrian can be set at about 1.5 m/s (see, e.g., [52]) therefore, assuming that a pedestrian can perform a single movement between a cell and an adjacent one (according to the Von Neumann neighborhood), the duration of a simulation turn is about 0.33 s in case of a 50 cm discretization and 0.27 in case of a finer 40 cm discretization.

Each cell can be either vacant, occupied by an obstacle or by a specific pedestrian. In order to support pedestrian navigation in the environment, each cell is also provided with specific floor fields [45]. In particular, each relevant final or intermediate target for a pedestrian is associated to a floor field, representing a sort of gradient indicating the most direct way towards the associated point of interest (e.g., see Fig. 5.1 in which a simple scenario and the relative floor field representation are shown). The GA-Ped (Group Aware Pedestrian model) model only comprises *static* floor fields, specifying the shortest path to destinations and targets. Interactions between pedestrians, that in other models are described by the use of *dynamic floor fields*, in this modeling approach are managed by the agent interpretation of the perceived situation.

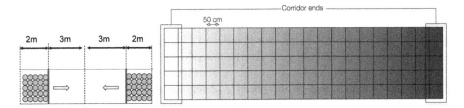

Fig. 5.1 Schematic representation of a simple scenario: a 2.5 by 10 m corridor, with exits on the short ends and 2 sets of 25 pedestrians. The discretization of 50 cm and the floor field directing towards the right end is shown on the *right*

5.4.2 Pedestrians

Pedestrians in the GA-PED model have a limited form of autonomy, meaning that they can choose were to move according to their perception of the environment and their goal, but their action is actually triggered by the simulation engine and they are not thus provided with a thread of control of their own. More precisely, the simulation turn activates every pedestrian once in every turn, adopting a random order in the agent selection: this agent activation strategy, also called *shuffled sequential updating* [30], is characterized by the fact that conflicts between pedestrians are prevented.

Each pedestrian is provided with a simple set of attributes: $pedestrian = \langle pedID, groupID \rangle$ with $pedID$ being an identifier for each pedestrian and $groupID$ (possibly null, in case of individuals) the group the pedestrian belongs to. For the applications presented in this paper, the agents have a single goal in the experimental scenario, but in more complex ones the environment could be endowed with multiple floor fields and the agent could be also characterized by a *schedule*, in terms of a sequence of floor fields and therefore intermediate destinations to be reached.

The behavior of a pedestrian is represented as a flow made up of three stages: *sleep, movement evaluation, movement*. When a new iteration starts each pedestrian is in a sleeping state. The system wakes up each pedestrian once per iteration and, then, the pedestrian passes to a new state of movement evaluation. In this stage, the pedestrian collects all the information necessary to obtain spatial awareness. In particular, every pedestrian has the capability to observe the environment around him, looking for other pedestrians (that could be part of his/her group), walls and other obstacles, according to the Von Neumann neighborhood. The choice of the actual movement destination between the set of potential movements (i.e. non empty cells are not considered) is based on the elaboration of an utility value, called *likability*, representing the desirability of moving into that position given the state of the pedestrian.

Formally, given a pedestrian belonging to a group g and reaching a goal t, the *likability* of a cell c is defined as:

$$li(c,g,t) = w_t \cdot goal(t,c) + w_g \cdot group(g,c) - w_o \cdot obs(c) - w_s \cdot others(g,c) + \varepsilon$$

where the functions *obst* counts the number of obstacles in the Von Neumann neighborhood of a given cell, *goal* returns the value of the floor field associated to the target t in a give cell, *group* and *other* respectively count the number of members and non-members of the group g, ε represents a random value. Group cohesion and floor field are positive components because the pedestrians wish to reach their destinations quickly, while staying close to other group members. On the contrary, the presence of obstacles and other pedestrians have a negative impact as a pedestrian usually tends to avoid them. A random factor is also added to the overall evaluation of the desirability of every cell.

In the usual floor field models, after a deterministic elaboration of the utility of each cell, not comprising thus any random factor, the utilities are translated into the probabilities that the related cell is selected as movement destination. This means that for a pedestrian generally there is a higher probability of moving towards his/her destination and according to proxemic considerations, but there is also the probability, for instance, to move away from his/her goal or to move far from his/her group. In this work, we decided to include a small random factor to the utility of each cell and to choose directly the movement that maximizes the agent utility. A more thorough comparison of the implications of this choice compared to the basic floor field approach is out of the scope of this chapter and it is object of future works.

5.5 Experimental Scenario

The GA-Ped model was adopted to realize a set of simulations in different starting conditions (mainly changing density of pedestrians in the environment, but also different configurations of groups present in the simulated pedestrian population) in a situation in which experiments focused at evaluating the impact of the presence of groups of different size was being investigated.

5.5.1 Experiments

The environment in which the experiments took place is represented in Fig. 5.1: a 2.5 by 10 m corridor, with exits on the short ends. The experiments were characterized by the presence of 2 sets of 25 pedestrians, respectively starting at the 2 ends of the corridor (in 2 by 2.5 m areas), moving towards the other end. Various cameras were positioned on the side of the corridor and the time required for the two sets of pedestrians to complete their movement was also measured (manually from the video footage).

Several experiments were conducted, some of which also considered the presence of groups of pedestrians, that were instructed on the fact that they had to behave as friends or relatives while moving during the experiment. In particular, the following scenarios have been investigated: (i) single pedestrians (three experiments); (ii) three couples of pedestrians for each direction (two experiments); (iii) two triples of pedestrians for each direction (three experiments); (iv) a group of six pedestrians for each direction (four experiments).

One of the observed phenomena was that the first experiment actually required more time for the pedestrians to complete the movement; the pedestrians actually learned how to move and how to perform the experiment very quickly, since the first experiment took them about 18 s while the average completion time over 12 experiments is about 15 s.

The number of performed experiments is probably too low to draw some definitive conclusions, but the total travel times of configurations including individuals and pairs were consistently lower than those not including groups. Qualitative analysis of the videos showed that pairs can easily form a line, and this reduces the friction with the facing group. Similar considerations can be done for large groups; on the other end, groups of three pedestrians sometimes had difficulties in forming a lane, retaining a triangular shape similar to the 'V' shaped observed and modeled in [34], and this caused a total travel times that were higher than average in two of the three experiments involving this type of group.

5.5.2 Simulation Results

We applied the model described in Sect. 5.4 to the previous scenario by means of an agent-based platform based on GA-Ped approach. A description of the platform can be found in [9]. We employed the gathered data and additional data available in the literature to perform a calibration of the parameters, essentially determining the relative importance of (a) the goal oriented, (b) general proxemics and (c) group proxemic components of the movement choice. In particular, we first identified a set of plausible values for the w_t and w_o parameters employing experimental data regarding a one-directional flow. Then we employed data from bidirectional flow situations to further tune these parameters as well as the value of the w_g parameter: the latter was set in order to achieve a balance between effectiveness in preserving group cohesion and preserving aggregated measures on the overall pedestrian flow (an excessive group cohesion value reduces the overall pedestrian flow and produces unrealistic behavior).

We investigated the capability of our model to fit the fundamental diagram proposed in the literature for characterising pedestrian simulations [46] and other traffic related phenomena. This kind of diagram shows how the average velocity of pedestrians varies according to the density of the simulated environment. Moreover, we wanted to distinguish the different performance of different agent types, and essentially individuals, members of pairs, groups of three and five pedestrians over a relatively wide spectrum of densities. To do so, we performed continuous

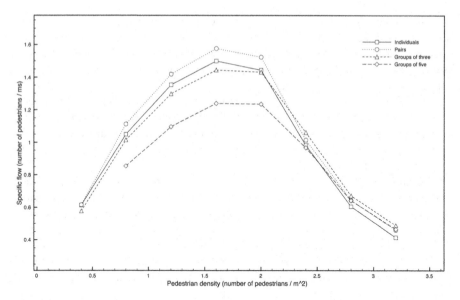

Fig. 5.2 Fundamental diagram for different pedestrian densities in the corridor scenario

simulations of the bidirectional pedestrian flows in the corridor with a changing number of pedestrians, to alter their density. For each density value displayed in the graph shown in Fig. 5.2 is related to at least 1 h of simulated time.

The achieved fundamental diagram represents in qualitatively correct way the nature of pedestrian dynamics: the flow of pedestrians increases with the growing of the density of the corridor unit a critical value is reached. If the system density is increased beyond that value, the flow begins to decrease significantly as the friction between pedestrians make movements more difficult.

The simulation results are in tune with the experimental data coming from observations: in particular, the flow of pairs of pedestrians is consistently above the curve of individuals. This means that the average speed of members of pairs is actually higher than the average speed of individuals. This is due to the fact that they easily tend to form a line, in which the first pedestrian has the same probability to be stuck as an individual, but the follower has a generally higher probability to move forward, following the path "opened" by the first member of the pair, as exemplified in Fig. 5.3b. The same does not happen for larger groups, since for them it is more difficult to form a line and therefore they offer a larger profile to the counter flow, as shown in Fig. 5.3a: the curves related to groups of three and five members are below the curve of individuals for most of the spectrum of densities, precisely until very high density values are reached. In this case, the advantage of followers overcomes the disadvantage of offering a larger profile to the counter flow and the combined average velocity is higher than that of individuals.

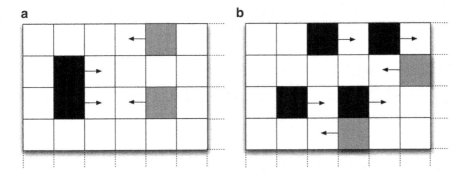

Fig. 5.3 In the *left figure*, the *black* pedestrians have not formed a line and they offer a larger profile to the counter flow. In the *right figure*, they formed a line and the follower has a lower probability to find an opposing pedestrian due to the presence of a sort of "emergent" leader

5.6 Real World Scenario

5.6.1 Environment and observations

The model was also adopted to elaborate different what-if scenarios in a real world case study. In particular, the simulated scenario is characterized by the presence of a station of the Mashaer line, a newly constructed rail line in the area of Makkah. The goal of this infrastructure is to reduce the congestion caused by the presence of other collective means of pilgrim transportation (i.e. buses) during the Hajj: the yearly pilgrimage to Mecca that involves over two millions of people coming from over 150 countries and some of its phase often result in congestions of massive proportions. In this work, we are focusing on a specific point of one of the newly constructed stations, Arafat I. One of the most demanding situations that the infrastructure of the Mashaer Rail line must be able to sustain is the one that takes place after the sunset of the second day of the pilgrimage, which involves the transport of pilgrims from Arafat to Muzdalifah. The pilgrims that employ the train to proceed to the next phase of the process must be able to move from the tents or other accommodation to the station in an organised flow that should be consistent with the movement of trains from Arafat to Muzdalifah stations. Since pilgrims must leave the Arafat area before midnight, the trains must continuously load pilgrims at Arafat, carry them to Muzdalifah, and come back empty to transport other pilgrims.

The size of the platforms was determined to allow hosting in a safe and comfortable way a number of pilgrims also exceeding the potential number of passengers of a whole train. Each train is made up of 12 wagons, each able to carry 250 passengers for a total of approximately 3,000 persons. In order to achieve an organized and manageable flow of people from outside the station area to the platforms, the departure process was structured around the idea of waiting–boxes: pilgrims are subdivided into groups of about 250 persons that are led by specific

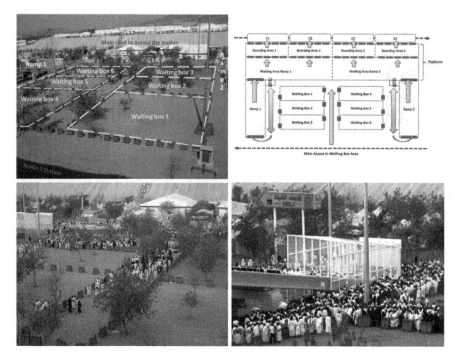

Fig. 5.4 Photos and a schematic representation of the real world scenario and the related phenomena: groups of pilgrims move from the tents area according to a precise schedule and flow into the waiting boxes, fenced queuing areas located in immediately outside the station, between the access ramps. The groups wait in these areas for an authorization by the station agents to move towards the ramps or elevators

leaders (generally carrying a pole with signs supporting group identification). The groups start from the tents area and flow into these fenced queuing areas located in immediately outside the station, between the access ramps. Groups of pilgrims wait in these areas, called waiting boxes, for an authorization by the station agents to move towards the ramps or elevators. In this way, it is possible to stop the flow of pilgrims whenever the number of persons on the platforms (or on their way to reach it using the ramps or elevators) is equal to the train capacity, supporting thus a smooth boarding operation.

Three photos and a schematic representation of the real world scenario and the related phenomena are shown in Fig. 5.4: the bottom right photo shows a situation in which the waiting-box principle, preventing the possibility of two flows simultaneously converging to a ramp, was not respected, causing a higher than average congestion around the ramp. This anomaly was plausibly due to the fact that it was the first time the station was actually used, therefore also the management personnel was not experienced in the crowd management procedures.

5.6.2 Simulation Results

Three different scenarios were realized adopting the previously defined model and using the parameters that were employed in the previous case study: (i) the flow of a group of pilgrims from one waiting box to the ramp; (ii) the simultaneous flow of two groups from two different waiting boxes to the same ramp; (iii) the simultaneous flow of three groups of pilgrims, two as in the previous situation, one coming directly from the tents area. Every group included 250 pilgrims. The goal of the analysis was to understand if the model is able to qualitatively reflect the increase in the waiting times and the space utilization when the waiting box principle was not respected.

The environment was discretized adopting 50 cm sided cells and the cell space was endowed with a floor field leading towards the platform, by means of the ramp. The floor field was generated according to well known techniques (essentially employing the Manhattan Distance [29] corrected introducing a minor effect of repulsion generated by obstacles, as in [37]). The different speed of pedestrians in the ramp was not considered: this scenario should be therefore considered as a best case situation, since pilgrims actually flow through the ramp more slowly than in our simulation. Consequently, we will not discuss here the changing of the travel time between the waiting boxes and the platform (that however increased with the growth of the number of pilgrims in the simulated scenario), but rather different metrics of *space utilization*. This kind of metric is tightly related to the so called *level of service* [18], a measure of the effectiveness of elements of a transportation infrastructure; it is also naturally related to proxemics, since a low level of service is related to a unpleasant perceived situation due to the invasion of the personal (or even intimate) space.

The diagrams shown in Fig. 5.5 report three metrics describing three different phenomena in the same area, including a ramp (on the left) and three waiting boxes (on the right). The three phenomena are related to (i) a situation in which an agent in a cell of the environment was willing to move but it was unable to perform the action due to the excessive space occupation; (ii) a situation in which an agent actually moved from a cell of the environment; (iii) the "set sum" of the previous situations, in other words, the situations in which a cell was occupied by agent, that either moved out of the cell or remained stuck in there. More precisely, diagrams show the relative frequency of the above events on the whole simulation time. The three metrics are depicted graphically following the same approach: the background color of the environment is black and obstacles are red (gray in a B&W rendering); each point associated to a walkable area (i.e. a cell of the model) is painted in a different shade of gray according to the value of the metric in that specific point. The black color is therefore associated to point if the environment in which the related metric is 0; the white color is associated to the point in which the metric assumes the highest value in the scenario (also shown in the legend). For instance, in all diagrams in the third row the points of space close to the ramp entrance are white or light gray,

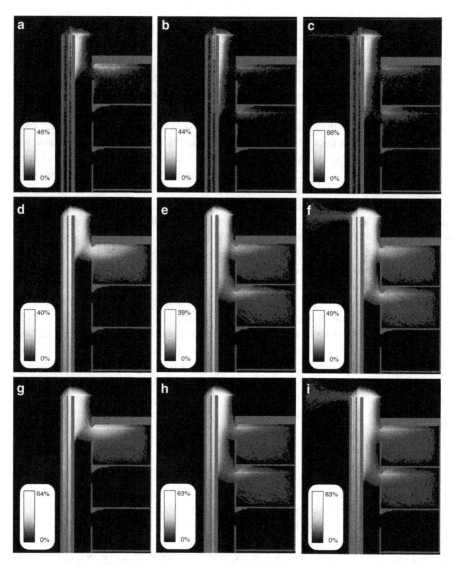

Fig. 5.5 Space utilization diagrams related to the three alternative simulated scenarios in the same area, including a ramp (on the *left*) and three waiting boxes (on the *right*). The different columns depict three space utilization metrics (described in the text) respectively in case of (i) a single group entering the station from one waiting box, (ii) two groups simultaneously approaching the ramp from two waiting boxes and (iii) two groups moving towards the station from waiting boxes and another one directly from the tents area. (**a**) One waiting box – block situations. (**b**) Two waiting boxes – blocks situations. (**c**) Two waiting boxes and external flow – block situation. (**d**) One waiting box – Flow from cell situation. (**e**) Two waiting boxes – flows from cell situation. (**f**) two waiting boxes and external flow – flow from cell situations. (**g**) One waiting box – total space utilization. (**h**) Two waiting boxes – total space utilization. (**i**) Two waiting boxes and external flow – total space utilization

while the space of the waiting area from which the second group starts is black in the first column, since the group is not present in the related situation and therefore that portion of space is not actually utilized.

The different columns depict three space utilization metrics respectively in case of (i) a single group entering the station from one waiting box, (ii) two groups simultaneously approaching the ramp from two waiting boxes and (iii) two groups moving towards the station from waiting boxes and another one directly from the tents area. The difference between the first and second scenario is not apparent in terms of different values for the maximum space utilization metrics (they are actually slightly lower in the second scenario), but the area characterized by a medium-high space utilization is actually wider in the second case. The third scenario is instead characterized by a noticeably worse performance not only from the perspective of the size of the area characterized by a medium-high space utilization, but also from the perspective of the highest value of space utilization. In particular, in the most utilized cell of the third scenario, an agent was stuck about 66% of the simulated time, compared to the 46 and 44% of the first and second scenarios.

This analysis therefore confirms that increasing the number of pilgrims that are simultaneously allowed to move towards the ramp highly increases the number of cases in which their movement is blocked because of overcrowding. Also the utilization of space increases significantly and, in the third situation, the whole side of the ramp becomes essentially a queue of pilgrims waiting to move towards the ramp. Another phenomenon that was not highlighted by the above diagrams is the fact that groups face a high pressure to mix when reaching the entrance of the ramp, which is a negative factor since crowd management procedures adopted in the scenario are based on the principle of preserving group cohesion and keeping different groups separated. According to these results, the management of the movement of group of pilgrims from the tents area to the ramps should try to avoid exceptions to the waiting box principle as much as possible.

5.7 Opportunities and Challenges for Crowd Analysis Methods

A comprehensive framework trying to put together different aspects and aims of pedestrians and crowd dynamics research has been defined in [28]. The central element of this schema is the mutually influencing (and possibly motivating) relationship between the above mentioned efforts aimed at synthesizing crowd behavior and other approaches that are instead aimed at analyzing field data about pedestrians and crowds in order to characterize it different ways. It must be noted, in fact, that some approaches have the goal of producing aggregate level quantities (e.g. people counting, density estimation), while others are aimed at producing finer-grained results (i.e. tracking people in scenes) and other ones are instead aimed at identifying some specific behavior in the scene (e.g. main directions, velocities, unusual events). The different approaches adopt different techniques, some performing a *pixel–level analysis*, others considering larger patches of the image, i.e. *texture–level analysis*; other techniques require instead the detection of proper objects in the scene, a real *object–level analysis*.

From the perspective of the requirements for the synthesis of quantitatively realistic pedestrian and crowd behavior, it must be stressed that both aggregate level quantities and granular data are of general interest: a very important way to characterize a simulated scenario is represented by the previously mentioned fundamental diagram [46], that is, the relationship in a given scenario between the flow of pedestrians in a section and their density. Qualitatively, a good model should be able to reproduce an empirically observed phenomenon characterized by the growth of the flow until a certain density value (also said *critical density*) is reached; then the flow should decrease. However, every specific situation is characterized by a different shape of this curve, the position of critical density point and the maximum flow level; therefore even relatively "basic" counting and density estimation techniques can provide useful information in case of observations in real world scenarios. Density estimation approaches can also help in evaluating qualitatively the patterns of space utilization generated by simulation models against real data. Tracking techniques instead can be adopted to support the estimation of travelling times (and length of the followed path) by pedestrians. Crowd behavior understanding techniques can help in determining main directions and the related velocities. In this perspective, some relevant and fruitful experiences can already be mentioned: in [41] the authors are able to essentially derive guidance fields, that is, significant elements of the modeling approach managing goal driven tendencies of pedestrian agents directly from video footage. In [19] an anticipative system integrating computer vision techniques and pedestrian simulation is used to suggest crowd management solutions (e.g. guidance signals) to avoid congestion situations in evacuation processes. In [31] a pedestrian model is instead exploited to improve the performance of a multiple–people tracker in semi–crowded conditions. Finally, in [43] the authors propose to employ the social force model to support the detection of abnormal crowd behavior in video sequences.

It is important to emphasize that anthropological considerations about human behavior [23] are growingly considered as crucial both in the computerized analysis of crowds [28] and in the synthesis of believable pedestrian and crowd behavior [32, 51]. They can also represent a useful source of considerations on the analyzed phenomenon and they can guide some relevant modeling choices.

One of the currently least investigated pathways in this articulated research context is characterized by new requirements coming from novel research questions that were defined in the area of synthesis of pedestrians and crowd behavior. In particular, we think that the first results in the modeling of the implications of groups of pedestrians in larger crowds, supported by empirical observations possibly deriving from the manual analysis of video footages of ad hoc experiments, can lead to the identification of patterns, particular shapes and morphologies, recurrent situations, that can represent a form of *contextual information* from which automated computer vision techniques can benefit [21]. The automatic identification, tracking and characterization of groups (e.g. shapes, estimation of the number of members) by means of computer vision techniques could lead, in turn, to a substantial improvement of the possibility to effectively calibrate and validate pedestrian models considering groups in challenging innovative scenarios.

5.8 Conclusions and Future Work

The chapter has introduced a recent and current trend of research in the modeling, simulation and visual analysis of pedestrians and crowds, that is, the study of the impact of groups on the overall crowd dynamics, and its implications of the aforementioned research activities as well as their outcomes. The chapter has described some relevant influential contributions from anthropological and sociological disciplines, and it has presented a brief state of the art of pedestrians and crowd modeling and simulation. An effort aimed at modeling crowds in terms of groups has been introduced and its results in an experimental and a real-world scenario have been discussed. Finally, the opportunities arising from a more systematic interaction between the efforts aimed at synthesizing and analyzing pedestrians and crowd behaviors have been discussed. Future works in this framework are naturally aimed at extending the modeling approach to allow the representation of more composite forms of groups and extending the range of analyzed scenarios with a validation against empirical data.

Acknowledgements This work is a result of the Crystal Project, funded by the Center of Research Excellence in Hajj and Omrah (Hajjcore), Umm Al-Qura University, Makkah, Saudi Arabia. Our acknowledgement for the common work in the project and for fruitful discussions goes to Katsuhiro Nishinari (RCAST – Research Center for Advanced Science and Technology, The University of Tokyo, Japan), our valuable partner within the Crystals Project. We also thank Ugo Fabietti (CREAM – University of Milano-Bicocca) for his contribution from the area of Anthropology.

References

1. Bandini, S., Federici, M.L., Vizzari, G.: Situated cellular agents approach to crowd modeling and simulation. Cybern. Syst. **38**(7), 729–753 (2007)
2. Bandini, S., Manenti, L., Manzoni, S., Sartori, F.: A knowledge-based approach to crowd classification. In: Proceedings of the 5th International Conference on Pedestrian and Evacuation Dynamics, March 8–10, Gaithersburg, MD, USA (2010)
3. Bandini, S., Manzoni, S., Redaelli, S.: Towards an ontology for crowds description: a proposal based on description logic. In: Umeo, H., Morishita, S., Nishinari, K., Komatsuzaki, T., Bandini, S. (eds.) ACRI. Lecture Notes in Computer Science, vol. 5191, pp. 538–541. Springer, Berlin, Germany (2008)

4. Bandini, S., Manzoni, S., Vizzari, G.: Agent based modeling and simulation: an informatics perspective. J. Artif. Soc. Soc. Simul. **12**(4), 4 (2009)
5. Bandini, S., Rubagotti, F., Vizzari, G., Shimura, K.: An agent model of pedestrian and group dynamics: experiments on group cohesion. In: Pirrone, R., Sorbello, F. (eds.) AI*IA. Lecture Notes in Computer Science, vol. 6934, pp. 104–116. Springer, Berlin, Germany (2011)
6. Batty, M.: Agent based pedestrian modeling (editorial). Environ. Plan. B: Plan. Des. **28**, 321–326 (2001)
7. Blue, V.J., Adler, J.L.: Cellular automata microsimulation of bi-directional pedestrian flows. Transp. Res. Rec. **1678**, 135–141 (1999)
8. Blue, V.J., Adler, J.L.: Modeling four-directional pedestrian flows. Trans. Res. Rec. **1710**, 20–27 (2000)
9. Bonomi, A., Manenti, L., Manzoni, S., Vizzari, G.: Makksim: dealing with pedestrian groups in MAS-based crowd simulation. In: Fortino, G., Garro, A., Palopoli, L., Russo, W., Spezzano, G. (eds.) WOA. CEUR Workshop Proceedings, Rende, vol. 741, pp. 166–170 (2011). http://CEUR-WS.org
10. Canetti, E.: Crowds and power. Farrar, Straus and Giroux, New York (1984)
11. Challenger, R., Clegg, C.W., Robinson, M.A.: Understanding crowd behaviours: Supporting evidence. Tech. rep., University of Leeds (2009)
12. Chattaraj, U., Seyfried, A., Chakroborty, P.: Comparison of pedestrian fundamental diagram across cultures. Adv. Complex Syst. **12**(3), 393–405 (2009)
13. Costa, M.: Interpersonal distances in group walking. J. Nonverbal Behav. **34**, 15–26 (2010). http://dx.doi.org/10.1007/s10919-009-0077-y, doi:10.1007/s10919-009-0077-y
14. Dijkstra, J., Jessurun, J., de Vries, B., Timmermans, H.J.P.: Agent architecture for simulating pedestrians in the built environment. In: International Workshop on Agents in Traffic and Transportation, pp. 8–15, Hakodate, Japan (2006)
15. Dopfer, K., Foster, J., Potts, J.: Micro-meso-macro. J. Evol. Econ. **14**, 263–279 (2004). http://dx.doi.org/10.1007/s00191-004-0193-0, doi:10.1007/s00191-004-0193-0
16. Fabietti, U.E.M.: Gruppi – Antropologia, vol. Enciclopedia delle Scienze Sociali, pp. 424–429. Treccani (1994)
17. Federici, M.L., Gorrini, A., Manenti, L., Vizzari, G.: An innovative scenario for pedestrian data collection: the observation of an admission test at the university of Milano-Bicocca. In: Proceedings of the 6th International Conference on Pedestrian and Evacuation Dynamics – PED 2012, Zurich, Switzerland (2012)
18. Fruin, J.J.: Pedestrian planning and design. Metropolitan Association of Urban Designers and Environmental Planners, New York (1971)
19. Georgoudas, I.G., Sirakoulis, G.C., Andreadis, I.: An anticipative crowd management system preventing clogging in exits during pedestrian evacuation processes. IEEE Syst. J. **5**(1), 129–141 (2011)
20. Gloor, C., Stucki, P., Nagel, K.: Hybrid techniques for pedestrian simulations. In: Sloot, P.M.A., Chopard, B., Hoekstra, A.G. (eds.) 6th International Conference on Cellular Automata for Research and Industry, ACRI 2004. Lecture Notes in Computer Science, vol. 3305, pp. 581–590. Springer, Berlin, Germany (2004)
21. Gualdi, G., Prati, A., Cucchiara, R.: Contextual information and covariance descriptors for people surveillance: An application for safety of construction workers. EURASIP J. Image Video Process. **2011** (2011)
22. Hall, E.T.: A system for the notation of proxemic behavior. Am. Anthropol. **65**(5), 1003–1026 (1963). http://www.jstor.org/stable/668580
23. Hall, E.T.: The Hidden Dimension. Anchor Books, New York (1966)
24. Helbing, D.: A fluid–dynamic model for the movement of pedestrians. Complex Syst. **6**(5), 391–415 (1992)
25. Helbing, D., Molnár, P.: Social force model for pedestrian dynamics. Phys. Rev. E **51**(5), 4282–4286 (1995)
26. Helbing, D., Schweitzer, F., Keltsch, J., Molnár, P.: Active walker model for the formation of human and animal trail systems. Phys. Rev. E **56**(3), 2527–2539 (1997)

27. Henein, C.M., White, T.: Agent-based modelling of forces in crowds. In: Davidsson, P., Logan, B., Takadama, K. (eds.) Joint Workshop on Multi-agent and Multi-agent-based Simulation, MABS 2004, New York, 19 July 2004, Revised Selected Papers. Lecture Notes in Computer Science, vol. 3415, pp. 173–184. Springer (2005)
28. Junior, J.C.J., Musse, S.R., Jung, C.R.: Crowd analysis using computer vision techniques. IEEE Signal Process. Mag. 27(5), 66–77 (2010)
29. Kirchner, A., Schadschneider, A.: Simulation of evacuation processes using a bionics-inspired cellular automaton model for pedestrian dynamics. Phys. A: Stat. Mech. Appl. 312(1–2), 260–276 (2002). http://www.sciencedirect.com/science/article/pii/S0378437102008579
30. Klüpfel, H.: A cellular automaton model for crowd movement and egress simulation. P.hd. thesis, University Duisburg-Essen (2003)
31. Leal-Taixé, L., Pons-Moll, G., Rosenhahn, B.: Everybody needs somebody: modeling social and grouping behavior on a linear programming multiple people tracker. In: ICCV Workshops, pp. 120–127. IEEE, Barcelona, Spain (2011)
32. Manenti, L., Manzoni, S., Vizzari, G., Ohtsuka, K., Shimura, K.: Towards an agent-based proxemic model for pedestrian and group dynamic. In: Omicini, A., Viroli, M. (eds.) WOA. CEUR Workshop Proceedings, vol. 621, Rimini, Italy (2010). http://CEUR-WS.org
33. Manenti, L., Manzoni, S., Vizzari, G., Ohtsuka, K., Shimura, K.: An agent-based proxemic model for pedestrian and group dynamics: motivations and first experiments. In: Villatoro, D., Sabater-Mir, J., Sichman, J.S. (eds.) MABS. Lecture Notes in Computer Science, vol. 7124, pp. 74–89. Springer Berlin, Germany (2011)
34. Moussaïd, M., Perozo, N., Garnier, S., Helbing, D., Theraulaz, G.: The walking behaviour of pedestrian social groups and its impact on crowd dynamics. PLoS ONE 5(4), e10047 (2010). http://dx.doi.org/10.1371%2Fjournal.pone.0010047
35. Musse, S.R., Thalmann, D.: Hierarchical model for real time simulation of virtual human crowds. IEEE Trans. Vis. Comput. Graph. 7(2), 152–164 (2001)
36. Nagel, K., Schreckenberg, M.: A cellular automaton model for freeway traffic. Journal de Physique I France 2(2221), 222–235 (1992)
37. Nishinari, K., Kirchner, A., Namazi, A., Schadschneider, A.: Extended floor field ca model for evacuation dynamics. IEICE Trans. Inf. syst. 87(3), 726–732 (2004)
38. Nishinari, K., Suma, Y., Yanagisawa, D., Tomoeda, A., Kimura, A., Nishi, R.: Toward smooth movement of crowds. In: Pedestrian and Evacuation Dynamics 2008, pp. 293–308. Springer, Berlin/Heidelberg (2008)
39. Okazaki, S.: A study of pedestrian movement in architectural space, part 1: pedestrian movement by the application of magnetic models. Trans. A.I.J. 283, 111–119 (1979)
40. Paris, S., Donikian, S.: Activity-driven populace: A cognitive approach to crowd simulation. IEEE Comput. Graph. Appl. 29(4), 34–43 (2009)
41. Patil, S., van den Berg, J.P., Curtis, S., Lin, M.C., Manocha, D.: Directing crowd simulations using navigation fields. IEEE Trans. Vis. Comput. Graph. 17(2), 244–254 (2011)
42. Qiu, F., Hu, X.: Modeling group structures in pedestrian crowd simulation. Simul. Model. Pract. Theory 18(2), 190–205 (2010)
43. Raghavendra, R., Bue, A.D., Cristani, M., Murino, V.: Abnormal crowd behavior detection by social force optimization. In: Salah, A.A., Lepri, B. (eds.) HBU. Lecture Notes in Computer Science, vol. 7065, pp. 134–145. Springer, Berlin, Germany (2011)
44. Sarmady, S., Haron, F., Talib, A.Z.H.: Modeling groups of pedestrians in least effort crowd movements using cellular automata. In: Al-Dabass, D., Triweko, R., Susanto, S., Abraham, A. (eds.) Asia International Conference on Modelling and Simulation, pp. 520–525. IEEE Computer Society, Bali, Indonesia (2009)
45. Schadschneider, A., Kirchner, A., Nishinari, K.: CA approach to collective phenomena in pedestrian dynamics. In: Bandini, S., Chopard, B., Tomassini, M. (eds.) 5th International Conference on Cellular Automata for Research and Industry, ACRI 2002. Lecture Notes in Computer Science, vol. 2493, pp. 239–248. Springer, Berlin, Germany (2002)

46. Schadschneider, A., Klingsch, W., Klüpfel, H., Kretz, T., Rogsch, C., Seyfried, A.: Evacuation dynamics: empirical results, modeling and applications. In: Meyers, R.A. (ed.) Encyclopedia of Complexity and Systems Science, pp. 3142–3176. Springer, New York (2009)
47. Schreckenberg, M., Sharma, S.D. (eds.): Pedestrian and Evacuation Dynamics. Springer, Berlin, Germany (2001)
48. Schultz, M., Schulz, C., Fricke, H.: Passenger dynamics at airport terminal environment. In: Klingsch, W.W.F., Rogsch, C., Schadschneider, A., Schreckenberg, M. (eds.) Pedestrian and Evacuation Dynamics 2008, pp. 381–396. Springer, Heidelberg/New York (2010)
49. Shao, W., Terzopoulos, D.: Autonomous pedestrians. Graph. Models 69(5–6), 246–274 (2007)
50. Toyama, M.C., Bazzan, A.L.C., da Silva, R.: An agent-based simulation of pedestrian dynamics: from lane formation to auditorium evacuation. In: Nakashima, H., Wellman, M.P., Weiss, G., Stone, P. (eds.) 5th International Joint Conference on Autonomous Agents and Multiagent Systems (AAMAS 2006), pp. 108–110. ACM, Hakodate, Japan (2006)
51. Was, J.: Crowd dynamics modeling in the light of proxemic theories. In: Rutkowski, L., Scherer, R., Tadeusiewicz, R., Zadeh, L.A., Zurada, J.M. (eds.) ICAISC (2). Lecture Notes in Computer Science, vol. 6114, pp. 683–688. Springer, Berlin, Germany (2010)
52. Willis, A., Gjersoe, N., Havard, C., Kerridge, J., Kukla, R.: Human movement behaviour in urban spaces: implications for the design and modelling of effective pedestrian environments. Environ. Plan. B 31(6), 805–828 (2004)
53. Xu, S., Duh, H.B.L.: A simulation of bonding effects and their impacts on pedestrian dynamics. IEEE Trans. Intell. Transp. Syst. 11(1), 153–161 (2010)

Chapter 6
Scalable Solutions for Simulating, Animating, and Rendering Real-Time Crowds of Diverse Virtual Humans

Daniel Thalmann, Helena Grillon, Jonathan Maïm, and Barbara Yersin

Abstract In this chapter, we describe how we can model crowds in real-time using dynamic meshes, static meshes and impostors.Techniques to introduce variety in crowds including colors, shapes, textures, individual animation, individualized path-planning, simple and complex accessories are explained. We also present a hybrid architecture to handle the path planning of thousands of pedestrians in real time, while ensuring dynamic collision avoidance. Several behavioral aspects are presented as gaze control, group behavior, as well as the specific technique of crowd patches. Several case-studies are shown in cultural heritage and social phobia.

6.1 Introduction

To simulate large crowds at high frame rates, it is necessary to use several levels of detail (LOD). Characters close to the camera are accurately rendered and animated with costly methods, while those farther away are represented with less detailed, faster representations. In this chapter, we describe how we can model crowds in real-time using dynamic meshes, static meshes and impostors. Techniques to introduce variety in crowds including individual animation, individualized path-planning, simple and complex accessories are explained.

D. Thalmann (✉)
Institute for Media Innovation, Nanyang Technological University,
Singapore and EPFL, Switzerland
e-mail: danielthalmann@ntu.edu.sg

H. Grillon
Centrale de Compensation, Geneva, Switzerland
e-mail: helena.grillon@gmail.com

J. Maïm • B. Yersin
Minsh.net, Bengaluru, Karnataka, India
e-mail: Jonathan.Maim@gmail.com; Barbara@minsh.net

S. Ali et al. (eds.), *Modeling, Simulation and Visual Analysis of Crowds*, The International
Series in Video Computing 11, DOI 10.1007/978-1-4614-8483-7__6,
© Springer Science+Business Media New York 2013

We also present a hybrid architecture to handle the path planning of thousands of pedestrians in real time, while ensuring dynamic collision avoidance. The scalability of our approach allows to interactively create and distribute regions of varied interest, where motion planning is ruled by different algorithms. Practically, regions of high interest are governed by a long-term potential field-based approach, while other zones exploit a graph of the environment and short-term avoidance techniques. Our method also ensures pedestrian motion continuity when switching between motion planning algorithms. Tests and comparisons show that our architecture is able to realistically plan motion for many groups of characters, for a total of several thousands of people in real time, and in varied environments. We finally introduce a method for populating large-scale interactive virtual environments with walking and idle humans, as well as animated and static objects. The key idea of our solution is to build environments from a set of blocks, the crowd patches, that describe periodic motion for a small local population, as well as other environment details. Periodicity in time allows endless replay.

Several case-studies are shown in cultural heritage, emergency situations, and social phobia.

The rest of this chapter is organized into sections. Section 6.2 explains the principles of multiple levels of detail and how they are processed during the rendering of large crowd simulation. Section 6.3 is dedicated to the variety aspects. We explain how to create various shapes, colors, textures, animation styles. We also emphasize the concept of accessories like bags, glasses, mobile phones, hats. In Sect. 6.4, we introduce path planning with three approaches: navigation graphs, continuum crowds, and into more details an hybrid method. Section 6.5 describes two aspects related to behavior: gaze control and group behavior. The notion of crowd patches is also presented in this section. Finally, Sect. 6.6 briefly discusses a few applications like simulation of ancient cities or treatment of agoraphobia.

Since 1997, when Thalmann and Musse [1] started their pioneering work on crowds of Virtual Humans, a lot of researchers have investigated into this field. Today, there is almost no graphics or animation conference without papers on crowd simulation. We cannot easily introduce a state of the art in this chapter, but a few selected works will be cited along the chapter. The reader can find an exhaustive state of the art in the second edition of the book "Crowd Simulation" [2].

6.2 Representing Virtual Humans Using Multiple Levels of Detail

Animation and rendering of virtual humans relies on a set of geometric, kinematic and appearance models. When dealing with crowds made of thousands of virtual humans (see Fig. 6.1), it is not possible to dispose of a unique model for each of them; this would result in huge memory consumption and computation times. A key solution is to use templates defining each type of virtual human, which can be continuously derived in order to generate an infinite variety of instances of virtual humans.

Fig. 6.1 Large crowd of diverse virtual humans

The first step is to create a sufficient combination of templates with different genders and ages, i.e., adults, elderly, children, males and females. In a second step, creating different textures for each template allows additional variation in age and appearance. Finally, the colors of skin, hair, and clothes are varied per texture, as described in [3]. To further vary appearance for virtual humans in the vicinity of the camera, it is possible to add fine details like make-up, freckles and beard. Also, by controlling shading parameters per body part, as proposed in [4], allows to vary the materials used for the clothes, as for example to model shiny clothes. Thus, with a small number of templates, it is possible to synthesize a large number of virtual human instances that seem unique. Under some conditions recently studied in [5], it is possible to make the use of templates practically undetectable for a spectator. Further varying the shape of instantiated virtual humans can be achieved using accessories, i.e., simple items like hats or more complex accessories like cell phones or shopping bags that require upper body variations only.

The goal of the real-time crowd visualizer is to render a large number of entities according to the current simulation state, which provides the position, orientation, and velocity for each individual. System constraints include believability, real-time updates (25 frames per second) and the number of digital actors ranging in the tens of thousands. We make the population believable by varying the appearance (textures and colors) and animation of the individuals. Their graphical representation is derived from a template, which holds all the possible variations.

Thus, with only a limited set of such templates, we can achieve a varied crowd, leading to considerable time savings for designers.

A template is defined as:

- A set of three meshes with decreasing complexity (LODs),
- A set of textures in gray scale (except for the skin) identifying color modulation areas (pants, shirt, hair etc.),
- A skeleton (kinematic structure),
- A corresponding animation database as skeletal orientations (here 1,000 different walk cycles are generated using a motion blending-based loco-motion engine [6, 7]).

Each human in the visualization system is called an instance and is derived from a template. Individualization comes from assigning a specific gray scale texture and a color combination for each identifiable region. Instances have individualized walk velocities and are animated by blending the available walk animations.

The rendering pipeline advances consecutively in four steps. The first one consists of culling, that is, determining visibility, and choosing the rendering fidelity for each simulated human. By re-using the information stored in the navigation graph of the simulation system, this task is not done for each individual but at the vertex level, thereby determining fidelities for a whole subset of characters at once.

During this phase, humans are distributed in three different groups according to their fidelity level, which ensures efficient batched rendering. The next step of the pipeline is the rendering of dynamic meshes. This is the most detailed fidelity where the animation is obtained by interpolation of skeletal postures. According to the current instance state (linear and angular walk velocities and time), animations are retrieved from the database and interpolated, yielding a smooth animation, with continuous variations of velocities, and no foot-sliding. The resulting skeletal posture is sent to a hardware vertex shader and fragment shader deforming and rendering the human on the graphics card.

Static meshes (also called baked or predeformed) constitute the second rendering fidelity, which keeps a pre-transformed set of animations using the lowest resolution mesh of the deformed mesh in the previous step. Pre-computing deformations allows substantial gains in speed, but constrains the animation variety and smoothness.

The final rendering fidelity is the billboard model which, compared to previous approaches, uses a simplified scheme of sampling and lighting. World-aligned billboards are used, with the assumption that the camera will never hover directly above the crowd. Thus, only sampled images around the waist level of the character are needed. In our case, the templates are sampled at 20 different angles, for each of the 25 key frames composing a walk animation. When constructing the resulting texture, the bounding box of each sampled frame is detected to pack them tightly together. When rendering bill-boarded pedestrians, a specificity of our technique is to apply cylindrical lighting instead of using normal maps: each vertex normal is set to point in the positive Z direction, plus a small offset on the X axis, so that it points slightly outside the frame. We then interpolate the light intensity for each pixel in the fragment shader. Figure 6.2 shows an example of the distribution of the three fidelities.

Fig. 6.2 The three different rendering levels of detail: deformed meshes in *front*, rigid meshes in the *middle*, and billboards *behind*

6.3 Adding Variety in Appearance and Animation of Virtual Humans

6.3.1 Introduction

Our main interest is focused on real-time applications where the visual uniqueness of the characters composing a crowd is paramount. On the one hand, it is required to display several thousands of virtual humans at high frame rates, using levels of detail. On the other hand, each character has to be different from all others, and its visual quality highly detailed. Instantiating many characters from a limited set of human templates lead to the presence of multiple similar characters everywhere in the scene. However, the creation of an individual mesh for each character is not feasible, for it would have too high requirements in terms of design and memory. Thus, methods have to be introduced to modify each instance, so that it is visually different from all the others. Such methods also need to be scalable for all LOD used in crowd simulations to avoid inconsistencies in the individual appearances. Our main contribution is the introduction of techniques to improve the variety of crowds in three domains: visual appearance, shape, and animation. More details may be found in [11].

6.3.2 Appearance Variety

6.3.2.1 Color Variety

Previous work on color variety is based on the idea of dividing a human template into several body parts, identified by specific intensities in the alpha channel of the template texture. At runtime, each body part of each character is assigned a color in order to modulate the texture. Tecchia et al. [8] used several passes to render each impostor body part. Dobbyn et al. [3] extended the method to 3D meshes and avoided multi-pass rendering with programmable graphics hardware. Although these methods offer nice results from a reasonable distance, they produce sharp transitions between body parts. Based on the same idea, Gosselin et al. [9] showed how to vary characters with the same texture by changing their tinting. They also presented a method to selectively add decals to the characters' uniforms. However, their approach is only applied to armies of similar characters, and the introduced differences are not sufficient when working with crowds of civilians.

Using a single alpha layer to segment body parts has several drawbacks. No bi-linear filtering can be used on the texture, because incorrect interpolated values would be fetched in the alpha channel at body part borders. Moreover, for individuals close to the camera, the method tends to produce too sharp transitions between body parts, e.g., between skin and hair, due to the impossibility of associating a texel to several body parts at the same time. Also, character close-ups bring the need for a new method capable of handling detailed color variety. Subtle make-up, or detailed patterns on clothes greatly increase the variety of a single human template. Furthermore, changing illumination parameters of materials, e.g., their specularity, provides more realistic results. Previous methods would require costly fragment shader branching to achieve such effects. We apply a versatile solution based on segmentation maps to overcome previous method drawbacks.

For each texture of a human template, we create a series of segmentation maps. Each of them is an RGBA image, delimiting four body parts, i.e., one per channel, and sharing the same parameterization as the human template texture. This method allows for each texel to partially belong to several body parts at the same time through its channel intensities. As a result, it is possible to design much smoother transitions between body parts than in previous approaches. Figure 6.3 shows the principles of color variety.

6.3.2.2 Height and Shape Variety

Magnenat-Thalmann et al. [10] classified the methodologies for modeling virtual people into three major categories: creative, reconstructive, and interpolated. Geometric models created by artists such as anatomically based models fall into the former approach. The second major category built 3D virtual human's geometry by capturing existing shape from 3D scanners, images and even video sequences.

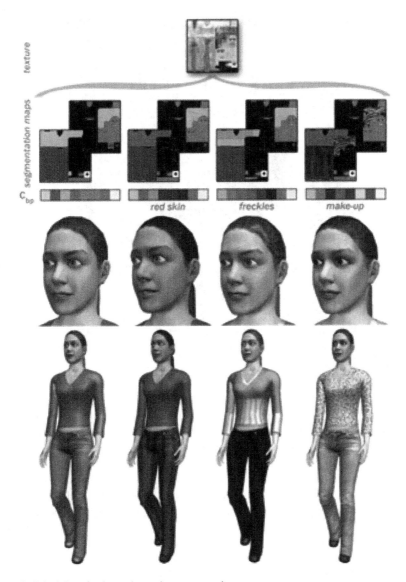

Fig. 6.3 Principles of color variety using segmentation maps

The interpolated modeling uses sets of example models with an interpolation scheme to reconstruct new geometric models. For crowds, the first approach is too expensive in terms of manual work. The second way is also prohibitive for large crowds and also presents a lack of flexibility. The last approach is the most convenient. For large crowds, another approach consists of modifying separately the height of the human body and its shape.

Fig. 6.4 Variation of height

We can modify the height of a human template, by scaling its skeleton. To help the designer in this task, we provide additional functionalities to the design tool presented above: for a given human template skeleton, the global space of height scaling can be defined. Fine-grained local tuning for each joint can also be specified, i.e., minimal and maximal scale parameters on the x, y, and z world axes. These data allow several different skeletons to be generated from a single template, which we call the meta-skeleton. Figure 6.4 shows an example. For each new skeleton, a global scale factor is randomly chosen within the given range. Then, the associated new scale for each of its bones is deduced. Short/tall skeletons mixed with broad/narrow shoulders are thus created. The skin of the various skeletons also needs adaptation. Each vertex of the original template is displaced by each joint that influences it. For the shape, the human mesh is modified using three steps:

1. Using a commercial $3D$ package like 3DSMax, it is possible for a designer to paint a FatMap for a given template, as seen in Fig. 6.5. The FatMap is an extra gray-scale UV texture that is used to emphasize body areas that store fat. Darker areas represent regions where the skin will be most deformed, e.g., on the belly, and lighter areas are much less deformed, like the head. When the creation of the FatMap is complete, the gray scale values at each texel are used to automatically infer one value for each vertex of the template's mesh. Each of these values, called a fatWeight, is attached to the vertex as an additional attribute.

Fig. 6.5 FatMaps designed
in 3DSMax. *Dark areas*
represent regions more
influenced by fat or muscles
modification, while *lighter*
parts are less affected

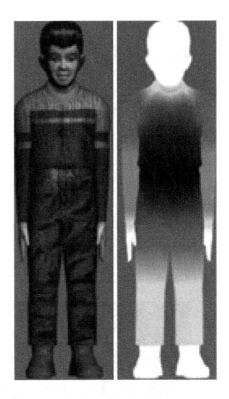

2. The next step is to compute in which direction the vertices are moved when scaled. For this, we compute the scaling direction of each vertex as the weighted normal of the bones influencing it.
3. Once the direction of the body scaling is computed for each vertex, the actual scaling can take place. The extent to which we scale the body is defined by a fatScale, randomly chosen within a pre-defined range.

Figure 6.6 shows the same character with various heights and shapes.

6.3.2.3 Accessories

Accessorizing crowds [21] offers a simple and efficient alternative to costly human template modeling. Accessories are small meshes representing elements that can easily be added to the human template original mesh. Their range is considerable, from subtle details, like watches, jewelry, or glasses, to larger items, such as hats, wigs, or backpacks, as illustrated in Fig. 6.7. Distributing accessories to a large crowd of a few human templates varies the shape of each instance, and thus makes it unique. Similar to deformable meshes, accessories are attached to a skeleton and follow its animation when moving.

Fig. 6.6 Various body shapes and heights

Fig. 6.7 Population with accessories: bags, hats, glasses

The first group of accessories does not necessitate any particular modification of the animation clips played. They simply need to be correctly placed on a virtual human. Each accessory can be represented as a simple mesh, independent from any virtual human. First, let us outline the problem for a single character. The issue is to render the accessory at the correct position and orientation, accordingly to the movement of the character. To achieve this, we can attach the accessory to a specific joint of the virtual human. Let us take a real example to illustrate our idea: imagine a walking person wearing a hat. Suppose that the hat has the correct size and does not slide, it basically has the same movement as the head of the person as he walks.

The second group of accessories we have identified is the one that requires slight modifications to the played animation sequences, e.g., the hand close to the ear to make a phone call, or a hindered arm sway due to carrying a heavy bag. Concerning the rendering of the accessory, we still keep the idea of attaching it to a specific joint of the virtual human. The additional difficulty is the modification of the animation clips to make the action realistic. We only focus on locomotion animation sequences. There are two options to modify an animation related to an accessory:

- If we want a virtual human to carry a bag for instance, the animation modifications are limited to the arm sway, and maybe a slight bend of the spine to counterweight the bag. The motion is restricted, in this case, we clamp the joints defining the limits of the joint angle (minimum angle, maximum angle).
- If it is a cell phone accessory that we want to add, we need to keep the hand of the character close to its ear and avoid any collision over the whole locomotion cycle. The motion is blocked and the angle is frozen to a certain value (Freezing Angle).

At runtime, the animation is updated as usual, the frozen joints are overwritten, and we use exponential maps to clamp joints.

6.3.2.4   How to Vary the Animation of Characters?

Animation is an another important factor that determines crowd heterogeneity. If they all perform the same animation, the results are not realistic enough [3]. We have implemented three techniques to vary the animation of characters, while remaining in the domain of navigating crowds, i.e., working with locomotion animations:

We introduce variety in the animation by generating a large amount of locomotion cycles (walking and running), and idle cycles (like standing, talking, sitting, etc.), that we morphologically adapt for each template. We take care to make these animations cyclic, and categorize them in the database, according to their type: sitting or standing, talking or listening, etc. For locomotion clips, walk and run cycles are generated from a locomotion engine based on motion capture data (see Sect. 6.3.2.5). We compute such locomotion clips for a set of speeds. Thus, during real-time animation, it is possible to directly obtain an adequate animation for a virtual human, given its current locomotion velocity, and its morphological parameters. Then, we designed the concept of motion kit, a data structure, that efficiently handles animations at all levels of detail (LOD).

We use a second technique of animation variety, i.e., how pre-computed animation cycles can be augmented with upper-body variations, like having a hand on the hip, or in a pocket.

Fig. 6.8 Locomotion generated using PCAs

Finally, we introduced procedural modifications applied at runtime on locomotion animations to allow crowds to wear complex accessories as mentioned in the previous section.

6.3.2.5 Locomotion Animation Clips

To generate our original set of walk and run cycles, we use the locomotion engine developed by Glardon et al. [6]; it is an integrated walking and running engine able to extrapolate data beyond the space described by the PCA basis. In this approach, the Principal Component Analysis (PCA) method is used to represent the motion capture data in a new, smaller space. As the first Principal Components contain the most variance of the data, an original methodology is used to extract essential parameters of a motion. This method decomposes the PCA in a hierarchical structure of sub-PCA spaces. At each level of the hierarchy, an important parameter of a motion is extracted and a related function is elaborated, allowing not only motion interpolation but also extrapolation. Figure 6.8 shows an example.

There are mainly three high-level parameters which allow to modulate these motions:

- Personification weights: several people, different in height and gait have been captured while walking and running. This variable allows the user to choose how he wishes to parametrize these different styles.

- Speed: the subjects have been captured at many different speeds. This parameter allows to choose at which velocity the walk/run cycle should be generated.
- Locomotion weights: this parameter defines whether the cycle is a walk or a run animation. Thus, the engine is able to generate a whole range of varied locomotion cycles for a given character. Each human template is also assigned a particular personification weight so that it has its own gait. With such a high number of animations, we are already able to per-ceive a sense of variety in the way the crowd is moving. Virtual humans walking together with different locomotion styles and speeds add to the realism of the simulation.

6.4   Motion Planning for Large-Scale Crowds

6.4.1   Introduction

Realistic real-time motion planning for crowds has become a fundamental research field in the Computer Graphics community. The simulation of urban scenes, epic battles, or other environments that show thousands of people in real time require fast and realistic crowd motion. Domains of application are vast: video games, psychological studies, and Architecture to name a few. In this section, we first focus on the Navigation Graph approach. A Navigation Graph is a simple structure that represents an environment topology by distinguishing navigable are-as from impassable obstacles. We then briefly discuss the continuum crowd concept. We finally present our motion planning architecture, offering a hybrid and scalable solution for real-time motion planning of thousands of characters in complex environments.

6.4.2   Navigation Graphs

Real-time crowd motion planning requires fast, realistic methods for path planning as well as obstacle avoidance. The difficulty to find a satisfying trade-off between efficiency and believability is particularly challenging, and prior techniques tend to focus on a single approach [12, 13]. We have presented [14, 15] a novel approach to automatically extract a topology from a scene geometry and handle path planning using a navigation graph. Figure 6.9 shows a crowd moving using navigation graphs. The main advantage of this technique is that it handles uneven and multi-layered terrains. Nevertheless, it does not treat inter-pedestrian collision avoidance. Given an environment geometry, a navigation graph can be computed [14, 21]: the vertices of the graph represent circular zones where a pedestrian can walk freely without the risk of colliding with any static object in the environment. Graph edges represent connections between vertices. In the environment, they are viewed as intersections

Fig. 6.9 Crowd moving

(or gates) between two circular zones (vertices). From a navigation graph, path requests can be issued from one vertex to another. Using an algorithm based on Dijkstra's, we are able to devise many different paths that join one point of the environment to another one. It is possible to provide the navigation graph with a second model of the environment, which usually has a much more simple geometry, annotated with information. This second model is automatically analyzed, its meta-information retrieved, and associated to the corresponding vertices. Figure 6.9 shows an example.

6.4.3   Scalable Motion Planning

More recently, Treuille et al. [16] proposed efficient motion planning for crowds. Their method produces a potential field that provides, for each pedestrian, the next suitable position in space (a waypoint) to avoid all obstacles. Compared to agent-based approaches, these techniques allow simulating thousands of pedestrians in real time, and are also able to show emergent behaviors. However, they produced less believable results, because they require assumptions that prevent treating each pedestrian with individual characteristics. For instance, only a limited number of goals can be defined and assigned to groups of pedestrians. The resulting performance depends on the size of the grid cells and the number of groups.

6.4.4 An Hybrid Architecture Based on Regions Of Interest (ROI)

We proposed a hybrid architecture [17] to handle the path planning of thousands of pedestrians in real time, while ensuring dynamic collision avoidance. The scalability of our approach allows to interactively create and distribute regions of varied interest, where motion planning is ruled by different algorithms. Practically, regions of high interest are governed by a long-term potential field-based approach, while other zones exploit a graph of the environment and short-term avoidance techniques. Our method also ensures pedestrian motion continuity when switching between motion planning algorithms. Tests and comparisons show that our architecture is able to realistically plan motion for many groups of characters, for a total of several thousands of people in real time, and in varied environments.

The goal of our architecture is to handle thousands of pedestrians in real time. We exploit the vertex structure described in Sect. 6.4.1 to divide the environment into regions ruled by different motion planning techniques. Regions of interest (ROI) can be defined in any number and anywhere in the walkable space with high-level parameters, modifiable at runtime.

By defining three different ROI, we obtain a simple and flexible architecture for realistic results: ROI 0 is composed of vertices of high interest, ROI 1 regroups vertices of low interest, and ROI 2 contains all other vertices, of no interest. For regions of no interest (ROI 2), path planning is ruled by the navigation graph. Pedestrians are linearly steered to the list of waypoints on their path edges. To use the minimal computation resources, obstacle avoidance is not handled. Path planning in regions of low interest (ROI 1) is also ruled by the navigation graph. To steer pedestrians to their waypoints, an approach similar to Reynolds' [18] is used, and obstacles are avoided with an agent-based short-term algorithm. Although agent-based, this algorithm works at low level, and thus stays simple and efficient. In the regions of high interest (ROI 0), path planning and obstacle avoidance are both ruled by a potential field-based algorithm, similar to Treuille et al. [16]. Figure 6.10 shows a crowd moving using the hybrid path planning algorithm.

6.5 Controlling Individual and Group Behavior in Crowds

6.5.1 Introduction

To make believable a crowd simulation is not just to generate many various individual characters. Modelling the behavior of the individuals, the groups, and even the crowd itself is very important. An important aspect is to give the feeling that people are aware of the environment and the other people. For this objective, the role of gaze control is essential. Group behavior is also a key issue. For example, in

Fig. 6.10 Crowd using hybrid path planning

real cities, many pedestrians are part of a group, whether they are sitting, standing, or walking toward their shared goal. They behave differently than if they were alone: they adapt their pace to the other members, wait for each other, may get separated in crowded places to avoid collisions, but regroup afterwards. Finally, we explain in this section a method to generate unlimited cities or streets sing crowd patches.

6.5.2   Gaze Control

We can improve the realism of a crowd simulation by allowing its pedestrians to be aware of their environment and of the other characters present in this environment. They can even seem to be aware of a user interacting with this environment. We introduced [19] the various setups which allow for crowd characters to gaze at environment objects, other characters or even a user. Finally, we developed a method to add these attention behaviors in order for crowd characters to seem more individual.

The first step is to define the interest points, i.e. the points in space which we consider interesting and which therefore attract the characters' attention. We use several different methods to do this depending on the result we want to obtain:

Fig. 6.11 An example depicting the types of possible gaze behaviors

- The interest points can be defined as regions in space which have been described as interesting. In this case, they will be static.
- They can be defined as characters evolving in space. All characters may then potentially attract the attention of other characters as long as they are in their field of view. In this case, we have dynamic constraints, since the characters move around.
- They can be defined as a user if we track a user interacting with the system. A coupled head- and eye-tracking setup allows us to define the position of the user in the 3D space. Characters may then look at the user.

The second step to obtain the desired attention behaviors consists of computing the displacement map which allows for the current character posture to achieve the gaze posture, i.e. to satisfy the gaze constraints. Once the displacement map has been computed, it is dispatched to the various joints composing the eyes, head, and spine in order for each to contribute to the final posture. Finally, this displacement is propagated in time in order for the looking or looking away motions to be smooth, natural, and human-like. Figure 6.11 shows virtual humans with gaze.

6.5.3 Group Behavior

The behavior of people in a crowd is a fascinating subject: crowds can be very calm but also rise to frenzy, they can lead to joy but also to sorrow. It is quite a common idea that people not only behave differently in crowd situations, but that they undergo some temporary personality change when they form part of a crowd. Most writers in the field of mass- or crowd-psychology agree that the most discriminating property of crowd situations is that normal cultural rules, norms and organization forms cease to be applicable For instance in a panic situation the normal rule of waiting for your turn, and the concomitant organization form of the queue, are violated and thus become obsolete.

In Musse et al. [20], the model presents a simple method for describing the crowd behavior through the group inter-relationships. Virtual actors only react in the presence of others, e.g., they meet another virtual human, evaluate their own emotional parameters with those of the other one and, if they are similar, they may walk together. The group parameters are specified by defining the goals (specific positions which each group must reach), number of autonomous virtual humans in the group and the level of dominance of each group. This is followed by the creation of virtual humans based on the groups' behavior information. The individual parameters are: a list of goals and individual interests for these goals (originated from the group goals), an emotional status (random number), the level of relationship with the other group members (based on the emotional status of the agents from a same group) and the level of dominance (which follows the group trend). With these rules, we can model the following sociological effects:

- *Grouping* of individuals depending on their inter-relationships and the *domination* effect;
- *Polarization* and the *sharing* effects as the influence of the emotional status and domination parameters; and finally,
- *Adding* in the relationship between autonomous virtual humans and groups.

The group behavior is formed by two behaviors: seek goal, that is the ability of each group to follow the direction of motion specified in its goals, e.g. in the case of a visit to a museum, the agents walk in the sense of its goals; and the flocking (ability to walk together), has been considered as a consequence of the group movement based on the specific goals during a specific time.

Generally, the available computational resources to trigger intelligent behaviors are very limited in crowds, because their navigation, animation, and rendering are already very expensive tasks that are absolutely paramount. Our approach to this problem is to find a trade-off that simulates intelligent behaviors, while remaining computationally cheap. In [10], we describe the various experiments performed to improve pedestrians behaviors. To make crowd movements more realistic, a first important step is to identify the main places where many people tend to go, i.e., places where there is a lot of pedestrian traffic. It can be a shopping mall, a park, a circus, etc. Adding meta-information to key places in an environment has been achieved in many different ways. Our approach is to use the navigation graph of an environment to hold this meta-information, which is a very advantageous solution: instead of tagging the meshes of an environment, or creating a new dedicated informational structure, we directly work on the structure that is already present, and which is used for path planning and pedestrian steering.

Fig. 6.12 *Left*: A procedurally computed pedestrian street, where patches are generated at runtime. *Right*: The same image revealing the patch borders and their trajectories. (**a**) Density profile. (**b**) Velocity profile

6.5.4 Crowd Patches

We break classical crowd simulation limitations on the environment dimensions: instead of pre-computing a global simulation dedicated to the whole environment, we independently pre-compute the simulation of small areas, called crowd patches [21]. To create virtual populations, the crowd patches are interconnected to infinity from the spectator's point of view. We also break limitations on the usual durations of pre-computed motions: by adapting our local simulation technique, we provide periodic trajectories that can be replayed seamlessly and endlessly in loops over time.

Our technique is based on a set of patch templates, having specific constraints on the patch content, e.g., the type of obstacles in the patch, the human trajectories there, etc. A large variety of different patches can be generated out of a same template, and then be assembled according to designers' directives.

Patches can be pre-computed to populate the empty areas of an existing virtual environment, or generated online with the scene model. In the latter case, some of the patches also contain large obstacles such as the buildings of a virtual city.

Patches are geometric areas with convex polygonal shapes. They may contain static and dynamic objects. Static objects are simple obstacles with its geometry is fully contained inside the patch. Figure 6.12 shows an example.

Larger obstacles, such as buildings, are handled differently. Dynamic objects are animated: they are moving in time according to a trajectory $\tau(t)$. In this context, we want all dynamic objects to have a periodic motion (of period π) in order to be seamlessly repeated in time.

Two categories of dynamic objects may be distinguished: endogenous and exogenous objects. The trajectory of endogenous objects remains inside the geometrical limits of the patch for the whole period. The point's trajectory is fully contained in the patch and respects the periodicity condition (1). If the animation is looped with a period π, the point appears to be moving endlessly inside the patch. Note that static objects can be considered as endogenous objects, with no animation.

Exogenous objects have a trajectory $\tau(t)$ that goes out of the patch borders at some time, and thus, does not meet the periodicity condition (1). In order to en-force this condition, we impose the presence of another instance of the same exogenous object whose trajectory is $\tau'(t)$. As the two objects are of the same type, i.e., they have an identical kinematics model, their trajectories can be directly compared. Different cases are then to be distinguished and are discussed in [17].

We build environments and their population by assembling patches. Thus, two adjacent patches have at least one common face. They also share identical limit conditions for exogenous objects' trajectories. Indeed, when an exogenous object goes from one patch to an adjacent one, it first follows the trajectory contained by the first patch, and then switches to the one described by the second patch. These two trajectories have to be at least continuous $C0$ to ensure a seamless transition from the first patch to the second one. The patterns between the two adjacent patches allow to share these limit conditions.

6.6   Applications

We can conclude this chapter with a few applications and the challenges associated to them.

6.6.1   Virtual Heritage

Based on archaeological data, we have presented the different steps of our work to generate the ancient city of Pompeii and populate it with virtual romans [22] (see Fig. 6.13). Thanks to the semantic data labeled in the geometry, crowds are able to exhibit particular behaviors relative to their location in the city. Our results and show that we are able to simulate several thousands of virtual characters in the reconstructed city in real-time. The use of a procedural technique for the creation of city models has proven to be very flexible and allows for quick variations and tests not possible with manual editing techniques.

One possible challenging work would be to make the Virtual Romans interact with the model, e.g., opening doors. This would allow creating more intelligent and varied behaviors for crowds.

Fig. 6.13 Roman crowd in Pompeii

6.6.2 Agoraphobia Treatment

We developed an application allowing for characters to perform gazing motions in a real-time virtual crowd in a CAVE environment [23]. Moreover, it allows for users to interact with those crowd characters. It is an adaptation of the model of visual attention described in [16] in order to integrate it in a crowd engine and al-low for the method to function online (in real-time). Certain aspects of the automatic interest point detection have been greatly simplified. The existing architecture has been also modified in order to abide with the limitations induced by the real-time implementation.

The final application consists in a city scene, projected in a CAVE setup, in which a crowd of characters walks around (see Fig. 6.14). The application uses a Phase-space optical motion capture device to evaluate where a user is looking and more specifically, which character he/she is looking at. Finally, we further enhance this setup with an RK-726PCI pupil/corneal reflection tracking device in order to evaluate more precisely where a user is looking. The system then allows for the crowd characters to react to user gaze. For example, since we can determine the user's position and orientation in the virtual world, the characters can look at the user.

6.6.3 Transportation and Urbanism

Virtual crowds are used for simulation of new train stations and airports. A challenge would be to introduce natural motivations to simulate more complex and realistic

Fig. 6.14 Agoraphobia
treatment

situations. For example, in an airport, people should not just check in, go to
the security then the gate, as in most simulations. They should be able to go
to restaurants, cafes, shops, toilets, according to their internal motivations. Such
models exist, but the problem is that it will be extremely CPU intensive to introduce
them.

Acknowledgements Most of this research has been performed at the VRlab in EPFL, directed by
the first author.

References

1. Musse, S.R., Thalmann, D.: A model of human crowd behavior. In: Proceedings of the
   Eurographics Workshop on Computer Animation and Simulation '97, Budapest, pp. 39–51.
   Springer, Wien (1997)
2. Thalmann, D., Musse, S.R.: Crowd Simulation, 2nd edn. Springer, London (2012)
3. Dobbyn, S., Hamill, J., O'Conor, K., O'Sullivan, C.: Geopostors: a realtime geometry/impostor
   crowd rendering system. In: SI3D '05: Proceedings of the 2005 Symposium on Interactive 3D
   Graphics and Games, New York, pp. 95–102. ACM (2005)
4. Maïm, J., Yersin, B., Pettré, J., Thalmann, D.: YaQ: an architecture for real-time navigation
   and rendering of varied crowds. IEEE Comput. Grap. Appl. **29**(4), 44–53 (2009)
5. McDonnell, R., Larkin, M., Dobbyn, S., Collins, S., O'Sullivan, C.: Clone attack! per-ception
   of crowd variety. ACM Trans. Graph. **27**(3), 1–8 (2008)
6. Glardon, P., Boulic, R.Thalmann, D.: Robust on-line adaptive footplant detection and enforce-
   ment for locomotion. Vis. Comput. **22**(3), 194–209 (2006)
7. Glardon, P., Boulic, R., Thalmann, D.: PCA-based walking engine using motion capture data.
   In: Proceedings of the Computer Graphics International. IEEE Computer Society, Washington,
   DC, USA (2004)

8. Tecchia, F., Loscos, C., Chrysanthou, Y.: Visualizing crowds in real-time. Comput. Graph. Forum **21**(4), 753–765 (2002)
9. Gosselin, D., Sander, P.V., Mitchell, J.L.: Drawing a crowd. In: Engel, W. (ed.) ShaderX3: Advanced Rendering Techniques in DirectX and OpenGL. Charles River Media, Cambridge (2004)
10. Magnenat-Thalmann, N., Seo, H., Cordier, F.: Automatic modeling of virtual humans and body clothing. In: Proceedings of SIGGRAPH ACM, New York, pp. 19–26 (2003)
11. Yersin, B., Maïm, J., Thalmann, D.: Unique instances for crowds. IEEE Comput. Graph. Appl. **29**(6), 82–90 (2009)
12. Lamarche, F., Donikian, S.: Crowd of virtual humans: a new approach for real time navigation in complex and structured environments. Comput. Graph. Forum **23**(3), 509–518 (2004)
13. Pelechano, N., Allbeck, J., Badler, N.: Controlling individual agents in high-density crowd simulation. In: SCA '07, ACM/Eurographics, NY and Geneva (2007)
14. Pettré, J., de Heras Ciechomski, P., Maïm, J., Yersin, B., Laumond, J.-P., Thalmann, D.: Real-time navigating crowds: scalable simulation and rendering. J. Vis. Comput. Animat. **17**(3–4), 445–455 (2006)
15. Pettre, J., Grillon, H., Thalmann, D.: Crowds of moving objects: navigation planning and simulation. In: Proceedings of IEEE International Conference on Robotics and Automation. IEEE Computer Society, Washington, DC, USA, pp. 3062–3067 (2007)
16. Treuille, A., Cooper, S., Popovic, Z.: Continuum crowds. In: Proceedings of the SIGGRAPH 2006, ACM, New York, USA, pp. 1160–1168 (2006)
17. Morini, F., Yersin, B., Maïm, J., Thalmann, D.: Real-time scalable motion planning for crowds. Vis. Comput. **24**(10), 859–870 (2008)
18. Reynolds, C.W.: Steering behaviors for autonomous characters. In: Proceedings of Game Developers Conference, San Jose, pp. 763–782 (1999)
19. Grillon, H., Thalmann, D.: Simulating gaze attention behaviors for crowds. Comput. Animat. Virtual Worlds **3–4**, 111–119 (2009)
20. Musse, S.R., Thalmann, D.: A hierarchical model for real time simulation of virtual human crowds. IEEE Trans. Vis. Comput. Graph. **7**(2), 152–164 (2001)
21. Yersin, B., Maïm, J., Pettré, J., Thalmann, D.: Crowd patches: populating large-scale virtual environments for real-time applications. Proceedings of I3D, ACM, New York (2009)
22. Maïm, J., Haegler, S., Yersin, B., Mueller, P., Thalmann, D., Van Gool, L.: Populating ancient pompeii with crowds of virtual romans. Proceedings of the VAST 2007, Eurographics Association, Geneva, pp. 109–116 (2007)
23. Peternier, A., Cardin, S., Vexo, F., Thalmann, D.: Practical design and implementation of a CAVE environment. In: Proceedings of the 2nd International Conference on Computer Graphics, Theory and Applications GRAPP 2007, Barcelona (2007)

Chapter 7
Authoring Multi-actor Behaviors in Crowds with Diverse Personalities

Mubbasir Kapadia, Alexander Shoulson, Funda Durupinar, and Norman I. Badler

Abstract Multi-actor simulation is critical to cinematic content creation, disaster and security simulation, and interactive entertainment. A key challenge is providing an appropriate interface for authoring high-fidelity virtual actors with feature-rich control mechanisms capable of complex interactions with the environment and other actors. In this chapter, we present work that addresses the problem of behavior authoring at three levels: Individual and group interactions are conducted in an event-centric manner using parameterized behavior trees, social crowd dynamics are captured using the OCEAN personality model, and a centralized automated planner is used to enforce global narrative constraints on the scale of the entire simulation. We demonstrate the benefits and limitations of each of these approaches and propose the need for a single unifying construct capable of authoring functional, purposeful, autonomous actors which conform to a global narrative in an interactive simulation.

7.1 Introduction

Multi-actor simulation is a critical component of cinematic content creation, disaster and security simulation, and interactive entertainment. Depending on the application, a simulation may involve two or three actors interacting in complex ways, a group of actors participating in an event, or a large crowd with hundreds and thousands of actors. For example, a user may want to author huge armies in movies, the repercussions of a car accident in a busy city street, the reactions of a crowd to

M. Kapadia (✉)
Center for Human Modeling and Simulation, University of Pennsylvania, Philadelphia, PA, USA
e-mail: mubbasir.kapadia@gmail.com

A. Shoulson • F. Durupinar • N.I. Badler
University of Pennsylvania, Philadelphia, PA, USA
e-mail: shoulson@seas.upenn.edu; fundad@seas.upenn.edu; badler@seas.upenn.edu

S. Ali et al. (eds.), *Modeling, Simulation and Visual Analysis of Crowds*, The International   147
Series in Video Computing 11, DOI 10.1007/978-1-4614-8483-7__7,
© Springer Science+Business Media New York 2013

a disturbance, a virtual marketplace with buyers and vendors haggling for prices, and thieves that are on the lookout for stealing opportunities. The existing baseline for multi-actor simulations consists of numerous relatively independent walking pedestrians. While their visual appearances may be quite varied, their behavioral repertoire is not, and their interactions are generally limited to attention control and collision avoidance. The next generation of interactive virtual world applications require functional, purposeful, heterogeneous actors with individual personalities and desires, while exhibiting complex group interactions, and conforming to global narrative constraints.

Readily authoring such complex multi-actor situations is an open problem. Existing techniques are often a bottleneck in the production process, requiring the author to either manually script every detail in an inflexible way or to provide a higher level description that lacks appropriate control to ensure correct or interesting behavior. The challenge is to provide a method of authoring that is intuitive, simple, automatic, yet has enough expressive power to control details at the appropriate level of abstraction. In this chapter, we present work that addresses the problem of behavior authoring at three levels. Our goal is to explain these levels and construct feasible and authorable computational models of when and how they interact.

First, we present a method to capture social crowd dynamics by mapping low-level simulation parameters to the OCEAN personality model. Each personality trait is associated with nominal behaviors – facilitating a plausible mapping of personality traits to existing behavior types. We validate our mapping by conducting a user study which assesses the perception of personality traits in a variety of crowd simulations demonstrating these behaviors [1].

Second, we describe a framework for authoring background characters using an event-centric control model, which shifts behavior authoring from writing complex reactive agents to defining particular activities. Interactions between groups of actors are defined using parameterized behavior trees, and a centralized *Group Coordinator* dispatches events to agents based on their situational and locational context, while satisfying a global distribution of events that is user specified [2–4].

Third, we present a multi-actor planning framework for generating complicated behaviors between interacting actors in a user-authored scenario. Users define the state and action space of actors and *specialize* existing actor definitions to add variety and purpose to their simulation. Actors with dependent goals are grouped together into a set of independent composite domains. For each of these domains, a multi-actor planner generates a trajectory of actions for all actors to meet the desired behavior. We author and demonstrate a simulation of more than 100 pedestrians and vehicles in a busy city street and inject heterogeneity and drama into our simulation using specializations [5].

The rest of this document is articulated as follows. Section 7.2 reviews prior work in behavior authoring for interactive virtual characters. Section 7.3 describes the use of the OCEAN personality model to capture social crowd dynamics. Section 7.4 presents an event-centric paradigm for authoring multi-actor interactions, and Section 7.5 proposes the use of domain-independent planning for behavior generation. Finally, Section 7.6 discusses the comparative benefits and limitations of each of these approaches, and proposes the need for a single unifying construct capable of authoring functional, purposeful, autonomous actors which conform to a global narrative in an interactive simulation.

7.2 Related Work

Behavioral animation in crowds has been studied extensively from many different perspectives [6,7] which can be broadly classified into three overlapping categories: (1) steering based approaches, (2) cognitively based approaches and, (3) narrative driven approaches. Many implementations blend aspects of these three categories – steering-based models in particular are often used in concert with one of the two other approaches. However, since each model has a very different approach to agent control and motivation it is difficult to evenly incorporate all three.

Steering based approaches. These techniques focus on agent movement with a focus on collision avoidance and trajectory planning. Centralized techniques [8–11] focus on the system as a whole, modeling flow characteristics rather than individual pedestrians. Particle based approaches [12, 13] simulate agents using particle dynamics. Social force based approaches [14–17] simulates physical as well as psychological forces between steering agents. Cellular Automata models [18–20] simulate agents defined as mathematical idealizations for physical systems in which space and time are discretized. Rule-based approaches [21–23] use carefully designed conditions and heuristics to define agent behavior. Data-driven methods [24, 25] use real world data to derive steering choices. The works of [26–28] use predictions in the space-time domain to perform steering in environments populated with dynamic threats. Local field methods [29, 30] uses egocentric fields to model agent affordances. The work in [31, 32] uses space-time planning in different problem domains to produce collision-free trajectories, and recent work [33] demonstrates a synthetic vision-based approach for steering.

Cognitively-based Approaches. These techniques populate virtual worlds with rich individual agents which sense the environment and other agents, and act based on personalized desires, motivations, and other attributes such as mood and emotions. Agent decision-making is simulated using a wide variety of cognitively based models such as decision networks [34], neural networks [35], partially-observable Markov decision problems [36], fuzzy logic [37], hierarchical state machines [38], scripts [39–41], and planners [42]. These models capture domain specific knowledge, effectual actions, and personal agent goals to simulate functional, purposeful autonomous agents [43]. Several studies represent individual differences through psychological states [17, 44]. The OCEAN personality model [45] and the OCC emotion model [46] are commonly used in the simulation of autonomous agents. Such models aid to improve believability of embodied conversational characters [47, 48] as well as agents in a crowd [49].

Narrative Driven Approaches. These systems orchestrate the behavior of actors in a scene from a global scope, dictating actions to participants based on the needs of the scenario constraints rather than the individual agents' motivations. Drama Managers [50] are used weave a story around the actions of a player and the principal actors in the environment. Director-based systems such as Facade [51], Thespian [52], Mimesis [53], and others act upon a small number of high-dimensional agents representing principal characters in the simulation. These systems can be controlled by a planning approach [54], using actions and preconditions as a dynamic script for the intended plot. Smart Events [55] externalize behavior logic to authored events that occur in the environment. Unlike cognitively-driven simulations, virtual actors respond to impulses sent by a central controller responsible for enforcing the constraints of a global narrative system.

7.3   The Impact of the OCEAN Personality Model on the Perception of Crowds

Personality is the sum of a person's behavioral, temperamental, emotional, and mental traits. A popular model that describes personality is the Five Factor, or OCEAN (openness, conscientiousness, extroversion, agreeableness, and neuroticism) model. The personality space is composed of these five orthogonal dimensions.

- *Openness* describes a dimension of personality that portrays the imaginative and creative aspect of human character. Appreciation of art, inclination towards going through new experiences and curiosity are characteristics of an open individual.
- *Conscientiousness* determines the extent to which an individual is organized, tidy and careful.
- *Extroversion* is related to the social aspect of human character.
- *Agreeableness* is a measure of friendliness, generosity and the tendency to get along with other people.
- *Neuroticism* refers to emotional instability and the tendency to experience negative emotions. Neurotic people tend to be too sensitive and they are prone to mood swings.

Each factor is bipolar and composed of several traits, which are essentially the adjectives that are used to describe people [56]. Some of the relevant adjectives describing each of the personality factors for each pole are given in Table 7.1.

We have mapped these trait terms to the low-level behavior parameters in the HiDAC (High-Density Autonomous Crowds) crowd simulation system. HiDAC models individual differences by assigning each person different psychological and physiological traits. Users normally set these parameters to model the non-uniformity and diversity of a crowd. Our approach frees users of the tedious task of low-level parameter tuning by combining all these behaviors in distinct personality factors.

Table 7.1 Trait-descriptive adjectives

O+ Curious, alert, informed, perceptive
O− Simple, narrow, ignorant
C+ Persistent, orderly, predictable, dependable, prompt
C− Messy, careless, rude, changeable
E+ Social, active, assertive, dominant, energetic
E− Distant, unsocial, lethargic, vigorless, shy
A+ Cooperative, tolerant, patient, kind
A− Bossy, negative, contrary, stubborn, harsh
N+ Oversensitive, fearful, dependent, submissive, unconfident
N− Calm, independent, confident

By incorporating a standard personality model to a high-density crowd simulation, our approach creates plausible variations in the crowd and enables novice users to dictate these variations. A crowd consists of subgroups with different personalities. Variations in the characteristics of subgroups influence emergent crowd behavior. The user can add any number of groups with shared personality traits and can edit these characteristics during the course of an animation.

In order to verify the plausibility of our mapping we have conducted tests that evaluate users' perception of the personality traits in the generated animations. We created several animations to examine how modifying the personality parameters of subgroups affects global crowd behavior. The animations exhibit the emergent behaviors of agents in scenarios in which the settings assigned according to the OCEAN model drive crowds behavior. In order to validate our system, we determined the correspondence between our mapping and the users' perception of these trait terms in the videos. The results indicate a high correlation between our parameters and the participants perception of them.

7.3.1 Personality-to-Behavior Mapping

A crowd is composed of subgroups with different personalities. Variations in the characteristics of the subgroups influence emergent crowd behavior. The user can add any number of groups with shared personality traits and can edit these characteristics during the course of an animation. An agent's personality π is a five-dimensional vector, where each dimension is represented by a personality factor, ψ_i. The distribution of the personality factors in a group of individuals is modeled by a Gaussian distribution function N with mean μ_i and standard deviation σ_i:

$$\pi = <\psi_O, \psi_C, \psi_E, \psi_A, \psi_N> \tag{7.1}$$

$$\psi_i = N(\mu_i, \sigma_i^2), \; for \; i \in \{O, C, E, A, N\}, \tag{7.2}$$

where $\mu \in [0,1]$ and $\sigma \in [-0.1, 0.1]$.

Table 7.2 Low-level parameters vs. trait-descriptive adjectives

Leadership	Dominant, assertive, bossy, dependable, confident, unconfident, submissive, dependent, social, unsocial	E, A−, C+, N
Trained/not trained	Informed, ignorant	O
Communication	Social, unsocial	E
Panic	Oversensitive, fearful, calm, orderly, predictable	N, C+
Impatience	Rude, assertive, patient, stubborn, tolerant, orderly	E+, C, A
Pushing	Rude, kind, harsh, assertive, shy	A, E
Right preference	Cooperative, predictable, negative, contrary, changeable	A, C
Avoidance/personal space	Social, distant	E
Waiting radius	Tolerant, patient, negative	A
Waiting timer	Kind, patient, negative	A
Exploring environment	Curious, narrow	O
Walking speed	Energetic, lethargic, vigorless	E
Gesturing	Social, unsocial, shy, energetic, lethargic	E

An individual's overall behavior β is a combination of different behaviors. Each behavior is a function of personality as:

$$\beta = (\beta_1, \beta_2, \ldots, \beta_n) \tag{7.3}$$

$$\beta_j = f(n), \; for \; j = 1, \ldots, n \tag{7.4}$$

Since each factor is bipolar, ψ can take both positive and negative values. For instance, a value of 1 for extroversion means that the individual has extroverted character; whereas a value of -1 means that the individual is highly introverted.

By analyzing the meaning and usage of each low-level parameter and built-in behavior in the HiDAC model, we characterize these by the adjectives that are used to describe personalities. Thus, we devise a mapping between the agents' personality factors (adjectives) and the HiDAC parameters, as shown in Table 7.2. A positive factor takes values in the range $[0.5, 1]$, whereas a negative factor takes values in the range $[0, 0.5)$. A factor given without any sign indicates that both poles apply to that behavior. For instance E+ for a behavior means that only extroversion is related to that behavior; introversion is not applicable. As indicated in Table 7.2, a behavior can be defined by more than one personality dimension. The more adjectives of a certain factor defined for a behavior, the stronger is the impact of that factor on that behavior. We assign a weight to the factor's impact on a specific behavior. The sum of the weights for a specific type of behavior is 1. In order to understand how the mapping from a personality dimension to a specific type of behavior is performed, we explain four representative mappings. The remaining ones are mathematically similar.

Right preference. When the crowd is dispersed, individuals tend to look for avoidance from far away and they prefer to move towards the right hand side of the obstacle they are about to face. This behavior shows the individual's level of conformity to the rules A disagreeable or non-conscientious agent makes a right or left preference with equal probability, while the probability of choosing the right side increases with increase in values of agreeableness and conscientiousness. Given $P_i(Right) \alpha A, C$ and $\beta_i^{Right} \in \{0,1\}$, right preference $P_i(Right)$ is computed as follows

$$P_i(Right) = \begin{cases} 0.5 & \text{if } \psi_i^A < 0 \text{ or } \psi_i^C < 0 \\ \omega_{AR}\psi_i^A + \omega_{CR}\psi_i^C & \text{otherwise} \end{cases} \tag{7.5}$$

$$\beta_i^{Right} = \begin{cases} 1 \text{ if } P_i(Right) \geq 0.5 \\ 0 \text{ otherwise} \end{cases} \tag{7.6}$$

Personal space. Personal space determines the territory in which an individual feels comfortable. Agents try to preserve their personal space when they approach other agents and when other agents approach from behind. However, these two values are not the same. According to the research on Western cultures, the average personal space of an individual is found to be 0.7 m in front and 0.4 m behind [57]. Given $\beta_i^{PersonalSpace} \alpha^{-1} E$ and $\beta_i^{PersonalSpace} \in \{0.5, 0.7, 0.8\}$, the personal space of an agent i with respect to an agent j is computed as follows

$$\beta_{i,j}^{PersonalSpace} = \begin{cases} 0.8 \, f(i,j) \text{ if } \psi_i^E \in [0, \frac{1}{3}) \\ 0.7 \, f(i,j) \text{ if } \psi_i^E \in [\frac{1}{3}, \frac{2}{3}] \\ 0.5 \, f(i,j) \text{ if } \psi_i^E \in (\frac{2}{3}, 1] \end{cases} \tag{7.7}$$

$$f(i,j) = \begin{cases} 1 & \text{if } i \text{ is behind } j \\ \frac{0.4}{0.7} & \text{otherwise} \end{cases} \tag{7.8}$$

Waiting radius. In an organized situation, individuals tend to wait for space available before moving. This waiting space is called the waiting radius and it depends on the kindness and consideration of an individual, i.e., the agreeableness dimension. Given $\beta_i^{WaitingRadius} \alpha A$ and $\beta_i^{WaitingRadius} \in \{0.25, 0.45, 0.65\}$, the waiting radius is computed as follows

$$\beta_{i,j}^{WaitingRadius} = \begin{cases} 0.25 \text{ if } \psi_i^A \in [0, \frac{1}{3}) \\ 0.45 \text{ if } \psi_i^A \in [\frac{1}{3}, \frac{2}{3}] \\ 0.65 \text{ if } \psi_i^A \in (\frac{2}{3}, 1] \end{cases} \tag{7.9}$$

Walking speed. The maximum walking speed is determined by an individual's energy level. As extroverts tend to be more energetic while introverts are more lethargic, this parameter is controlled by the extroversion trait. Given $\beta_i^{WalkingSpeed} \alpha E$ and $\beta_i^{WalkingSpeed} \in [1,2]$, the walking speed is computed as follows

$$\beta_i^{WalkingSpeed} = \psi_i^E + 1. \tag{7.10}$$

7.3.2   User Studies on Personality

In order to evaluate if the suggested mappings are correctly perceived, we conducted user studies. We created several animations to see how global crowd behavior is affected by modifying the personality parameters of subgroups. Some of these animations can be found at http://cg.cis.upenn.edu/hms/research/Ocean/.

7.3.2.1   Experiment Design

We created 15 videos presenting the emergent behaviors of people in various scenarios where the crowds' behavior is driven by the settings assigned through the OCEAN model. We performed the mapping from HiDAC parameters to OCEAN factors by using trait-descriptive adjectives. We determined the correspondence between our mapping and the users' perception of these trait terms in the videos in order to validate our system. Seventy subjects (21 female, 49 male, ages 18–30) participated in the experiment. We showed the videos to the participants through a projected display and asked them to fill out a questionnaire consisting of 123 questions– about 8 questions per video. The videos were shown one by one; after each video, participants were given some time to answer the questions related to the video. The participants did not have any prior knowledge about the experiment. Questions assessed how much a person agreed with statements such as "I think the people in this video are kind." or "I think the people with black suits are calm." We asked questions that included the adjectives describing each OCEAN factor instead of asking directly about the factors because we assumed that the general public might be unfamiliar with the OCEAN model. Participants chose answers on a scale from 0 to 10, where 0 = totally disagree, 5 = neither agree nor disagree, and 10 = totally agree. We omitted the antonyms from the list of adjectives for the sake of conciseness. The remaining adjectives were *assertive,calm, changeable, contrary, cooperative, curious, distant, energetic, harsh, ignorant, kind, orderly, patient, predictable, rude, shy, social, stubborn,* and *tolerant.*

7.3.2.2   Sample Scenarios

A sample scenario testing the impact of openness took place in a museum setting as one of the key factors determining openness is the belief in the importance of art.

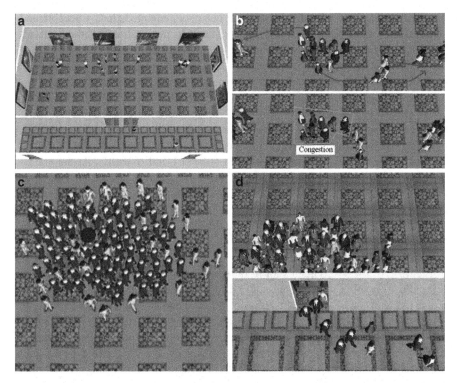

Fig. 7.1 Snapshots of a crowd simulation authored using our framework: (**a**) Openness tested in a museum. The most open people stay the longest, whereas the least open people leave the earliest. (**b**) People with low conscientiousness and agreeableness values cause congestion. (**c**) Ring formation where extroverts are inside and introverts are outside. (**d**) Neurotic, non-conscientious and disagreeable agents show panic behavior

Figure 7.1a shows a screen-shot from the sample animation. We tested the adjectives curiosity and ignorance with this scenario. There were three groups of people, with openness values of 0, 0.5, and 1. We mapped the number of tasks that each agent must perform to openness, with each task requiring looking at a painting. The least open agents (with blue hair) left the museum first, followed by the agents with openness values of 0.5 (with black hair). The most open agents (with red hair) stayed the longest.

In order to test whether the personalities of people creating congestion are distinguished, we showed the participants two videos of same duration and asked them to compare the characteristics of the agents in each video. Each video consisted of two groups of people moving through each other. The first video showed people with high agreeableness and conscientiousness values ($\mu = 0.9$ and $\sigma = 0.1$ for both traits), whereas the second video showed people with low agreeableness and conscientiousness values ($\mu = 0.1$ and $\sigma = 0.1$ for both traits). In the first video, groups managed to cross each other while in the second video congestion occurred after a fixed period of time. Such behaviors emerged since agreeable and conscientious individuals are more patient; they do not push each other and are always predictable, as they prefer to move on the right side. Figure 7.1b shows how congestion occurred due to low conscientiousness and agreeableness. People were stuck at the center and refused to let other people move. They were also *stubborn*, *negative*, and not *cooperative*.

Another video assessed how extroverts and introverts were perceived according to their distribution around a point of attraction. Figure 7.1c shows a screen-shot from the video in which the agents in blue suits are extroverted ($\mu = 0.9$ and $\sigma = 0.1$) and those in grey suits are introverted ($\mu = 0.1$ and $\sigma = 0.1$). At the end of the animation, introverts were left out of the ring structure around the object of attraction. Because extroverts are faster, they approached the attraction point in less time. In addition, when other agents blocked their way, they tended to push them to reach their goal. The figure also shows the difference between the personal spaces of extroverts and introverts. This animation tested the adjectives, *social*, *distant*, *assertive*, *energetic*, and *shy*.

Figure 7.1d shows a screen-shot from the animation demonstrating the effect of neuroticism, non-conscientiousness and disagreeableness on panic behavior. Five of the 13 agents had neuroticism values of $\mu = 0.9$ and $\sigma = 0.1$, conscientiousness values of $\mu = 0.1$ and $\sigma = 0.1$ and agreeableness values of $\mu = 0.1$ and $\sigma = 0.1$. The other agents, which are psychologically stable, have neuroticism values of $\mu = 0.1$ and $\sigma = 0.1$, conscientiousness values of $\mu = 0.9$ and $\sigma = 0.1$ and agreeableness values of $\mu = 0.9$ and $\sigma = 0.1$. The agents in black suits are neurotic, less conscientious, and disagreeable. The figure shows that they tend to panic more, push other agents, force their way through the crowd, and rush to the door. They are not *predictable*, *cooperative*, *patient*, or *calm* but they are *rude*, *changeable*, *negative*, and *stubborn*.

7.3.2.3 Analysis

After collecting the participants' answers for all the videos, we first organized the data for the adjectives. Each adjective is classified by its question number, the actual simulation parameter and the participants' answers for the corresponding question. We calculated the Pearson correlation (r) between the simulation parameters and the average of the subjects' answers for each question.

We grouped the relevant adjectives for each OCEAN factor to assess the perception of personality traits. The evaluation process is similar to the evaluation of adjectives; this time considering the questions for all the adjectives corresponding to an OCEAN factor. For instance, as openness is related to curiosity and ignorance, we took into account the adjectives *curious* and *ignorant*. Again, we averaged the subjects' answers for each question. Then, we computed the correlation with the parameters and the mean throughout all the questions inquiring *curious* and *ignorant*.

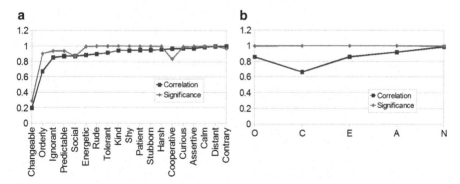

Fig. 7.2 (**a**) The correlation coefficients between the parameters and the subjects answers for the descriptive adjectives, and the significance values for the corresponding correlation coefficients. Significance is low (<0.95) for changeable, orderly, ignorant, predictable, social, and cooperative. (**b**) The correlation coefficients between actual parameters and subjects answers for the OCEAN factors, and the two-tailed probability values for the corresponding correlation coefficients. All the coefficients have high significance

We computed the significance of the correlation coefficients as $1 - p$, where p is the two-tailed probability that is calculated considering the sample size and the correlation value. Higher correlation and significance values suggest more accurate user perception.

7.3.2.4   Results

Figure 7.2a depicts the correlation coefficients and significance values for the adjectives. Significance is low (<0.95) for *changeable, orderly, ignorant, predictable, social* and *cooperative*. Low significance is caused by low correlation values for *changeable* and *orderly*. However, although the correlation coefficients are found to be high for *predictable, ignorant, social* and *cooperative*, low significance can be explained due to small sample size. From the participants' comments, we determined that *changeable* is especially confusing because the participants identified non-conscientious agents as rude but perceived them as persistent in their rudeness.

Orderly is another weakly correlated adjective. Analyzing the results for each video, we found that agents in the evacuation drill scenario were perceived to be orderly although they displayed panic behavior. In these videos, even if the agents pushed each other and moved fast, some kind of order could be observed. This was due to the smooth flow of the crowd during building evacuation. Although people were impatient and rude, the overall crowd behavior appeared orderly. On the other hand, in a scenario showing queuing behavior in front of a water dispenser, the participants could easily distinguish orderly agents from disorderly ones. Orderly agents waited at the end of the queue, whereas disorderly agents rushed to the front.

In this scenario, although the main goal was the same for all the agents (drinking water), there were two distinguishable groups that acted differently.

Figure 7.2b shows the correlation coefficients and their significance for the OCEAN parameters. These values are computed by taking into account all the relevant adjectives for each OCEAN factor. All the coefficients have high significance, with a probability of less than 0.5% of occurring by chance ($p < 0.005$). The significance is high because all the adjectives describing a personality factor are taken into account, achieving sufficiently large sample size.

The correlation coefficient for conscientiousness is comparatively low, showing that the participants correctly perceived only approximately 44% of the traits ($r^2 \approx 0.44$). Low correlation values for *orderly* and *changeable* reduce the overall correlation. If we consider only *rude* and *predictable* for conscientiousness, correlation increases by 18.6%. The results suggest that people can observe the politeness aspect in short-term crowd behavior settings more easily than the organizational aspects. This observation also explains why the perception of agreeableness is highly correlated with the actual parameters.

Figure 7.2 also shows that the participants perceived neuroticism the best. In this study, we have only considered the calmness aspect of neuroticism, which is tested in emergency settings and building evacuation scenarios.

7.4   Coordinating Agent Interactions with Behavior Events

When two actors interact in a virtual world, they must coordinate tasks and exchange information. Managing this complexity is difficult when designing the behavior of each actor in isolation. For sophisticated cooperative or competitive behavior, an actor must constantly communicate and perform actions dependent on both its own and other actors' current state. This call-and-response type of interchange is traditionally authored in pieces across multiple actors, with certain steps anticipating the behavior of another actor involved in the interaction. If a cooperative or competitive behavior involves actors taking on certain roles (such as "leader/follower"), participating actors must negotiate the nature of their participation in the interaction, which further complicates the behavior authoring process.

Consider two agents participating in a transaction involving bargaining over a piece of merchandize. In a localized agent-centric model, the interaction begins when the buyer, B, has a desire for an item that is sold by the seller, S. B approaches S's market stall and displays a greeting animation. S is notified that B played a greeting animation and has to recognizes that B wants to buy something that he (S) has for sale in order to begin the bargaining process. Throughout the interaction, the agents transmit notifications to one another about which animations were played and the current price negotiated. They must regularly receive these notifications and interpret the mode and mood of the interaction, maintaining the state of the conversation and the currently negotiated price, as well as the item in question when deciding on an appropriate response. For a simple sequence of animations, this exhibits a high level of complexity that must be duplicated and maintained in both agents' state. In contrast, suppose a centralized data structure could coordinate these agents instead. The centralized structure selects B and S out of a pool of available agents, instructs B to approach S, dispatches the appropriate animations to each in sequence, and maintains centralized information such as the current mood of the conversation, the item being haggled over, and the current offers from both parties. Moreover, this centralized structure can be invoked immediately to involve the two agents in this kind of interaction, rather than waiting for the whim of actor B to decide that he suddenly wants to purchase an item from S.

This centralized data structure simplifies the process of designing interactions between actors, using a behavior paradigm we call event-centric authoring. Rather than requiring multiple actors to handle the responsibilities of message passing and stimuli response, a behavior event consists as a body of centralized control logic with its own state and unrestricted access to the actor(s) involved in the event. Participating actors temporarily suspend their own autonomy and are controlled entirely by the event structure, which treats them as limbs of a central entity for the duration of its execution. A conversation could be conducted with a central event instructing the two actors to approach one another and then take turns playing the requisite sounds and gesture animations. When the event is completed, or fails, the involved actors resume their own autonomy until co-opted to participate in another event. Events also define roles for their participants that must be filled on instantiation by a particular actor type. For example, an event for a transaction between two actors could may stipulate that the seller be of type "Merchant Actor". Events exist in a system to augment the richness of available behavior, and need not detract from an actor's own individuality. Actors can still retain rich autonomous behaviors and only occasionally be involved in higher-order events.

7.4.1 Parameterized Behavior Trees

Though they could be designed with any suitable method, we create events using Parameterized Behavior Trees (PBTs) [4], an expansion on standard behavior tree models [58, 59] with a specialized data flow architecture to handle multiple agents and shared state data. PBTs, and behavior trees in general, represent a flexible graphical programming language with explicit goal direction that is easy to visualize for complex behavior structures. They contain an inherent hierarchical structure that makes them easy to visualize and conceptualize at a macroscopic level. In general, a subtree represents a goal at its root, and the means by which that goal can be achieved with its leaves. This allows for implicit documentation within the tree structure, so that certain branches of the tree can be understood by the goals they attempt to accomplish, without the need to necessarily expose any of their children.

The core mechanic of behavior trees in general is the success or failure of each node in the tree. Each node attempts to execute an action and reports success or failure to its parent node. The parent can use that information for selecting the

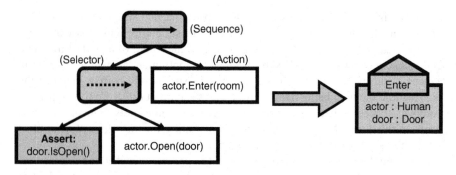

Fig. 7.3 An encapsulated behavior for entering a room through a door

next node to execute. Sequence nodes execute each child in order. If one child of a sequence node fails, that sequence reports failure to its parent and ceases execution. Sequence nodes succeed when all of their children succeed. Conversely, selector nodes cease execution and report success if any one of their children succeeds. A selector node reports failure only if all of its children fail. Figure 7.3 represents a simple behavior tree for opening a door and entering a room. If the assertion succeeds, the selector will propagate success and skip over the door opening action. Otherwise, if the assertion fails, the selector will attempt to perform its next child, which consists of an action opening a door. Assuming one of the two actions is successful, the selector will report success to the root sequence node, which will then execute its next action directing the actor to enter the opened room.

Unlike traditional behavior trees, PBTs are designed with data fields that take on values at runtime and propagate information through the structure of the tree, eventually moving through the tree's action or assertion leaf nodes to the underlying functions those leaves invoke. These data can be targets for low-level character controller, or flags that affect branching decisions within the tree itself. With parameterization, subtrees can be encapsulated for reuse with parameters by multiple PBTs, not unlike a subroutine in a traditional programming language. The right side of Fig. 7.3 shows how the simple door behavior tree can be encapsulated into a generic PBT with the typed parameters "actor" and "door" that take on values at runtime. The single node created by this encapsulation can be accessed by means of a lookup node, which acts as a placeholder for the entire subtree and fills the parameter fields with object references at runtime. Encapsulated subtrees can also act polymorphically, so that different actor types can "implement" a given subtree signature in multiple different ways. When an event invokes an actor's capabilities, or tells an actor to execute a specific subtree with certain parameters, the actual implementation of those actions can still be personalized to the actor or that actor's type.

Parameterized subtrees that take multiple actors as parameters define the logic for behavior events. Figure 7.4 displays an event for two actors playing a game of catch at different levels of the hierarchy. First we define a simple sub-event taking two

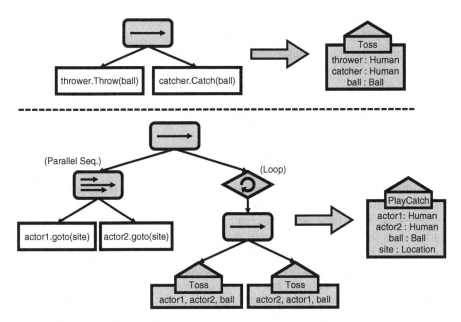

Fig. 7.4 Composing an event for two actors to play catch

actors and animating one tossing a ball to another. This sub-event is encapsulated as a "Toss" tree, with two actor parameters and an object parameter. Next, we define the actual "PlayCatch" event. This event directs the two actors to approach a location (given as a parameter when this tree is instantiated), and loops the actors playing the "Toss" sub-event. Observe that the event reverses the actor roles in the "Toss" subtree so that the ball passes back and forth between actors.

The "PlayCatch" subtree is also encapsulated with parameters, so the "Play-Catch" tree can be instantiated at runtime using entities in the environment for its actor or object roles. At any point during the simulation, a virtual director can select two actors and a location, provide those actors with a ball, and instantiate this event using those objects as parameters. The event will then completely manage the actors' behavior in a centralized manner as they perform the complicated interactions needed for playing a game of catch. This authoring approach avoids issues such as shared ownership of the ball item, since all state information for the event is maintained in the event's own state space, rather than in the state space of one of the actors. Note that the event does not own the ball or the actors involved, but does acquire control over them for the duration of the event. The ball, or any prop, may persist in the world or may be destroyed depending on the virtual director's decision.

In general, we use events for all complex interactive behavior, but actors are also independently controlled by separate PBTs for their autonomous actions. When an actor is not involved in an event, it executes its own internal behavior tree that locally invokes that actor's capabilities and modifies that actor's traits. When an actor is selected to participate in an event, that actor's internal tree is halted, and the actor's traits and capabilities are exposed to be modified and invoked by the PBT contained in the event. In general, events are designed to terminate after the implicit goal of the event is accomplished or fails, while actors' internal trees never terminate.

7.4.2 Selecting Actors for Events

Since events contain centralized behavior for interactions between actors, we expect scenario authors to design a library of events for a variety of interactions between actors of different types within an environment. Events accept parameters for the actors involved and any other modifiers that may affect the progression of the interaction between those actors, including the location in which it takes place. Since events work only on actor types, they are not designed for any one specific actor within the world. Instead, events are designed to operate generically on any selection of actors matching the interfaces expected by the event. At runtime, over the course of the simulation, we use a global director construct to determine which events should be performed in the world, and what actors should perform them.

When the global director decides that a specific event should be performed at a location in the world (a process we will discuss later), that director must select the actors to participate in that interaction. There are several details that factor into this decision. The first filter is the type of the actor. The structure of the event itself specifies that each of its roles can only be filled by an actor of a given hierarchical type, so the director can only select actors of that type when filling that role. Other filters for actor candidates may be more detailed. The director may select an actor for a role based on its traits, its relationships with other characters, or the history of events in which it has participated. These qualities of an actor may be fixed at initialization, or modifiable at runtime.

If a scenario has many events that select actors based on the actions they have performed during that simulation, then the simulation's actors exhibit a quality we call *progressive differentiation*. That is, actors begin as largely homogeneous characters in the world and are slowly differentiated by one another based on the actions they perform at the behest of the director [3]. This kind of differentiation is important because it matches the perception of a human user in the environment. To a user observing the simulation, actors are already undifferentiated because the user has just encountered them for the first time. The user only learns to distinguish actors' personalities based on the actions they perform and the manner in which they are performed. Much in the same way, the director personifies actors at runtime based on the actions it selects for them, and maintains that history to inform selection of actors for further events. An actor selected to tell a joke to a crowd may have that history recorded, making the director more likely to select that actor again for another "joke" event, and progressively characterize that actor as a "comedian" character. Because this differentiation experience happens simultaneously for the director and a human user, selective differentiation accomplishes a key goal of facilitating the user's internal narrative for rationalizing the chain of events presented in the scenario.

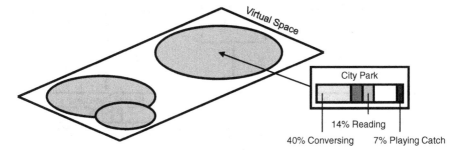

Fig. 7.5 Assigning regions in a virtual space, each with a desired actor activity distribution

7.4.3 Selecting Events to Perform

After the behavior authors for a given scenario create a library of potential events that can be performed in the environment, the centralized director needs a method to pick which events to execute, when, and where in the virtual world. There are many ways a director could be designed to do this, including planning approaches to capture causality in narrative, but we present a simple statistical model suitable for maintaining a variety of actions in the background of a scene. This mostly applies to undifferentiated characters acting as "extras" to add atmosphere to a location [2].

Areas in the environment are annotated as regions, which keep track of the actors currently present and the events in which they are involved. The scenario designer specifies a distribution of event classes that should be enforced within that given region for the actors present. Event classes may be broad categories such as "conversation", or "playing a game", and members of these classes may be more specific, such as "arguing over politics", or "playing catch with a ball". A desired event distribution specifies what percentage of the actors in that environment should be involved in events of a certain class. For instance, in a park, the author may designate that 15% of all actors in the park should be playing games with one another, while another 25% should be conversing. Figure 7.5 illustrates the process of assigning regions over a virtual space and specifying the event distribution for each.

We use a specialized director called a *Group Coordinator* to enforce these distributions. At a given point in time, the Group Coordinator inspects a region and determines the distribution density of the events being performed in that region at that instant. If the instantaneous calculated distribution differs significantly from the expected distribution specified by the scenario author, then the Group Coordinator must correct the error. The director accomplishes this by identifying the most under-represented event category, finding an event from the event library in that category, and invoking that new event on some suitable actors in or near that region. So long as there are suitable actors available to perform events in or near that region, the Group Coordinator will ensure that the actual distribution of events being performed will match the ideal distribution specified by the author within

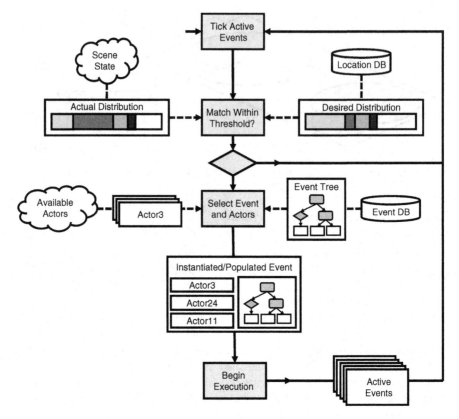

Fig. 7.6 The Group Coordinator's decision process for maintaining the distribution of events in a region

some threshold. This decision process is illustrated in Fig. 7.6. The desired event distribution for a region can also be changed over time, reflecting a gradual shift caused by some phenomenon such as a day/night cycle.

To achieve the greatest event variety, events in this framework should not be causally linked. This technique is suitable for ambient activities performed by characters on which the user won't focus, but prominent actors in the environment must account for causality. If the user does begin to focus on one of these background characters, then that character can be promoted to a principal character and managed by an event dispatch algorithm that does account for causality and differentiation. The goal of this algorithm is a simple interface to greatly affect the perceived variety of the activities being performed in a virtual space. Different regions also allow for different types of activity, where events suitable for a city park would not be appropriate in a movie theater.

We use the behavior trees from this Group Coordinator control process to implement a virtual middle eastern marketplace, as displayed in Fig. 7.7. To accurately

Fig. 7.7 A virtual Middle Eastern marketplace for demonstrating multi-actor interactions simulated using the ADAPT platform [60]

implement this setting, actors must be capable of interacting with one another in situations such as haggling, negotiating, and conversation. By default, actors maintain an individual "shopping list" of items they wish to procure from the stalls in the market. An actor's own tree directs that actor to move between stalls looking for items. When a stall has an item that an actor desires, the buyer and seller are placed in a negotiation event that directs them through gestures and movements representing two individuals haggling over an item. After the negotiation event ends, the buyer has either acquired the item or failed, and moves to the next stall on its shopping list. Periodically, actors are also selected by the Group Coordinator to stop and converse with one another, which suspends whatever else they were doing at the time (unless they were involved in a negotiation). This simple system captures the ambience of characters moving in the background and maintaining a baseline of activity in what should be a busy scene. We leverage the ADAPT platform [60] for generating the simulations described here, which provides the tools for navigation, full-body character animation, and multi-actor behavior using PBTs.

7.5 Authoring Complex Multi-actor Interactions using Domain Independent Planning

This section presents a multi-actor planning framework [5, 61] for generating complicated behaviors between interacting actors in a user-authored scenario. Users define the state and action space of actors and *specialize* existing actor definitions to add variety and purpose to their simulation. Actors with dependent goals are grouped together into a set of independent composite domains. For each of these domains, a multi-actor planner generates a trajectory of actions for all actors to meet the desired behavior. We author and demonstrate a simulation of more than 100 actors (pedestrians and vehicles) in a busy city street and inject heterogeneity and drama into our simulation using specializations.

Figure 7.8 presents an overview of our framework. First, a domain expert defines the problem domain of the actors in the scenario (domain specification). Next, a director specializes the actors using modifiers, constraints, and behaviors (domain

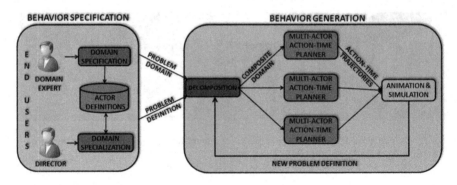

Fig. 7.8 An overview of the framework

specialization). Actors with dependent goals or constraints enforcing their interaction are grouped into a composite domain, forming a set of independent domains. For each domain, a multi-actor planner generates a trajectory of actions that satisfies the composite goal while optimizing each actors objective. Each searchs results become part of a global plan, which generates the resulting simulation.

7.5.1 Behavior Specification

Domain Specification. Domain specification is the lowest abstraction level for authoring behaviors. It involves defining the state space and action space for all actors in a scenario. Each actor has a state and can affect the state of itself or others through actions. Different actors in the same scenario might have different domain specifications. For example, we can define a traditional actor to simulate a pedestrian and define the environment as an actor that can be used to trigger global events, such as natural disasters (Fig. 7.9).

1. **The state space.** We represent an actors state space using metrics - physical or abstract properties that are affected by actions. Users can extend metrics by applying operators to existing metrics to provide an intuitive understanding of the simulations properties. We denote the space of metrics for all actors in the scenario as mi.
2. **The action space.** The action space is a set of actions an actor can perform to modify its state and the state of other actors. An action has three properties: preconditions that determine whether the action is possible in a given state, the actions effect on the state of the actor and target actors, and the cost of executing the action.
3. **Costs.** Costs are a numerical measure of executing an action. Different actions can affect different cost metrics by different amounts. Examples of cost metrics include distance and energy. We denote the space of costs as ci.

Fig. 7.9 Domain Specification and Specialization. Actions, modifiers, constraints and, behaviors defined using our framework

Action *actionName* (*parameters*) {
    Precondition: *conditions on elements of* $\{m_i\}$ *or* $\{c_i\}$
    Effect: *effects on elements of* $\{m_i\}$
    Cost Effect: *effects on elements of* $\{c_i\}$
}

Modifier *modifierName* {
    Precondition: *conditions on elements of* $\{m_i\}$ *or* $\{c_i\}$
    Effect: *effects on elements of* $\{m_i\}$ *or* $\{c_i\}$
}

Constraint *modifierName* {
    Precondition: *conditions on elements of* $\{m_i\}$ *or* $\{c_i\}$
    Constraint: *conditions on elements of* $\{m_i\}$
}

Behavior *behaviorName* {
    Precondition: *conditions on elements of* $\{m_i\}$
    Goal: *conditions on elements of* $\{m_i\}$
    Objective Function: *values of* $\{w_i\}$
}

Domain Specialization. Users can reuse actor definitions across different scenarios by specializing actors in a state-dependent manner without modifying the original definition. This allows authors to specify and generate vastly different, purposeful simulations intuitively, with minimal specification. We provide three ways to specialize actors: effect modifiers, cost modifiers, and constraints.

1. **Modifiers.** Modifiers specialize the effects and costs of actions in a state-dependent manner. For example, users can place an effect modifier on elderly actors to reduce their normal speed of movement. Cost modifiers indicate what actions are in an actors best interest at a particular state. For example, users can author a cautious actor by increasing the cost of actions that might place the actor in danger (for example, entering a burning building). Here, the notion of danger would be a user-specified metric in the actors state space.
2. **Constraints.** Constraints enforce strict requirements on actors; they can prune the choices of an actor in a particular state. For example, constraints can prevent pedestrians from walking on the road or from disobeying traffic signals. Users can also place constraints on the simulations trajectory to author specific events (for example, two cars must collide), generate complex interactions between actors, and direct high-level stories.

Behavior State Machine Specification. A behavior state defines an actors current goal and objective function. The goal is a desired state the actor must reach; the objective function is a weighted sum of costs the actor must optimize. Users can define multiple behaviors for an actor that depend on its current state.

7.5.2 Behavior Generation

The entire problem domain is decomposed into a set of independent composite domains where actors with dependent goals or constraints enforcing their interaction are part of one composite domain. The composite state space is the cartesian product of the states of each actor and the composite action space is the union of the actions of each actor in the composite domain. The composite domain is denoted by Σ.

We define a particular problem instance $P = (\Sigma, s_0, g, \{o_i\})$ by determining the initial state s_0, composite goal g, and objectives $\{o_i\}$ of each actor in the composite domain. A composite goal can be single or multiple objectives for an actor, common or conflicting objectives between actors, as well as global constraints specified for the entire scene. The composite goal, g, is the logical combination of the goals for all actors in the composite domain. We combine common goals using the $\wedge$ operator, indicating that all actors must satisfy these goals. We combine contradicting goals using the $\vee$ operator, indicating that any actor must satisfy its goal. The problem definition $P = (\Sigma, s_0, g, \{o_i\})$ becomes the input for the planner.

7.5.2.1 Multiactor Action-Time Planner

During planning, the heuristic search generates a trajectory of actions for all actors in the composite space that satisfies g while optimizing $\{o_i\}$. This facilitates the generation of complicated interactions between actors, without needing centralized planning across all actors in the scenario. Even though an actors action affect only the state space of its composite domain, the planner determines an actions possibility by considering the global state space of all actors in the scenario. This ensures collision-free trajectories between two independent plans. So, we can overlay the action trajectories for actors in different groups to generate a complete simulation.

Our planner builds on traditional planning approaches in three ways. First, it works in the composite space of multiple actors with competitive or collaborative goals. Second, it explicitly takes into account that different actions take various amounts of time and that actors actions overlap. Finally, it uses an automatically derived heuristic estimate to speed up the search.

Overview. For the current state, our heuristic planner generates a set of possible transitions. Each transition represents the forward simulation of the actions by one time step in the composite space in which actors are simultaneously executing actions. The planner chooses a transition by minimizing the sum of the transitions total cost and the heuristic estimate of reaching the composite goal. It computes a transitions cost such that an action optimizes its own objective function. When the planner reaches a state that satisfies g, it returns the generated plan.

Transitions. A transition represents the simultaneous execution of actions chosen by all actors in the composite domain by one time step. A transition is valid and ready to simulate if all actors have a valid action theyre executing or ready to execute. An action is possible if three conditions are met: (1) the actor is currently not executing an action, (2) the preconditions of the action are satisfied and, (3) no constraints prohibit the action.

For a valid transition, all actions are simulated for one time step in a random order. The explicit modeling of time in the action definition results in overlapping actions, partially executed actions (action failure), and actors performing new actions while other actors are still performing their current action. After a transition, an actor might be in one of three states: (1) *Success*. The actor successfully completed the action. (2) *Executing*. The actor partially executed the action at that time step. (3) *Failure*. Other actors actions negated the actions preconditions. For both the success and failure states, the actor must choose a new action in the next time step.

Cost and Heuristic Function. The cost of simulating a transition $\{a_i\}$, at state s, in the time interval $(t, t+1)$ where a_i is the action chosen by actor i in the composite domain is $\sum_i o_i(\{c_j\})$. The heuristic function is used to provide a cost estimate from the current state to the goal state. Our design of a heuristic function is straightforward and efficient. We first relax the preconditions on the actions (all actions are deemed possible at any given instant of time) and do a fast greedy search for a trajectory of actions that takes the planner from the current state to the goal. The sum of the cost of all actions is the heuristic, h for that particular state, s.

7.5.2.2 The Animation and Simulation Engine

Once we've generated trajectories for the actors, we use a simple steering algorithm to simulate coin-shaped agents to accurately follow paths. Then, the animation system animates models of virtual humans and vehicles along the simulated paths. It animates characters by transitioning between walk, run, and stop animations on the basis of the movement speed. It also employs animations to visualize actors current actions, such as a thief stealing a hot dog.

7.5.2.3 Behavior Generation Algorithm

The algorithm used to generate multi-actor behaviors is described below:

1. Define Actors, $\text{Ac}_i = \langle S_i, A_i, C_i, B_i \rangle$, where S_i is the state space, A_i is the action space, C_i is the set of constraints and modifiers, and B_i is the set of behaviors defined for actor i.
2. Determine Composite Domains, $\text{CD}_j = \langle S_j^c, A_j^c, C_j^c, B_j^c \rangle$, where $S_j^c = \{S_1 \times S_2 \times \ldots S_n\}$ is the composite state space, $A_j^c = \bigcup_{i=1}^n \{A_i\}$ is the composite action space, $C_i^c = \{C_i\}$ is the set of specializations, and $B_j^c = \{B_i\}$ is the set of behaviors defined for all actors $i = 1$ to n in the composite domain CD_j.

3. For each Composite Domain, CD_j

    a. Define Search Domain, $\Sigma = (S^c, A^c, C^c)$.

    b. Determine initial state in the composite space of all agents, $s^0 = \bigcup_{i=1}^{n} s_i^0$

    c. Determine active behaviors, b_i for each actor, i in composite domain, CD_j. The active behavior for each actor determines the goal, g_i and the objective function o_i.

    d. The composite goal, g is the logical combination of the goals, $\{g_i\}$ for all actors in the composite domain. Common goals are combined using an $\wedge$ operator, indicating that all actors must satisfy their goal. Contradicting goals are combined using an $\vee$ operator, indicating that any one of the actors must satisfy their goal.

    e. If no behavior is active for actors, **Return**.

    f. Solve for sequence of actions π by performing a search, $\pi = \textbf{Search}(\Sigma, s^0, g, \{o_i\})$, where Σ is the search domain, s^0 is the composite start state, g is the composite goal, and $\{o_i\}$ are the objective functions for each actor.

4. Combine plans for all domains, $\Pi = \pi_1 \cup \pi_2 \cup \ldots \pi_n$.
5. Execute Global Plan, Π.
6. Determine new states of all actors.
7. Repeat Steps 2–6.

7.5.3 City Simulation

We demonstrate the effectiveness of our framework by authoring a car accident in a busy city street and observing the repercussions of the event on other actors that are part of the simulation, such as the old man and his son, whose behaviors are automatically generated using our framework.

Actor Specification. We first define the state space and action space of three actors in the scenario: (1) a generic pedestrian, (2) a vehicle and, (3) a traffic signal.

1. **Pedestrian:** The state of a pedestrian comprises its position, orientation, speed of movement, mass, and, a collision radius. In addition, pedestrians have the following abstract metrics: hunger, safety, amount of money. These metrics are variables whose values are modified by actions. The *Move* action (Fig. 7.10a) kinematically translates an actor and has an associated distance and energy cost. The routine CheckCollisions(..) returns false if *Move* causes the pedestrian to enter a state of collision. Additional actions (e.g. Eat) can be associated with different metrics (e.g. hunger). A pedestrian is given a simple behavior to move towards a specified goal position (Fig. 7.10b) while minimizing distance and energy cost. The goal positions are randomly generated to produce a realistic city simulation with wandering pedestrians. Additionally, the pedestrians monitor the state of a traffic signal which coordinates the movement of pedestrians and vehicles at an intersection (Fig. 7.10c).

a

```
Action Move(Velocity : v, TStep: dt) {
  Precondition:
    CheckCollisions(self.position + v · dt) == false;
  Effect:
    self.position = self.position + v · dt;
  Cost Effect:
    self.energyCost = ½ (self.mass)|v|²;
    self.distanceCost = |v|dt;
}
```

b

```
Behavior GoalBehavior {
  Precondition: self.goalPosition ≠ 0;
  Goal: self.goalPosition;
  Objective Function:
    min(self.distanceCost + self.energyCost);
}
```

c

```
Constraint PedSignalConstraint {
  Precondition: true;
  Constraint:
    if ((signal.signalState == 0
    ∧CrossingRoad(self.position,C))
    ∨(signal.signalState == 1
    ∧CrossingRoad(self.position,A))
    ∨(trafficSignal.signalState == 2
    ∧CrossingRoad(self.position,B)))
      true;
    else false;
}
```

Fig. 7.10 A generic pedestrian with a simple *Move* action (**a**), a behavior to go to a specified goal position (**b**), and a constraint to follow the traffic signals (**c**)

2. **Vehicles:** The state and action space of vehicles is defined similarly to simulate their movement. In addition, they have a metric damage which increases if a vehicle collides with another vehicle. Vehicles are constrained to stay on the roads, give right of way to pedestrians, and obey the traffic lights.
3. **Traffic Signals:** A traffic signal represents an environment actor that models the simulation of the traffic signals at the intersection. It has a single metric signal state which is the current state that the traffic signals at the intersection are in. An action, ChangeTrafficSignal (Table 7.3(a)) determines the state of the traffic signal based upon the current simulation time. The pedestrian and the vehicles query the signal state in order to follow the traffic signals.

Table 7.3 Scripts used to author the city simulation

```
Action ChangeTrafficSignal {
Precondition:
true;
Effect:
timeMod = currentTime % 100;
if (timeMode <= 35)
self.signalState = 0;
else if (timeMode <= 70)
self.signalState = 1;
else self.signalState = 2;
}
(a)
```

```
EffectModifier RecklessVehicleEM {
Precondition:
true;
Effect:
self.collisionRadius = MIN;
self.followSignals = FALSE;
}

(f)
```

```
CostModifier DaringCM {
Precondition:
∃ a: a.danger > 0 ;
Cost Effect:
self.safetyCost =
MAX_COST - max(a.danger);
}

(b)
```

```
CostModifier RecklessVehicleCM {
Precondition:
true;
Effect:
// low cost for traveling
// at MAX_SPEED
self.speedCost
= MAX_SPEED-self.speed;
}

(g)
```

```
EffectModifier DaringEM {
Precondition:
true;
Effect:
a = argmax(a.danger) ;
self.goalPosition =
a.position;
}

(c)
```

```
Behavior CooperativeVendorB {
Precondition:
true;
Goal:
self.money >= 100 ∧
otherVendor.money >= 100;
Objective Function:
min(self.stolenCost
+ otherVendor.stolenCost);
}

(h)
```

```
Behavior FireFighterB {
Precondition:
∃ a ∈ Actors: a.fire > 0;
Goal:
∀ a ∈ Actors a.fire = 0;
Objective Function:
min(0.3·self.safetyCost
+ self.distanceCost
+ self.energyCost);
}

(d)
```

```
Behavior ThiefB {
Precondition:
true;
Goal:
self.money >= 100
Objective Function:
min(self.distanceCost
+ self.energyCost);
}

(i)
```

```
Constraint AccidentC {
Precondition:
true;
Constraint:
// Two vehicles must collide
// at some point in time
∃ a1,a2 :
IsAVehicle(a1) ∧
IsAVehicle(a2) ∧
Distance(a1,a2) < 5.0;
}

(e)
```

```
Action Steal(Actor a, Amount: m){
Precondition:
m <= a.money ∧
DistanceBetween(self,a) < 1.0;
Effect:
a.money = a.money - m;
self.money = self.money + m;
Cost:
self.stealCost = m;
a.stolenCost = m;
}
(j)
```

Actor Specialization We can easily and intuitively specialize actors using our framework. Fire-fighter actors are specialized pedestrians with lower weights to the safety cost metric and with a common goal to extinguish fires. A grandfather pedestrian is specialized by reducing the walking speed and specializing them to follow their grandson. The objective of the grandson is to escort his grandson at all times and to keep him away from danger (e.g. car accidents, oncoming traffic and other pedestrians) which is achieved by incorporating the safety cost of the grandfather in the grandsons objective function.

Cautious and daring actors are authored by affecting the cost of actions that place them in danger (Table 7.3(b)). A street vendor is given the behavior of manning his hot dog stand and ensuring that his money is not stolen. A thief is authored with a goal to steal money using a *Steal* action (Table 7.3 (i),(j)) while minimizing the risk of getting caught (Table 7.3(j)). A cost modifier assigns a high cost to stealing in the presence of other actors. A reckless vehicle is modeled by introducing a high cost to moving at slower speeds and relaxing the constraints of obeying traffic signals and collisions with other vehicles (Table 7.3 (f),(g)).

7.5.4 Results

We populate a city block with pedestrians and vehicles using our framework. Actor specializations provide an easy and intuitive way to add variety and purpose to the virtual world. We observe pedestrians walking along the sidewalks in the city in a goal-oriented manner (satisfying hunger by getting a hot dog, going to the park to meet a friend, stopping to take a look at objects of interest) while obeying constraints and modifications (obey traffic lights, avoid collisions, stay off the streets etc.). A video demonstrating the results can be found at http://cg.cis.upenn.edu/hms/research/MultiActorPlanning/.

The accident scenario. To add drama to the simulation, we introduce constraints on the trajectory of the entire simulation. First, we introduce a constraint, *AccidentC* (Table 7.3(e)), that an accident must happen (i.e. two vehicles must collide). A simulation is generated where two reckless vehicles collide with one another, resulting in a fire that stops the traffic at the intersection (Fig. 7.11d). Cautious pedestrians who are near the accident run away to a safe distance in panic or walk away calmly (depending on their specialization) while daring actors approach the scene of the accident. The car accident triggers the activation of the behaviors in the fire-fighters who run to the location of the fires. They work together collaboratively to extinguish both fires (a result of the planner working in the composite domain). Upon noticing the accident, the vendor runs to a place of safety (high cost modifier on safety). As soon as the thief notices that the vendor has left his stand, he slowly approaches the stand, steals the money and runs to a place of safety.

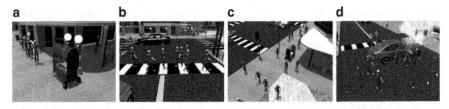

Fig. 7.11 Snapshots of a city simulation authored using our framework: (**a**) Actors queue up at a hot dog stand while the vendors talk to one another. In the meantime, a thief lies in the shadows waiting for an opportunity to steal the money from the stand. (**b**) Cars giving right of way to pedestrians. (**c**) Cautious actors run to a place of safety in the event of an accident. (**d**) Fire-fighters extinguish the fire while daring actors look on

Fig. 7.12 Interaction between thief and the vendors: (**a**) The thief steals money from the hot dog stand when the vendors walk away (because of the accident). (**b**)–(**d**) The vendors collaboratively work together to surround the thief in the alley and manage to catch him

Varying the simulation. We vary the simulation result by introducing other specializations or modifying existing ones. In a first take, we define the objectives of the two vendors to minimize safety cost as well as the cost of being robbed as individuals. When the accident happens, they run to a place of safety while keeping the stand in eyesight. As soon as they see the thief stealing the money, they both chase after him. However, the thief has a head-start and runs away. This is because the planner generates solutions that tries to achieve the objective of each vendor independently. Hence, we observe that in the composite domain of the thief and two vendors, the thief succeeds. In a second take, we modify the objectives of the vendors to minimize the cost of both being robbed (Table 7.3(h)). The common goal of the vendors implies that the planner searches for a solution that optimizes their combined objectives. As a result, the two vendors cooperate to corner the thief in an alley (Fig. 7.12a–d).

Performance and Implementation Details. We demonstrate 106 actors in the city simulation, with 15 cars and 91 pedestrians. Based on constraints, goal definitions and spatial locality, the following composite domains are defined: (1) 15 cars and 4 fire-fighters, (2) old man and son and, (3) generic pedestrians grouped together based on spatial locality. Dividing the problem domain into smaller composite domains reduces the branching factor of the search by two orders of magnitude, reducing the search problem to smaller, more feasible searches. The plans for each of these domains is then overlayed to form the complete solution. The performance

Fig. 7.13 Performance
Results

Number of actors	106
Number of composite domains	12
Max # of actors in a composite domain	19
Total generation time	219 sec
Max generation time for one domain	76 sec
Min generation time for one domain	8 sec
Generation time per actor	2.06 sec
Length of output simulation	95 sec
Amortized time per actor per second	0.02 sec

results are provided in Fig. 7.13. The amortized performance of our behavior generation framework for the results shown in the video is 0.02 s per actor per second of simulation generated.

7.6   Discussion

A key challenge in multi-actor simulations is to provide an appropriate interface for authoring high-fidelity virtual actors with feature-rich control mechanisms capable of complex interactions, while satisfying global scenario constraints. This chapter presents work that addresses the problem of behavior authoring at three levels.

Modeling Agent Personality. The OCEAN personality model defines five orthogonal axes to intuitively describe the personality of an agent. We describe a mapping of these personality traits to low-level simulation parameters, facilitating the control of agent and group personality in order to observe the emergence of different crowd behaviors. We validate our mapping by conducting a user study which assesses the perception of personality traits in a variety of crowd simulations demonstrating these behaviors. The personality traits provide an intuitive and flexible interface for authors to control social crowd dynamics.

Event Centric Authoring of Multi-actor Interactions. We describe an event-centric behavior authoring paradigm where a user-authored centralized controller, defined using parameterized behavior trees (PBTs) [4] is used to coordinate behaviors between multiple interacting actors. Actors participating in an event temporarily suspend autonomy and are controlled by the event structure which treats them as limbs of a central entity for the duration of the event. PBTs provide a flexible, graphical programming language for authoring behaviors, and event-centric authoring alleviates the burden or coordinating the interactions between virtual actors. However, there is greater burden on the author to define events, and to stitch together event sequences to create more complex narratives.

Automated Planning for Simulating Multi-actor Interactions. The use of domain independent planners automates the behavior generation process where users only need to specify goals and objectives for actors. In order to inject

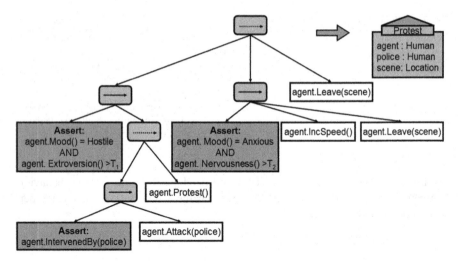

Fig. 7.14 A *Protest* event that takes into account the personality and mood of actors participating in the event

heterogeneity into the simulation, generic actor definitions can be specialized by modifying the cost and effect of actions. Furthermore, the use of composite domains where actors with common or conflicting objectives are part of a single planning domain, facilitates the generation of complex interactions between actors. However, this method has the following limitations:

- The behaviors generated by our framework are heavily dependent on the manner in which actors are grouped together and the weights of the objectives.
- One of the major design choices of this framework was the use of a multi-actor planner which prohibits its use in interactive applications such as games.
- It is non-intuitive to define interactions between multiple actors which is accomplished indirectly by specifying common objectives and global constraints.

Towards A Unified Framework for Authoring Diversity in Personality and Behavior Multi-actor Simulations. The traits of an actor that define personality, mood, and emotion provide an intuitive and high-level interface for specializing individual actors as well as groups of actors. Incorporating these traits into the behavior authoring process, especially at the event-centric level, facilitates the emergence of different actor interactions as a parametrization of these traits. For example, a behavior to describe a *Protest* event would differ based on the personality and mood of actors participating in the protest. A hostile and extroverted actor is more likely be an active participant while an anxious and nervous actor is more likely to steer away from the scene of the protest. Figure 7.14 illustrates the use of personality and mood traits in an authored event.

Authoring complex multi-actor interactions becomes prohibitive when designing the behavior of each agent in isolation. In order to cooperate or compete, actors must constantly communicate information and account for the current state of neighboring actors while performing an action. Planning based approaches provide flexibility of automation at higher performance, while event-centric authoring mitigates the need for treating every actor as an individual at greater burden to the author. A promising direction of exploration is to develop a planning based framework that automatically triggers sequences of user-authored events that satisfy global narrative constraints, while conforming to the roles of individual actors in the scenario. A planner can also be used at the discretion of the centralized controller to act alongside the event level as another tool for coordinating small narrative-driven actor interactions.

Events provide the facility to pre-author and tune actor interactions, while planners allow the generation of more dynamic activities in an automatic fashion. Using events, we can build a library of pre-authored collaborative action sequences of narrative and interactive significance. After preparing these actor interactions, we can apply a multi-actor planner to influence more local behaviors among groups of actors related by activity or domain in addition to selecting and applying the events we have designed. The planner may even be invoked at the discretion of the centralized controller, allowing various degrees of autonomy and control as needed by computational load, actor characteristics, or interactive user actions.

Acknowledgements We would like to thank the following people for their significant contributions in the research projects reported here. Francisco Garcia, Matthew Jones, Robert Mead, and Daniel Garcia helped define a parameterization of behavior trees for use in event-centric authoring. Nuria Pelechano, Jan Allbeck, and Ugur Gudukbay were involved in defining personality parameters for crowd simulations. Shawn Singh, Petros Faloutsos, and Glenn Reinman were instrumental in proposing the use of domain-independent planning for behavior authoring.

Parts of this research were supported by U.S. Army MURI "SUBTLE" and U.S. Army Robotics Collaborative Technology Alliance. The opinions expressed are solely those of the authors and not the sponsors.

References

1. Durupinar, F., Pelechano, N., Allbeck, J.M., Güdükbay, U., Badler, N.I.: How the ocean personality model affects the perception of crowds. IEEE Comput. Graph. Appl. **31**(3), 22–31 (2011)
2. Shoulson, A., Badler, N.I.: Event-centric control for background agents. In: Proceedings of the 4th International Conference on Interactive Digital Storytelling (ICIDS '11), pp. 193–198. Springer, Vancouver, Canada (2011)
3. Shoulson, A., Garcia, D., Badler, N.I.: Selecting agents for narrative roles. In: Proceedings of the 4th Workshop on Intelligent Narrative Technologies (INT4). AAAI, Stanford, Palo Alto, California (2011)
4. Shoulson, A., Garcia, F., Jones, M., Mead, R., Badler, N.I.: Parameterizing behavior trees. In: Proceedings of the 4th International Conference on Motion in Games (MIG '11), pp. 144–155. Springer, Edinburgh, UK (2011)
5. Kapadia, M., Singh, S., Reinman, G., Faloutsos, P.: A behavior-authoring framework for multiactor simulations. Comput. Graph. Appl. IEEE **31**(6), 45–55 (2011)

6. Badler, N.: Virtual Crowds: Methods, Simulation, and Control. Synthesis Lectures on Computer Graphics and Animation. Morgan and Claypool, San Rafael (2008)
7. Kapadia, M., Badler, N.I.: Navigation and steering for autonomous virtual humans. In: Wiley Interdisciplinary Reviews: Cognitive Science. Wiley, Hoboken, NJ (2013)
8. Henderson, L.F.: The statistics of crowd fluids. Nature **229**(5284), 381–383 (1971). (Online). Available: http://dx.doi.org/10.1038/229381a0
9. Lovas, G.: Modeling and simulation of pedestrian traffic flow. Transp. Res. Rec. **28**(6), 429–443 (1994)
10. Milazzo, J., Rouphail, N., Hummer, J., Allen, D.: The effect of pedestrians on the capacity of signalized intersections. Transp. Res. Rec. **1646**(1), 37–46 (1998)
11. Hoogendoorn, S.P., Hoogendoorn, S.P.: Pedestrian travel behavior modeling. In: In 10th International Conference on Travel Behavior Research, Lucerne, pp. 507–535 (2003)
12. Reynolds, C.W.: Flocks, herds and schools: a distributed behavioral model. In: SIGGRAPH '87: Proceedings of the 14th Annual Conference on Computer Graphics and Interactive Techniques, pp. 25–34. ACM, Anaheim, California (1987)
13. Reynolds, C.: Steering behaviors for autonomous characters (1999). (Online). Available: citeseer.ist.psu.edu/reynolds99steering.html
14. Helbing, D., Molnar, P.: Social force model for pedestrian dynamics. Phys. Rev. E **51**, 4282 (1995). (Online). Available: doi:10.1103/PhysRevE.51.4282
15. Helbing, D., Farkas, I., Vicsek, T.: Simulating dynamical features of escape panic. Nature **407**(6803), 487–90 (2000). (Online). Available: http://www.ncbi.nlm.nih.gov/pubmed/11028994
16. Braun, A., Musse, S.R., Oliveira, L.P.L.d., Bodmann, B.E.J.: Modeling individual behaviors in crowd simulation. In: CASA '03: Proceedings of the 16th International Conference on Computer Animation and Social Agents (CASA 2003), p. 143. IEEE Computer Society, Washington, DC, Rutgers University, New-Brunswick, New Jersey, USA (2003)
17. Pelechano, N., Allbeck, J.M., Badler, N.I.: Controlling individual agents in high-density crowd simulation. In: Proceedings of the 2007 ACM SIGGRAPH/Eurographics Symposium on Computer Animation. Series SCA '07, pp. 99–108. Eurographics Association, Aire-la-Ville (2007). (Online). Available: http://dl.acm.org/citation.cfm?id=1272690.1272705
18. Dijkstra, J., Timmermans, H.J.P., Jessurun, A.J.: A multi-agent cellular automata system for visualising simulated pedestrian activity. In: Bandini, S., Worsch, T. (eds.) Theoretical and Practical Issues on Cellular Automata – Proceedings on the 4th International Conference on Cellular Automata for research and Industry, pp. 29–36. Springer, Karlsruhe, Germany (2000)
19. Chenney, S. Flow tiles. In: Proceedings of the 2004 ACM SIGGRAPH/Eurographics Symposium on Computer Animation (SCA 04), pp. 233–242. Grenoble, France (2004)
20. Nishinari, K., Kirchner, A., Namazi, A., Schadschneider, A.: Extended floor field ca model for evacuation dynamics. IEICE Trans. Inf. Syst. **E84-D**(1), 7 (2003) (Online). Available: http://arxiv.org/abs/cond-mat/0306262
21. Lamarche, F., Donikian, S.: Crowd of virtual humans: a new approach for real time navigation in complex and structured environments. Comput. Graph. Forum **23**(2), 128 (2004)
22. Loscos, C., Marchal, D., Meyer, A.: Intuitive crowd behaviour in dense urban environments using local laws. In: TPCG '03: Proceedings of the Theory and Practice of Computer Graphics 2003, p. 122. IEEE Computer Society, Washington, DC, University of Birmingham, Edgbaston, Birmingham, United Kingdom (2003)
23. van den Berg, J., Patil, S., Sewall, J., Manocha, D., Lin, M.: Interactive navigation of multiple agents in crowded environments. In: SI3D '08: Proceedings of the 2008 Symposium on Interactive 3D Graphics and Games, pp. 139–147. ACM (2008)
24. Lee, K.H., Choi, M.G., Hong, Q., Lee, J.: Group behavior from video: a data-driven approach to crowd simulation. In: SCA '07: Proceedings of the 2007 ACM SIGGRAPH/Eurographics symposium on Computer animation, pp. 109–118. Eurographics Association, Aire-la-Ville, San Diego, USA (2007)
25. Lerner, A., Chrysanthou, Y., Lischinski, D.: Crowds by example. Comput. Graph. Forum **26**(3), 655–664 (2007)

26. Paris, S., Pettré, J., Donikian, S.: Pedestrian reactive navigation for crowd simulation: a predictive approach. In: EUROGRAPHICS 2007, Prague, Czech Republic, **26**, 665–674 (2007). Prague, Czech Republic
27. van den Berg, J., Lin, M.C., Manocha, D.: Reciprocal velocity obstacles for real-time multi-agent navigation. In: IEEE International Conference on Robotics and Automation, pp. 1928–1935. IEEE, Pasadena, California (2008)
28. Singh, S., Kapadia, M., Hewlett, W., Faloutsos, P.: A modular framework for adaptive agent-based steering. In: Proceedings of the 2011 Symposium on Interactive 3D Graphics and Games. Series I3D '11. ACM, San Francisco, CA (2011)
29. Kapadia, M., Singh, S., Hewlett, W., Faloutsos, P.: Egocentric affordance fields in pedestrian steering. In: I3D '09: Proceedings of the 2009 Symposium on Interactive 3D Graphics and Games pp. 215–223. ACM, Boston, MA (2009)
30. Kapadia, M., Singh, S., Hewlett, W., Reinman, G., Faloutsos, P.: Parallelized egocentric fields for autonomous navigation. Vis. Comput. Int. J. Comput. Graph. **28**(12), 1209–1227 (2012)
31. Singh, S., Kapadia, M., Reinman, G., Faloutsos, P.: Footstep navigation for dynamic crowds. Comput. Anim. Virtual Worlds **2**(2–3) (2011). Wiley
32. Kapadia, M., Porres, A., Garcia, F., Reddy, V., Pelechano, N., Badler, N.I.: Multi-domain Real-Time Planning in Dynamic Environments. In: Proceedings of the 2013 ACM SIG-GRAPH/EUROGRAPHICS Symposium on Computer Animation, Anaheim, CA (2013)
33. Ondřej, J., Pettré, J., Olivier, A.-H., Donikian, S.: A synthetic-vision based steering approach for crowd simulation. In: ACM SIGGRAPH 2010 Papers. Series SIGGRAPH '10, pp. 123:1–123:9. ACM, New York (2010). (Online). Available: http://doi.acm.org/10.1145/1833349.1778860
34. Yu, Q., Terzopoulos, D.: A decision network framework for the behavioral animation of virtual humans. In: Proceedings of the 2007 ACM SIGGRAPH/Eurographics Symposium on Computer Animation. Series SCA '07, pp. 119–128. Eurographics Association, Aire-la-Ville (2007). (Online). Available: http://dl.acm.org/citation.cfm?id=1272690.1272707
35. Blumberg, B.M.: Old tricks, new dogs: ethology and interactive creatures. Ph.D. dissertation, Supervisor-Maes, Pattie (1997)
36. Pynadath, D.V., Marsella, S.C.: Psychsim: modeling theory of mind with decision-theoretic agents. In: Proceedings of the International Joint Conference on Artificial Intelligence, pp. 1181–1186. Edinburgh, Scotland, UK (2005)
37. Massive Software Inc.: Massive: Simulating Life (2010). www.massivesofware.com.
38. Menou, E.: Real-time character animation using multi-layered scripts and spacetime optimization. In: Proceedings of ICVS '01, pp. 135–144. Springer, London, Avignon, France (2001)
39. Perlin, K., Goldberg, A.: Improv: a system for scripting interactive actors in virtual worlds. In: Proceedings of ACM SIGGRAPH, pp. 205–216. ACM, New York, New Orleans, Lousiana, USA (1996)
40. Vilhjálmsson, H., Cantelmo, N., Cassell, J., Chafai, N.E., Kipp, M., Kopp, S., Mancini, M., Marsella, S., Marshall, A.N., Pelachaud, C., Ruttkay, Z., Thórisson, K.R., Welbergen, H., Werf, R.J.: The behavior markup language: recent developments and challenges. In: Proceedings of IVA '07, pp. 99–111. Paris, France (2007)
41. Loyall, A.B.: Believable agents: building interactive personalities. Ph.D. dissertation, Pittsburgh (1997)
42. Funge, J., Tu, X., Terzopoulos, D.: Cognitive modeling: knowledge, reasoning and planning for intelligent characters. In: proceedings of the 26th annual conference on Computer graphics and interactive techniques, SIGGRAPH '99, pp. 29–38. ACM Press/Addison-Wesley Publishing Co., New York, Los Angeles, California, USA (1999). Available: http://dx.doi.org/10.1145/311535.311538.doi:10.1145/311535.311538
43. Shao, W., Terzopoulos, D.: Autonomous pedestrians. Graph. Models **69**, 246–274 (2007). (Online). Available: http://dl.acm.org/citation.cfm?id=1323742.1323926
44. Allbeck, J., Badler, N.: Toward representing agent behaviors modified by personality and emotion. Workshop Embodied conversational agents - let's specify and evaluate them! at AAMAS, Bologna, Italy (2002)

45. Wiggins, J.: The Five-Factor Model of Personality: Theoretical Perspectives. Guilford (1996). (Online). Available: http://books.google.com/books?id=UzXvvAsESjEC
46. Ortony, A., Clore, G., Collins, A.: The Cognitive Structure of Emotions. Cambridge University Press, Cambridge (1988)
47. Gebhard, P.: Alma: a layered model of affect. In: Proceedings of the Fourth International Joint Conference on Autonomous Agents and Multiagent Systems. Series AAMAS '05, pp. 29–36. ACM, New York (2005). (Online). Available: http://doi.acm.org/10.1145/1082473.1082478
48. Egges, A., Kshirsagar, S., Magnenat-Thalmann, N.: Generic personality and emotion simulation for conversational agents: research articles. Comput. Animat. Virtual Worlds 15, 1–13 (2004). (Online). Available: http://dl.acm.org/citation.cfm?id=1071195.1071196
49. Guy, S.J., Kim, S., Lin, M.C., Manocha, D.: Simulating heterogeneous crowd behaviors using personality trait theory. In: Proceedings of the 2011 ACM SIGGRAPH/Eurographics Symposium on Computer Animation. Series SCA '11, pp. 43–52. ACM, New York (2011). (Online). Available: http://doi.acm.org/10.1145/2019406.2019413
50. Magerko, B., Laird, J.E., Assanie, M., Kerfoot, A., Stokes, D.: Ai characters and directors for interactive computer games. Artif. Intell. 1001, 877–883 (2004)
51. Mateas, M., Stern, A.: Integrating Plot, Character and Natural Language Processing in the Interactive Drama Faade, vol. 2 (2003). (Online). Available: http://www.cs.ucsc.edu/~michaelm/tenurereview/publications/mateas-tidse2003.pdf
52. Si, M., Marsella, S.C., Pynadath, D.V.: THESPIAN: An Architecture for Interactive Pedagogical Drama. In: proceedings of the 2005 conference on Artificial Intelligence in Education: Supporting Learning through Intelligent and Socially Informed Technology. pp. 595–602. IOS Press, Amsterdam, The Netherlands (2005). http://dl.acm.org/citation.cfm?id=1562524.1562605
53. Riedl, M.O., Saretto, C.J., Young, R.M.: Managing Interaction Between Users and Agents in a Multi-agent Storytelling Environment, vol. 34, pp. 186–193. ACM (2003). (Online). Available: http://portal.acm.org/citation.cfm?doid=860575.860694
54. Li, B., Riedl, M.: Creating Customized Game Experiences by Leveraging Human Creative Effort: A Planning Approach, p. 99–116. Springer (2011). (Online). Available: http://www.springerlink.com/index/428G8512624879U6.pdf
55. Stocker, C., Sun, L., Huang, P., Qin, W., Allbeck, J.M., Badler, N.I.: Smart events and primed agents. In: Proceedings of the 10th International Conference on Intelligent Virtual Agents (IVA 10), 6356, 15–27. Philadelphia, PA, USA (2010)
56. Goldberg, L.R.: An alternative "description of personality": the big-five factor structure. J. Pers. Soc. Psychol. 59, 1216–1229 (1990). Washington, DC
57. Hall, E.T.: The Hidden Dimension. Anchor Books, (1966)
58. Isla, D.: Handling complexity in the Halo 2 ai. In: Game Developers Conference (2005). (Online). Available: http://www.gamasutra.com/gdc2005/features/20050311/isla_01.shtml
59. Knalfa, B.: Introduction to behavior trees (2011). http://www.altdevblogaday.com/2011/02/24/introduction-to-behavior-trees/.
60. Shoulson, A., Marshak, N., Kapadia, M., Badler, N.I.: ADAPT: the agent development and prototyping testbed. In: Proceedings of the ACM SIGGRAPH Symposium on Interactive 3D Graphics and Games. I3D '13, Orlando, pp. 111–116 (2013)
61. Kapadia, M., Singh, S., Reinman, G., Faloutsos, P.: Multi-actor planning for directable simulations. In Proceedings of the 2011 Workshop on Digital Media and Digital Content Management, pp. 111–116. IEEE Computer Society, Chengdu, China (2011)

Chapter 8
Virtual Tawaf: A Velocity-Space-Based Solution for Simulating Heterogeneous Behavior in Dense Crowds

Sean Curtis, Stephen J. Guy, Basim Zafar, and Dinesh Manocha

Abstract We present a system to simulate the movement of individual agents in large-scale crowds performing the Tawaf. The Tawaf serves as a unique test case; the large crowd consists of a heterogeneous set of pilgrims, varying in both physical capacity and activity. Furthermore, the density of the crowd reaches extremely high levels (up to 8 people/m^2). This extreme density can place impractical constraints on simulation parameters. We use a velocity-space-based pedestrian model which exhibits consistent results even under extreme density: reciprocal velocity obstacles (RVO). Furthermore, we extend RVO to include *priority* and *right of way*—agents respond to potential collisions asymmetrically depending on context; one agent may yield, to varying degrees, to another. Our system uses a finite state machine to specify the behavior of the agents at each time step, to model the varied behaviors seen during the Tawaf. The finite-state machine, used in conjunction with RVO, generates collision-free trajectories for tens of thousands of agents in the performance of the Tawaf. The overall system can model agents with varying age, gender and behaviors, supporting the heterogeneity observed in the performance of the Tawaf, even at high densities.

8.1 Introduction

The Tawaf is one of the Islamic rituals of pilgrimage performed by Muslims when they visit Al-Masjid al Harām. Located in Makkah, Saudi Arabia, Al-Masjid al Harām surrounds the Kaaba, the site Muslims around the world turn towards

S. Curtis (✉) • S. J. Guy • D. Manocha
University of North Carolina at Chapel Hill, Chapel Hill, NC, USA
e-mail: seanc@cs.unc.edu; sjguy@cs.unc.edu; dm@cs.unc.edu

B. Zafar
Hajj Research Institute, Umm al-Qura University, Makkah, Saudi Arabia
e-mail: Bjzafar@uqu.edu.sa

S. Ali et al. (eds.), *Modeling, Simulation and Visual Analysis of Crowds*, The International   181
Series in Video Computing 11, DOI 10.1007/978-1-4614-8483-7_8,
© Springer Science+Business Media New York 2013

while performing daily prayers. Al-Masjid al Harām is the largest mosque in the world and is regarded as Islam's holiest place. During the Tawaf, Muslim pilgrims circumambulate the Kaaba seven times in a counterclockwise direction, while in supplication to God.

The Tawaf is performed both during the Umrah and the Hajj. Performing the Hajj is one of the five pillars of Islam and every Muslim aspires to visit Makkah at least once in his or her life. Annually, more than two million Muslims perform the Hajj. While the Hajj has several stages and takes place over several days, all pilgrims move through the various stages of the Hajj on the same days which creates limitations in both time and space resulting in very high crowd densities during the Tawaf, especially on the Mataf, the marble floor of the mosque, in the center of which stands the Kaaba. During the Hajj season, or the last few days of the month of Ramadan, as many as 35,000 pilgrims perform Tawaf at the same time in the Mataf area in Al-Masjid al Harām. Given the large scale of the gathering, it is important to understand and model the behavior and movement of the crowd to provide insight which may improve crowd management techniques and help ensure the safety of the pilgrims.

The Tawaf has several properties which make simulating it particularly challenging:

Heterogeneous Population: The population of pilgrims varies significantly, spanning gender, a wide range of ages and physical capacity, and representing different cultures from all over the world.

High Density: The crowd density throughout the Mataf often varies considerably. It can become as high as eight pilgrims per square meter near the Kaaba [32]. The extremely high density greatly restricts the movement of the pilgrims.

Varying Velocities: The velocity of the pilgrims in Mataf can vary depending on many factors such as their distance from the Kaaba and the proximity of structures on the floor or congestion caused by other agents due to the capacity saturation and geometry of the mosque; the irregular shape of the Mataf is not well suited to the inherently elliptical movement around the Kaaba.

Complex Motion Flows: Different types of crowd flows have been observed during the Tawaf. These flows arise out of the sometimes contradictory intentions of the many pilgrims; at any given time, pilgrims will be simultaneously trying to stand still to kiss the Black Stone at the corner of the Kaaba, circumambulate the Kaaba, or attempt to move orthogonally to the circular flow, inwards, toward the Kaaba, or outwards, towards the exit, preventing purely circular flow.

Simulating the Tawaf will afford those who administer Al-Masjid al Harām the ability to evaluate alternative crowd flow control systems or architectural changes to improve the comfort and safety of the pilgrims and increase the capacity of the Mataf. But creating a practical simulator for such a complex scenario is challenging. The crowd simulator must account for the heterogeneous population, allowing for large variance in the capabilities and actions of the agents. Furthermore, to capture the acts of the ritual, the simulator must provide a mechanism in which the activities and strategies of each agent change with respect to time. These features must be embedded in a computationally efficient simulator. It should scale well with respect to both the number of virtual pilgrims and increasing density. The greater the computation time, the less flexibility the simulator provides in evaluating scenario variations or producing stochastic studies in which multiple runs with randomly perturbed initial conditions are analyzed in aggregate. Satisfying these challenges and producing an accurate simulation of complex and dynamic interactions between pedestrians of this sort remains an open problem.

Main Results In this chapter, we describe a system to model the movement of individual agents in a large-scale crowd performing the Tawaf. To address the above challenges, we present an agent-based model which combines a velocity-based pedestrian model to control local interactions between the agents with a finite state machine (FSM) to model the intentions of each pilgrim. We extend the pedestrian model with a parameter called *priority* which governs how the agents divide the effort to avoid collision between them. In some cases, agents act cooperatively to avoid collision. In other cases greater priority gives one agent *right of way* over another agent, causing the agent with less priority to yield. This enables us to model the asymmetric relationships observed between pilgrims in the Tawaf such as when pilgrims stop to kiss the Black Stone while others move around them. To model the changing goals of agents, each state in the FSM encodes a particular behavior which defines both strategy and tactics for navigating the shared space. The state provides a function which defines time-dependent values for a sub-space of the agent configuration space, including, but not limited to, such agent properties as preferred velocity and priority. The pedestrian model computes a collision-free trajectory by computing a new velocity based on the agent's time-dependent state. We use several criteria to transition between the states based on spatial, agent-property, temporal and stochastic conditions.

The resulting system allows us to simulate crowds of heterogenous individuals, including variations in age and gender, performing the Tawaf. Each agent is associated with a unique instance of the pilgrim FSM. The FSM defines the general form of the Tawaf ritual, but each individual instance can allow for individual variance in the particular performance. For example, some pilgrims may possess a strong desire to approach the Kaaba, while others avoid the dense region near the Kaaba and maintain a greater distance.

From these simulations, we measure aggregate behavior such as density and velocity. We also measure Tawaf-specific metrics, such as the time to complete the Tawaf, and the overall throughput, in terms of the number of pilgrims that can complete the Tawaf per hour and show correlation with empirical observations.

Chapter Organization: The rest of the chapter is organized as follows. In Sect. 8.2, we survey related work on crowd simulation, behavior modeling, and simulation of the Tawaf. We discuss the full simulation pipeline in Sect. 8.3, paying particular attention to the formulation of the high-level behavioral finite state machine. In Sect. 8.4 we present four models for pedestrian simulation and discuss their particular suitability for modeling the Tawaf. In addition, we discuss the details of *priority* and *right of way* and illustrate its effect on pedestrian relationships.

Section 8.5 contains specific details on the actual performance of the Tawaf by living pilgrims and the mapping to a particular finite state machine. Finally, in Sect. 8.6 we provide the results of our system.

8.2 Related Work

In this section, we discuss related work in crowd simulation and behavior modeling for crowds. We also highlight some prior crowd simulation systems designed for simulating the Tawaf.

8.2.1 Crowd Simulation

There is extensive literature on crowd simulation and many techniques have been proposed.

Cellular automata (CA) are some of the oldest approaches for crowd simulation. In CA the workspace of agents is divided into discrete grid cells which can be occupied by zero or one agent. Agents then follow simple rules to move towards their goals through adjacent grid cells [25].

Continuum methods such as [26] and [19] treat the crowd as a whole and model the motion and interactions of agents based on equations that represent aggregate flow.

Agent-based approaches model each individual in the crowd and the interactions between them. Different techniques have been proposed to model these interactions. Reynolds [23] proposed Boids, which is a simple method based on rules for avoiding collisions while preserving flock cohesion. The rules are often implemented as forces. Other well known force-based methods including the social force model [11] (and its many variations), generalized centrifugal force model [4] and HiDAC [22]. These approaches use more complex forces between agents to model a larger domain of local interactions. Ondřej et al. [20] proposed a vision-based model in which agents respond to nearby obstacles based on the angle to the obstacle and the estimated "time to interaction". Recently, velocity-space methods have been proposed to model human pedestrians. These geometric formulations are often based on velocity obstacles [8, 10, 28] and have been shown to exhibit many emergent crowd phenomena.

8.2.2 Behavior Modeling

Many researchers have proposed approaches to simulate various aspects of human and crowd behaviors. Funge et al. [9] proposed using a cognitive model to allow agents to plan and perform high-level tasks. Yu and Terzopoulos [31] introduced a decision network framework that is capable of simulating interactions between multiple agents. Ulincy and Thalmann [27] used a modular behavioral architecture to allow a mixture of automated and scripted behavior in multi-agent simulations. Durupinar et al. [7] modeled the effects personality factors have on local behavior. Yersin et al. [30] used spatial patches to direct motion and behavior of agents. Bandini et al. [3] applied a state machine to an underlying CA model to create scenarios with more complex behaviors. Yeh et al. [29] employed a physical collision avoidance mechanism to model abstract factors in pedestrian interactions such as aggression, priority and authority.

Data-driven approaches have also been used to capture crowd behaviors, often by training models of agent motion based on video data. Lee et al. [15] used data-driven methods to create group behavior such as queueing and clustering. Ju et al. [13] proposed a data-driven method which attempts to match the style of simulated crowds to those in a reference video. Patil et al. [21] proposed a method of directing crowd simulations with flow fields extracted from video or specified by a user. Video data has also been used to analyze and interpret real-world crowd behavior. Mehran et al. [16] proposed a method to detect abnormal crowd behavior from video using the social-force model. Johansson et al. [12] used video to study crowd behavior during portions of the Hajj.

8.2.3 Tawaf Simulation

There is some prior work on simulating crowd movement during the Tawaf and other Hajj related rituals. Algadhi and Mahmassani [2] simulated crowd flows in the Jamarat area of the Hajj using continuum models. Mulyana and Gunawam [18] performed agent based simulations of various rituals of the Hajj including a 500-agent simulation of the Tawaf. Zainuddin et al. [33] used the commercial software SimWalk to perform a social force-based simulation of up to 1,000 agents performing the Tawaf ritual. Sarmady et al. [24] performed a large crowd simulation of the Tawaf using CA techniques combined with a discrete-event simulator.

A few studies have also been performed on crowd flow in the Mataf area in the Al-Masjid al Harām. Al-Haboubi and Selim [1] proposed a potential spiral movement path to increase safety and throughput of pilgrims during the Tawaf. Koshak and Fouda [14] collected trajectories of actual pilgrims performing the Tawaf during the Hajj using GPS devices. The Crystals project currently studies how to incorporate cultural differences into simulations of Hajj pilgrims [5].

8.3 Modeling Crowd Behaviors

In this section, we give an overview of our method for modeling the crowd behaviors during the Tawaf. Human behavior arises from the confluence of many factors, including culture, psychology, environment and physiology. Generally, human

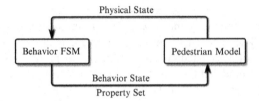

Fig. 8.1 We simulate crowd behaviors appropriate to the Tawaf by coupling a high-level finite-state machine (*FSM*) with a low-level pedestrian model. The *FSM* computes behavior-state and property-set values for each agent. The pedestrian model, in turn, updates the agent's physical state

behavior spans a wide range of activity. When discussing *crowd behavior* we limit our discussion to those human behaviors which affect how humans share space. For example, two people standing and discussing current events are functionally equivalent to those same people negotiating a business deal; the topic is unimportant, but the fact that they are stationary at a fixed distance away from each other is the behavioral detail which most influences crowd simulation. We characterize behaviors which affect the crowd with two concepts: where does an individual wish to be and how do they interact with those around them in reaching their goal? The first deals with the agent's intention—the general strategy, such as what path to take through an environment. The latter addresses the immediate tactics applied to execute the strategy under the dynamic constraints of a populated environment.

8.3.1   Agent-based Simulations

To simulate the Tawaf, with its heterogeneous population and widely varying activities, we need an approach which can accommodate a high-level of per-agent variability. To that end, we model the crowd with individual agents. Each agent is characterized by its *physical state* (position, velocity, size, etc.), its *behavior state* (preferred velocity, priority, its FSM, etc.), and its *property set* (a collection of associated data appropriate to the scenario). For example, a simulated pilgrim's property set includes a counter indicating how many circles the pilgrim has already completed around the Kaaba. The counter doesn't directly affect the computation of an agent's preferred velocity or how it interacts with other agents, but it is used in the behavior mechanism to know when the Tawaf is complete.

We model the behaviors of agents by coupling together a high-level finite-state machine (FSM) with a pedestrian model. The FSM evaluates the agent's physical state and defines the agent's behavior state and, optionally, changes values in the agent's property set. The behavior state is used by the pedestrian model, in conjunction with an agent's physical state, to compute a new velocity and update the agent's physical state. Figure 8.1 illustrates the two components of our system and how they interact.

8.3.2   The Behavior Finite State Machine

A finite state machine (FSM) defines the behavior of an agent at every time step. Each state in the FSM defines the behavior state, and, optionally, the property set for the agent. By providing unique definitions, each state can impart a distinct, observable behavior on the agent. We do not require any particular method for a state to use to compute the behavior state for an agent. The choice is arbitrary. All that matters is that the values for the agent's behavior state produce the desired behavior. For example, one element of the behavior state is the preferred velocity. It may be computed in any number of ways: by a simple rule, or as the result of a complex algorithm using techniques as varied as guidance fields or roadmaps. We give specific examples of this in Sect. 8.5.3

We have classified the FSM's transitions into four categories based on the types of conditions which cause the transition to become active: spatial, property, temporal, and stochastic. A *spatial transition* will cause the agent's current state to change when the agent's position achieves some pre-defined spatial configuration, such as entering an area, leaving an area, etc. For example, this transition will signal the start or end of a circumambulation. The *property transition* moves the FSM from the current state to a new state if some element of the agent's property set conforms to a particular condition. In the Tawaf, this transition causes an agent to exit when it has completed seven circles. The *temporal transition* acts as a timer for the state. The transition is activated when the agent has been in the current state for some pre-defined amount of time. For example, some agents in the Tawaf will stop and pray for a few seconds when completing a circumambulation. Finally, the *stochastic transition* becomes active according to a user-defined probability distribution. In the Tawaf, we expect that only a fraction of the participants stop to pray. We use the stochastic transition to model this distribution. Finally, we prioritize the transitions such that if two transitions conditions are both true, the transition with the higher priority is taken.

8.4   Pedestrian Modeling

With a mechanism in place to alter agent behavior over time, we need to select a pedestrian model to execute the high-level strategy. In this section, we discuss various types of pedestrian simulation algorithms and their suitability for a scenario such as the Tawaf. Finally, we present the reciprocal velocity obstacle pedestrian model and describe *priority* and *right of way*—an extension which increases the space of interactions between virtual agents.

8.4.1 Models of Pedestrian Simulation

There are numerous algorithms for simulating pedestrians. Each has its own unique set of advantages and disadvantages. It has yet to be shown that any single algorithm perfectly models pedestrian dynamics in arbitrary scenarios. For our purposes, we are most interested in those algorithms well suited to a specific scenario: the Tawaf. We are interested in simulators which will provide a mechanism to model the physical and behavioral heterogeneity observed in the Tawaf. Furthermore, to be useful, we require the simulator to be efficient (a hypothetical "perfect" algorithm which took hours to simulate seconds of data would be impractical).

We divide pedestrian simulators into two categories: macroscopic and microscopic. Macroscopic approaches model a crowd of pedestrians as an aggregate phenomenon (e.g. [19, 26]). Microscopic simulators deal with individual agents, trusting that the aggregate behavior will naturally arise from basic principles (e.g. [4, 8, 11, 20, 23, 28]).

Macroscopic models typically assume a relatively high density; they operate on the principle that the choices of an individual pedestrian are strongly constrained by its local conditions. In dense crowds this is a reasonable assumption. As the crowd becomes sparser, the effect of one pedestrian on its distant neighbors is significantly reduced. In some cases, these approaches can be quite efficient. Narain et al. [19] were able to simulate 100K agents at 450 ms per simulation step. These approaches usually treat the crowd as a continuous, homogenous medium. Assuming continuity and homogeneity precludes the variation in physical attributes and behaviors observed in the performance of the Tawaf. For example, such approaches would be unable to create a dense simulation in which some agents remain stationary while other agents move next to them. For these reasons, we consider macroscopic simulation algorithms to be inappropriate for simulating the Tawaf.

Microscopic models provide greater potential to realize the kind of per-agent heterogeneity we require. Each agent is individually simulated and, as such, can be assigned arbitrary properties to model varying physical capacity. Furthermore, their behaviors can be individually specified; one agent's motion is not explicitly constrained by its neighbors. It can try to pursue a goal that stands in direct opposition to its neighbors (its success is dependent on the pedestrian model). We consider three major categories of microscopic models: cellular automata (CA), social forces (SF) and velocity obstacles (VO).

As previously indicated, CA approaches decompose the simulation domain into a uniform grid. Each agent occupies a single cell and a single cell can contain at most one agent. Probabilities are applied to neighboring cells based on a movement protocol and each agent's position is updated according to the probability distribution and a set of rules for resolving conflict. CA approaches are typically simple to implement. A CA approach has even been applied to simulating the Tawaf before [24]. However, the authors indicate that while CA can generate emergent phenomena (such as lane formation, etc.), the individual microscopic trajectories are "unrealistic" [24]. Furthermore, the grid decomposition imparts a homogeneity on the agents as well. Agents can only move an integer number of cells in a single time step. Finally, the authors indicate that CA has limitations with respect to density. The maximum possible density is simply a function of the size of the cells. This maximum is only theoretical. In practice, that density is impossible to achieve because if all cells are filled, agents cannot resolve conflicts and the simulation reaches a deadlock. We feel that, despite its simplicity, the cost of the spatial discretization leads to too many undesirable artifacts to simulate the Tawaf effectively.

SF-based and VO-based approaches both operate in continuous space, obviating the artifacts observed in CA. SF-based models treat pedestrians as mass-particles. Various forces applied to a particle draw the particle towards a goal position and prevent collisions with obstacles and other agents. The many variations of SF-based models generally differ in how the forces are formulated.

VO-based approaches consider the relative velocities and positions of agents to select a feasible velocity—a velocity which will remain collision free for a specified window of time. For each neighboring agent, it computes a set of infeasible velocities—velocities which will lead to collision within a specified time window. The selected feasible velocity is the velocity which lies outside the union of all infeasible velocity regions but which minimizes some cost function. There are multiple variations on VO-based algorithms, which may differ in how they model the space of inadmissible velocities, define the cost function, and how they solve the optimization problem.

Generally, both SF- and VO-based algorithms appear to be viable candidates for simulating the Tawaf to the level of fidelity we seek. More detailed investigation is required to differentiate their suitability. We provide summaries of a recent SF-based model [4] and a recent VO-based model [28]. For simplicity, we limit the summary to agent-agent interactions and refer the reader to the original papers for details on agent-obstacle interactions.

Generalized Centrifugal Force: The Generalized Centrifugal Force (GCF) model is a SF-based model which formulates inter-agent repulsive forces in terms of the agents' positions and velocities.[1] The agent is modeled with the state vector: $[m \ \mathbf{p} \ \mathbf{v} \ \mathbf{v}^0]^T \in \mathbb{R}^7$, where $m \in \mathbb{R}^1$ is the agent's mass, $\mathbf{p}, \mathbf{v}, \mathbf{v}^0 \in \mathbb{R}^2$ are the agent's current position, current velocity, and preferred velocity, respectively. Preferred speed $v^0 = \|\mathbf{v}^0\|$ is simply the magnitude of preferred velocity.

At each time step, agent i's acceleration is computed as:

$$a_i = \frac{\mathbf{F}_i}{m_i} = \frac{\mathbf{F}_i^{drv} + \sum \mathbf{F}_{ij}^{rep}}{m_i}, \tag{8.1}$$

---

[1]The velocity term is the inspiration for the name. The original SF model considered only agent positions [11].

where $\mathbf{F}_i^{drv}$ is a "driving" force and $\mathbf{F}_{ij}^{rep}$ is the repulsive force applied to agent i by agent j. Given this acceleration, the agent's velocity and position are updated by integrating with respect to time using an explicit integrator.[2]

The driving force is what causes the agent to move toward its goal. These systems assume that for each time step, a "preferred velocity" ($\mathbf{v}^0$) is computed. How this velocity is defined is arbitrary, but we assume it represents the velocity the agent would "prefer" to take in the absence of dynamic constraints. The driving force is defined such that it imparts an acceleration on the agent sufficient to reach its preferred velocity in τ seconds:

$$\mathbf{F}_i^{drv} = m_i \frac{\mathbf{v}_i^0 - \mathbf{v}_i}{\tau}. \tag{8.2}$$

The presence of other agents may prevent an agent from following its preferred velocity. This interference is modeled by repulsive forces. Each nearby agent j applies a repulsive force to the agent i of the form:

$$\mathbf{F}_{ij}^{rep} = -m_i k_{ij} \frac{(\eta v_i^0 + v_{ij})^2}{d_{ij}} \hat{\mathbf{e}}_{ij}, \tag{8.3}$$

$$v_{ij} = \max(0, (\mathbf{v}_i - \mathbf{v}_j) \cdot \hat{\mathbf{e}}_{ij}), \tag{8.4}$$

$$k_{ij} = \max\left(0, \frac{\mathbf{v}_i \cdot \hat{\mathbf{e}}_{ij}}{v_i}\right) \tag{8.5}$$

where $d_{ij} = \|\mathbf{p}_j - \mathbf{p}_j\|$ is the distance between agents i and j, $\hat{\mathbf{e}}_{ij} = \frac{\mathbf{p}_j - \mathbf{p}_j}{d_{ij}}$ is the normalized *direction* vector from agent i to agent j, v_{ij} is the amount of agent i's and j's relative velocity that lies in the direction of $\hat{\mathbf{e}}_{ij}$, clamped to the range $[0, \infty]$, η is a simulation variable used to tune the behavior of the simulation, and k_{ij} is a field-of-view weight—the strongest response is to agents in the direction of travel with decreasing weight as the angle increases to $90°$ on either side of that direction.

According to the authors, the formulation of the repulsive force has several desirable properties:

1. Repulsion is a local effect because the magnitude of the force is dependent on inverse distance. effect.
2. The v_{ij} term accounts for relative velocity so that a slow moving agent will not be affected by a fast moving agent in front of it.
3. The k_{ij} term gives the agent an active field of view. Agents will not be repulsed by agents behind them.

According to (8.3), the repulsive force between agents has infinite support; no matter how far the distance between two agents, some small contribution to one

---

[2] While the formula doesn't preclude using an implicit integration scheme, the common practice has been to use a low-order explicit integrator such as forward Euler.

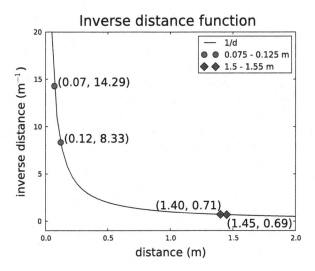

Fig. 8.2 (Color online) The inverse distance function. At small distances, a small perturbation in distance leads to a large change in the function value (*red circles*). The same sized perturbation at large distances leads to a correspondingly small change in function value (*blue diamonds*)

agent's acceleration will be due to an unreasonably distant agent. Conversely, when the agents overlap, their distance converges to zero and the repulsive force can grow infinitely large. The authors combat both of these undesirable artifacts by approximating (8.3) with a spline which bounds the growth at small distances and limits the functions domain to a user-defined maximum distance.

Unfortunately, this formulation still exhibits some undesirable properties as well. The combination of how the forces are defined and the integration scheme can lead to very "jittery" agent behavior, especially under high densities; an agent's trajectory may exhibit high-frequency oscillations because of numerical integration error which can only be addressed through taking extraordinarily small simulation time steps.

The full analysis of this behavior is beyond the scope of this work. However, we feel a brief intuitive discussion of the causes will illustrate why we deem a social-force-based model impractical for simulating the Tawaf. We leave the full, formal analysis for future work. We focus on two particular properties of the formulation as the cause of the undesirable oscillatory behavior: the explicit integration of a "stiff" physical system and the out-of-phase nature of driving and repulsive forces.

The inter-agent repulsive force (8.3) is essentially a function of the inverse distance between the agents (see Fig. 8.2). The function has relatively compact support; the force is greatest at near distances when collision is most imminent. As distance increases, the magnitude quickly decreases. When simulating a relatively sparse environment, where distance between agents is high, the magnitude of the repulsive force is quite small. A small perturbation in distances between agents produces forces with only slightly different magnitude. But when agents are close, small changes in distance lead to very large changes in forces—in other words, the slope of the force function is quite steep. This is a classic characteristic of a stiff system. When performing explicit integration, the common practice for SF-based pedestrian models, small time steps must be taken to prevent oscillatory behavior and unbounded error in the undamped system.

Furthermore, the driving force (8.2) and repulsive forces (8.3) are solved out of phase with each other. Imagine two agents moving towards each other at their preferred velocity. Because they are moving at their preferred velocity, their driving force is zero. They continue on their trajectories until they are close enough for the repulsive forces to be non-zero. At that time step when the repulsive forces are first non-zero, the driving force is still zero. Thus, the repulsive forces are the sole influence on the agent and the current velocity is accelerated accordingly. At the next time step, the repulsive force will be significantly reduced (because the relatively velocity has been reduced), but the driving force now increases due to the deviation between preferred and current velocities. This alternating dominance can eventually converge to a steady-state where they will be in balance, but it requires a small time step.

We seek to simulate the Tawaf during its peak performance, when tens of thousands of pilgrims pack into a small area reaching densities as high as 8 people/m^2. There is an unavoidable computational cost in increasing the size of the simulation to a 35,000 agents. If we also had to significantly reduce the simulation time step to an extremely small time step, the simulation would no longer be tractable in reasonable time frames. For this reason, we consider SF-based models impractical for simulating the Tawaf.

Reciprocal Velocity Obstacle: The Reciprocal Velocity Obstacle (RVO) updates an agent's velocity by performing geometric calculations in velocity space. Agents in RVO are modeled with a state vector similar to that of GCF: $[\mathbf{p} \ \mathbf{v} \ \mathbf{v}^0]^T \in \mathbb{R}^6$, where $\mathbf{p}$, $\mathbf{v}$, and $\mathbf{v}^0$ are defined as before.

The *velocity obstacle* lies at the core of these approaches. As the name implies, it is an obstacle, but rather than lying in workspace or configuration space, it lies in velocity space. For agents i and j, agent j induces a velocity obstacle on i, VO_{ji}, and i induces a symmetric velocity obstacle on j, VO_{ij}. The velocity obstacle is a cone, originating at $\mathbf{p}_i$, which tightly bounds the Minkowski sum of agent i's geometry with j's. Figure 8.3a illustrates VO_{ji} for two agents with circular geometry. If the relative velocity between agents i and j remains within this cone, there will be an inevitable collision. In practice, we are only concerned with collisions that can occur within the next τ seconds. Including this term truncates the cone (as illustrated in Fig. 8.3b).

This obstacle is the space of *relative* velocities that lead to collision. A single agent cannot exert unilateral control over the relative velocity. If agent i assumes that j will not change velocity, i must take full responsibility for avoiding the collision. This is accomplished by translating VO_{ji} by j's velocity, as shown in Fig. 8.3c. This is the original velocity obstacle formulation, in which each agent assumes that every other agent is a non-responsive, dynamic obstacle [8].

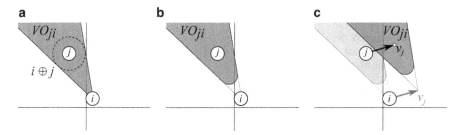

Fig. 8.3 Velocity obstacles. (**a**) The velocity obstacle formed by agent j on i. (**b**) The truncated velocity obstacle for time window τ formed by j on i. (**c**) Agent i assumes the full burden to avoid collision, assuming j's velocity will be unchanged; VO_{ji} is displaced by j's velocity

However, this model is a poor representation of people because agents do, in fact, react to each other. This leads to two significant issues. First, the velocity obstacle is only valid based on the assumption that the other agent maintains a constant velocity. It is possible for both agents to pick a velocity outside of their respective velocity obstacles but the resulting relative velocity may place them on a collision course. Secondly, both agents overreact to their neighbors (because they falsely assume that the other will make no effort to avoid collision). This can easily lead to oscillatory motion as the agents overreact in successive steps.

Van den Berg et al. proposed an alternate formulation to VO which addresses these issues: Optimal Reciprocal Collision Avoidance (ORCA) [28]. The truncated cone, VO_{ij}, is replaced with a half plane, $ORCA_{ij}$. This solves the first issue by defining the half planes of $ORCA_{ij}$ and $ORCA_{ji}$ to contain "mutually reciprocal" sets of velocities. That means there is no pair of velocities, selected from each agents admissible region, which will lead to a collision within τ seconds. Furthermore, the planes are defined such that the amount of change to the relative velocity required to avoid collision is evenly distributed between the two agents, removing the danger of overcompensation and oscillation.

Finally, ORCA provides an additional advantage. When an agent has multiple neighbors, the inadmissible velocities is the union of all velocity obstacles. For truncated cones, computing this region, and finding the best admissible velocity outside is complex and expensive. With half planes, the admissible velocities form a convex polygon. For a convex cost function, the optimal velocity can be computed in $O(n)$ time, for n ORCA half planes.

The ORCA half-plane can be constructed geometrically in the following manner. Assume agents i and j adopt velocities $\mathbf{v}_i^{opt}$ and $\mathbf{v}_j^{opt}$, respectively, and that these velocities place them on a collision course (i.e. $\mathbf{v}_i^{opt} - \mathbf{v}_j^{opt} \in VO_{ji}$). Let $\mathbf{u}$ be the vector from $\mathbf{v}_i^{opt} - \mathbf{v}_j^{opt}$ to the closest point on the boundary of the velocity obstacle (see Fig. 8.4). More formally,

$$\mathbf{u} = \left(\operatorname*{argmin}_{\mathbf{v} \in \delta VO_{ji}} \|\mathbf{v} - (\mathbf{v}_i^{opt} - \mathbf{v}_j^{opt})\|\right) - (\mathbf{v}_i^{opt} - \mathbf{v}_j^{opt}) \qquad (8.6)$$

Fig. 8.4 The formulation of
the ORCA half-plane. The
various components of the
definition are illustrated. The
minimum change in relative
velocity, **u**, the direction of
minimum change, **n̂**, and the
resultant ORCA velocity
obstacles, ORCA$_{ji}$ and the
symmetric half plane ORCA$_{ij}$

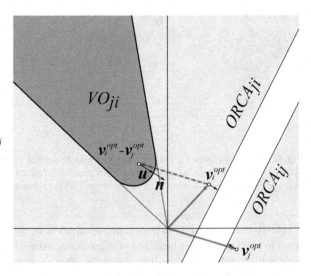

is the minimum change in relative velocity between i and j necessary to guarantee
no collision within τ seconds. To model the reciprocity, half of the minimum change
is applied to each agent. So, we can define the ORCA velocity obstacle induced by
agent j on agent i as:

$$ORCA_{ji} = \left\{ \mathbf{v} \mid \left(\mathbf{v} - \left(\mathbf{v}_i^{opt} + \frac{1}{2}\mathbf{u} \right) \right) \cdot \hat{\mathbf{n}} < 0 \right\}, \tag{8.7}$$

where $\hat{\mathbf{n}}$ is the normalized direction of **u**. Proof of the guarantees can be seen in the
original paper [28]. Henceforth, we will refer to this model as RVO.

RVO is less prone to jitter oscillations than GCF. Whereas GCF computes a
new velocity by integrating a stiff physical system with explicit integration, RVO
computes a new velocity directly in velocity space. Unlike GCF, for a given agent
state, RVO will produce the same feasible velocity, regardless of time step size.
Like GCF, agent position is still integrated using explicit integration and has limits
on the size of the time step. Theoretically, the time step must be strictly larger than
the time window τ. In practice, RVO has been shown to be stable for time steps as
large as 0.2 s [6].

8.4.2 Priority and Right of Way

One of the appealing properties of RVO for pedestrian simulation is its reciprocal
nature. The idea that moving pedestrians will each make an effort to avoid collisions
with others is consistent with anecdotal evidence. However, the model's exactitude
in defining the reciprocity to be precisely half implies a precision that does not exist
in nature. While, generally, the equal division of effort is a reasonable model of the

most generic behavior, there are scenarios in which effort is not shared equally and the model becomes highly dissatisfying.

On a subway platform, pedestrians enter the platform, find a location to await the train and then stop. In navigating the platform, moving pedestrians typically move around those already waiting. After they've stopped in their chosen position, those following behind, must move likewise around them. At that moment, the pedestrian shifts paradigms from an expectation of full responsibility for avoiding collisions, to the expectation that other moving pedestrians will assume the responsibility to avoid collision with them.

In a more subtle vein, even when all pedestrians are moving, the burden isn't necessarily shared equally. It may be that some pedestrians are more conservative and more willing to give way to others. Some pedestrians may seem more aggressive or determined. Subtle social and psychological clues affect how people react to each other and shifts the distribution of responsibility for avoiding collision.

The ability to model asymmetry plays an important role in the simulation of the Tawaf as well. There are several instances in which asymmetrical responses are vital to reproduce observed behaviors. When pilgrims queue up to kiss the Black Stone, their relationship with other pilgrims in the queue is different than with those still circling; they should not yield position to those still circling, but must cooperate, to some degree, with those in the queue. When actually kissing the Black stone, a pilgrim must ignore all other pilgrims, holding their position in front of the stone. Finally, when exiting the Mataf, the agents must work outwards when the rest of the agents are moving tangentially or even spiraling inwards. At any given moment, the exiting pilgrims are in the minority. The ability of the minority to move counter to the flow of the majority is predicated on their ability to enforce their will on the majority.

In the study of traffic, there is a concept that perfectly captures this phenomenon: *right of way*. Right of way is the set of rules which define when one entity must yield to another entity. When moving pedestrians walk around standing commuters on a train platform, the stationary people have right of way. When an aggressive person moves through a crowd and those around him part to let him through, it is because he implicitly has right of way.

Unlike with vehicles, where right of way has a very discrete, exclusionary interpretation (i.e. between two cars, right of way belongs entirely to one vehicle), between pedestrians it can be considered a continuous quantity. Right of way can be absolute, when one pedestrian completely yields to another or it can be shared such that each pedestrian partially yields, albeit to different degrees, to avoid collision.

RVO's formulation provides a simple mechanism by which we can model continuous right of way. We introduce a new agent state parameter, p, called *priority*—a non-negative, real number. An agent with higher priority has right of way. We define the right of way of agent i over agent j as:

$$R_{ij} = \begin{cases} \max(1, p_i - p_j) & \text{if } p_i \geq p_j \\ 0 & \text{otherwise} \end{cases}. \tag{8.8}$$

As implied by (8.8), the value of R_{ij} lies in the range $[0, 1]$, regardless of what the relative priorities of the two agents are. Furthermore, $R_{ij} > 0$ implies $R_{ji} = 0$. Right of way can only be held by a single agent and an agent cannot have more than 100% right of way. Another implication of this formulation is that agents can be assigned tiered priorities—an aggressive agent may acquire full right of way over a passive agent, but it may still be required to yield right of way to a stationary agent. This is easily achieved by assigning priority values to the shy, aggressive and stationary agents of 0, 1, and 2, respectively (or any sequence of monotonically increasing values such that each value is at least one greater than the previous value).

In the formulation of RVO, the velocity obstacle is defined with respect to an abstract relative velocity between agents i and j. The definition uses $\mathbf{v}^{opt}$ to compute the relative velocity. Van den Berg et al. refer to this as the "optimization" velocity and suggest that this is typically the agent's current velocity because it minimizes the amount of change to the current agent state required to avoid collision, but it need not necessarily be the current velocity [28].

We redefine $\mathbf{v}^{opt}$ in terms of right of way. This new definition will affect the definition of $ORCA_{ji}$ (8.7) in the following manner:

$$ORCA_{ji} = \{\mathbf{v} | (\mathbf{v} - (\mathbf{v}_{ij}^{opt} + \alpha_{ij}\mathbf{u}_{ij}) \cdot \hat{\mathbf{n}} < 0\}, \tag{8.9}$$

$$\mathbf{v}_{ij}^{opt} = (1 - R_{ij})\mathbf{v}_i + R_{ij}\mathbf{v}_i^0, \tag{8.10}$$

$$\alpha_{ij} = \begin{cases} 0.5 & \text{if } R_{ij} = R_{ji} = 0 \\ \frac{1-R_{ij}}{2} & \text{if } R_{ij} > 0 \\ \frac{1+R_{ji}}{2} & \text{if } R_{ji} > 0 \end{cases}. \tag{8.11}$$

The effect of right of way is as follows. If both agents have the same priority, no agent has right of way and the new formulation or RVO is equivalent to the old; both agents optimize with respect to their current velocities and share an equal burden in avoiding collision. As one agent's priority increases, its right of way also increases. The increased right of way affects the computation in two ways. First, the higher-priority agent's optimization velocity becomes a linear interpolation between its current and its preferred velocity. Second, the higher-priority agent's share of the burden linearly decreases. When an agent has full right of way, its optimization velocity is its preferred velocity and it bears no responsibility for avoiding collision.

We illustrate the impact of priority and right of way in four experiments (see Fig. 8.5). We apply the following methodology for each experiment. We construct a group of grey agents consisting of eight rows with 28 agents on each row. The rows are vertically offset to increase the average density. The priority of the grey agents always remains zero. We vary the priority of the white subject agent over the range $[0, 1]$. For each priority value, we run 20 iterations with a small random noise applied to the initial positions of the grey agents. In addition, for experiments 1, 2, and 3, we repeat the set of iterations while changing the average density of the grey agents over the values: 2, 3, 4, and 5 agents/m². Experiment 4 has a single density, 8 agents/m² (the maximum possible density when all agents converge in the center

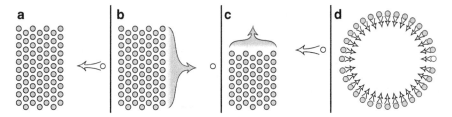

Fig. 8.5 Four experiments for evaluating right of way. In each experiment, the subject agent's (*white circle*) progress is measured. (**a**) Experiment 1: A single agent moves through a stationary group of agents. (**b**) Experiment 2: A single agent holds position against a moving group of agents. (**c**) Experiment 3: A single agent moves perpendicularly to a moving group of agents. (**d**) Experiment 4: A circle of 100 agents, each trying to move to its antipodal position

of the circle). For experiments 1, 3, and 4, the subject agent travels from an initial position to a goal position. For these experiments, we measure the impact of priority by examining the travel time to its goal. More particularly, given its preferred speed (v^0) and the straight-line distance (d) to its goal, we compute the baseline travel time ($t_b = d/v^0$) and report the travel time as a multiple of the baseline. In experiment 2, the agent tries to maintain its position, so we examine the impact of priority by measuring the total distance it travels in the course of the simulation.[3] The results of these experiments can be seen in Figs. 8.6 and 8.7.

There are several salient points to be made about the results of these experiments. First, in experiments 1, 2, and 3, as the subject agent's priority and the corresponding right of way increases, the subject agent's performance quickly converges to the baseline. This can be seen in Fig. 8.6a–c. The performance curves, at all densities, converge to the baseline value (bottom of the figure) at a priority value ranging between 0.4 and 0.6. This phenomenon becomes clearer when we observe the trajectory of the subject agent as shown in Fig. 8.7. The subject agent starts at the right in each figure and seeks to move in a straight line to its mirrored position on the left. The baseline trajectory would be a straight, horizontal line. With low priority, the agent is forced to deviate from the straight line. But for all priority levels, when the agent reaches the mid-point, it is able to travel directly toward its goal position.

We conjecture this quick convergence is due to two reasons. First, it has been shown that, like other pedestrian simulators, RVO exhibits emergent phenomena such as lane formation [10]. We conjecture that experiments 1, 2, and 3 benefit from this property. The experiments are orderly scenarios featuring simple bi-direction flows—an ideal circumstance for lane formation. The subject agent moves contrary to the large contingent of grey agents and as its priority increases, those agents nearest it begin to move out of its way. The following grey agents implicitly follow the divergent paths of the lead agents, forming lanes around the subject. Once those

---

[3]If the agent were perfectly capable of maintaining its position, it would travel no distance at all.

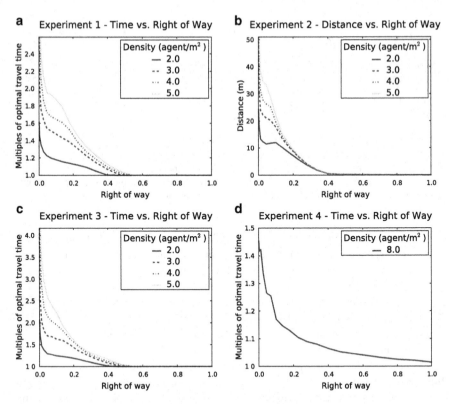

Fig. 8.6 The impact of priority on the experiment scenarios. (**a**), (**c**), and (**d**) report a multiple of the baseline travel time based on right-of-way value and density. (**b**) Shows the absolute distance traveled

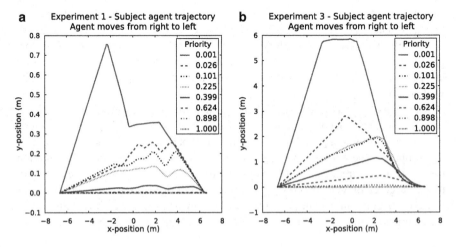

Fig. 8.7 The trajectory of the subject agent at varying priority levels. (**a**) Experiment 1. (**b**) Experiment 3

lanes have formed, the path for the subject agent remains clear. Second, the agents are arranged in a hexagonal lattice. Moving diagonally through the lattice is the clearest path possible. So, as the agent is pushed off of the horizontal, baseline trajectory, the most direct path to its goal eventually becomes a diagonal path which can exploit the greater clearance in the hexagonal lattice. So, for such orderly scenarios, a right-of-way value as little as 0.5 is sufficient for the subject agent to achieve baseline performance.

In comparison, experiment 4 represents a far more chaotic scenario. Agents moving to their antipodal positions do not share a preferred velocity with any of their neighbors. This significantly reduces the formation of lanes. The subject agent must contest with every agent in its path to achieve its goal. The experimental results support this idea. Figure 8.6d shows increasing priority values contribute to the subject agent's performance over the entire range of possible right-of-way values.

In addition, the impact of priority and right of way are dependent on the density around the subject agent. This is as expected. When the region around the subject agent is densely populated, taking any trajectory counter to its neighbors is significantly more difficult. The cause is two-fold. First, because the neighbors are near, the amount they interfere with the subject agent's intentions is much higher; the subject is in danger of colliding with its neighbors in a very small time frame. Also, nearby agents have very little flexibility in responding to the subject agent. So, the agent with right of way needs more priority to successfully influence its neighbors. But in a sparsely populated areas, neighboring agents are more distant, interfering less with the subject agent, and have a great deal more space to respond to the higher priority agent which leads to fast convergence to the baseline value. For the sake of visual clarity, we have vertically clipped the data shown in Fig. 8.6; the performance of the subject agent without right of way in high density scenarios was extraordinarily bad. Including those complete curves would have rendered the lower-density curves undifferentiable. At a density of 5 agents/m^2, the subject agent required 4.1× as much time for experiment 1, traveled 71.6 m in experiment 2, and took 7.9× as much time for experiment 3.

It is worth underscoring, that we are not modeling specific psychological factors nor advocating specific values which map human personality traits to priority values. That is a question for sociologists and psychologists to address. We simply provide a mathematical model which reproduces the phenomenon of asymmetric responses between pedestrians. Whence this asymmetry springs is an open question and we would hope that fellow scientists, better qualified to study these issues, will provide for us suitable characterizations for when such asymmetric responses occur and to what degree.

8.5 Simulating The Tawaf

In this section we give specific details on how the observed behaviors for performing the Tawaf are modeled.

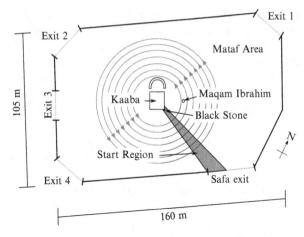

Fig. 8.8 The layout of the Mataf area in the Al-Masjid al Harām. Pilgrims walk seven counter-clockwise circles around the Kaaba and Hateem. Each circle starts in front of the black stone (indicated as the start region)

Figure 8.8 shows the layout of the Mataf area, the location where the Tawaf takes place including the Kaaba, Hateem and Maqam Ibrahim. The Hateem is a semi-circular structure which was originally part of the Kaaba when the Kaaba was rebuilt in A.D. 692. The Maqam Ibrahim is a structure of religious significance, to the northeast of the Kaaba.

8.5.1   The Rite

The Tawaf is performed in the following manner:

1. Pilgrims enter the Mataf area and proceed towards the Black Stone. The Black Stone is located at the Kaaba's eastern corner. This landmark serves as the start and finish point of each circumambulation.
2. After reaching the region in front of the black stone, pilgrims perform Istilam, which can consist of kissing the Black Stone, touching the Black Stone with hands, or raising hands towards the Black Stone, all while saying *Tekbir*, "God is Great". On crowded days, only a small number of pilgrims will attempt to approach the Black Stone to kiss it. Those desirous to kiss the Black Stone will queue up near the southeast wall of the Kaaba. A pilgrim typically will only seek to kiss the Black Stone once, if at all.
3. The pilgrims walk, in a counter-clockwise direction, around the Kaaba and Hateem.
4. At the completion of each circumambulation, the pilgrims perform Istilam again.
5. At the end of the seventh circle, the pilgrims perform a short prayer outside the Mataf area, preferably in front of the Maqam Ibrahim or any convenient location in the mosque. A small number approach to kiss the Black Stone upon completion of the Tawaf.

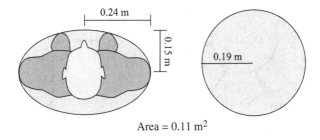

Area = 0.11 m^2

Fig. 8.9 A circle of radius 0.19 m has the same area as an ellipse with major and minor axes of 0.24 and 0.15 m, respectively

6. Pilgrims exit the Mataf area. A recent study [32] has shown that 61% of the pilgrims exit the Mataf through the Safa exit in preparation for the next ritual.

8.5.2 Population Characteristics

One of the parameters of our simulation is the composition of the population. To that end we specify agent characteristics using population *classes*. Each population class defines a numerical distribution of values for a set of agent parameters. These values represent the physical capacity of the virtual pilgrims. The classes we use in simulating the Tawaf include the following parameters:

1. **Preferred speed**: a normal distribution.
2. **Maximum speed**: a normal distribution.

Properties not enumerated in a class (such as agent radius) are the same for all agents. We defined four agent classes to model both genders in two age categories ("old" and "young"). Agents are assigned a population class based on a user-defined distribution. The initial position of the agents is uniformly distributed in a circular area around the Kaaba. To achieve "steady-state" as quickly as possible, we set the agents randomly to have already completed some number of circumambulations (a uniformly distributed integer in the range [0, 7]). Finally, we force the flow into the Mataf to be equal to the flow out of the Tawaf by reintroducing each exiting agent into the system at a random entrance.

The space occupied by the human body can reasonably be bound with an ellipse with major and minor radii of 0.24 and 0.15 m, respectively, with an area of 0.11 m^2. RVO uses circles to represent agents. A circle with a 0.19-m radius has the same total area as the ellipse (as shown in Fig. 8.9). We use this circle to model the pilgrims. Circles of this size can be optimally packed to yield a maximum density of 8 agents/m^2.

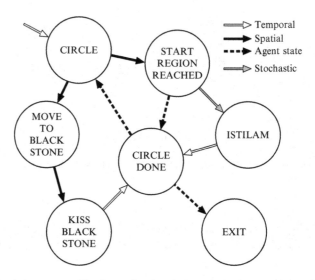

Fig. 8.10 The finite state machine for performing the Tawaf. Pilgrims start in the CIRCLE state. At the end of each circle, they either attempt to move to the black stone or perform Istilam and then perform another circle. After seven circles, they begin movement towards an exit

8.5.3 The Tawaf FSM

We have mapped the above behavior description to an FSM as shown in Fig. 8.10. Here we will enumerate the states and their transitions.

CIRCLE: The circle state is the main circumambulation state. It contains two velocity components represented as guidance fields (a 2D vector field defined over the simulation domain specifying velocity directions). The first is a radial guidance field with directions pointing towards the center of the Kaaba and the second is a tangential guidance field representing the direction of travel around the Kaaba. The tangential field causes the pilgrims to circle around the Kaaba and the radial field draws them toward it. Although it is desirable to approach and kiss the Black Stone, on crowded days it can prove too difficult and many pilgrims choose not to attempt it. We model a variable degree of desire to approach the Kaaba and Black Stone by normally varying the weight of the radial velocity component. Agents with a large radial weight model those pilgrims with a greater desire to approach and put themselves in a better position to kiss the Black Stone.

There are two transitions out of this state. The first transition determines if an agent will queue up to kiss the Black Stone. The transition is a combination of spatial and property transitions. If the agent has not yet kissed the Black Stone and enters into a region near the southern corner of the Kaaba, the condition of the transition is met and the agent enters the MOVE TO BLACK STONE state.

The second transition is a spatial transition. If the agent reaches the start region in front of the Black Stone, the agent enters the START REGION REACHED state.

START REGION REACHED: This state is a decision point. It contains no velocity components. When an agent reaches this state, the state's transitions are evaluated and the agent immediately advances to the corresponding state.

This state contains two transitions. The first transition is a stochastic transition. This is the likelihood that a given agent will attempt to perform Istilam by stopping while turning to face the Kaaba. Anecdotal evidence suggests that this probability is about 15%. We generate a uniformly distributed random value in the range $[0, 1]$. If the value is in the range $[0, p]$, where p is the probability of stopping for Istilam, then the transition is active, moving the FSM to the ISTILAM state.

If the transition to ISTILAM is not taken, then the second transition is taken. This transition is, by definition, active. It moves the FSM to the CIRCLE DONE state.

CIRCLE DONE: This state is another decision point. Like START REGION REACHED, it contains no velocity components. At this state, we determine whether the agent has completed the Tawaf or not.

This state contains two transitions. The first transition is a property transition. If the agent has completed seven circles around the Kaaba, the FSM transitions to the EXIT state. Otherwise, the FSM transitions back to the circle state for the next circle.

MOVE TO BLACK STONE: This state controls the queue for those agents waiting to kiss the Black Stone. Upon entering this state, the agent is marked as having kissed the black stone. Subsequently, the transition from CIRCLE to MOVE TO BLACK STONE cannot be active for this agent. The velocity is computed as follows: the direction of the preferred velocity is towards the Black Stone. If there is another agent in the queue between the agent and the Black Stone, the speed is the lesser of two speeds: the agent's preferred speed and the speed that will guarantee the agent reaches the other agent's position in 1 s. If the space in front of the agent is clear, the preferred velocity's magnitude is simply the agent's preferred speed.

This state has a single spatial transition. It activates when the agent reaches the stone and moves the FSM to the KISS BLACK STONE state.

KISS BLACK STONE: This state contains a single velocity component and a single transition. Upon reaching the area directly in front of the Black Stone, the velocity is computed to hold the agent in that position. To aid in this purpose, the agent's determination property is set to one. The single transition is a temporal transition. After a randomly determined duration the agent enters the CIRCLE DONE stage.

ISTILAM: This state, like the KISS BLACK STONE state, has a single velocity component and transition. It likewise computes a velocity to keep the agent fixed in the position at which the agent was when entering this state. However, this is a softer constraint and the determination is set to zero. The single transition is a temporal transition. After a randomly determined duration (1–2 s), the agent enters the CIRCLE DONE stage.

EXIT As pilgrims complete the Tawaf and exit the Mataf floor, they do so in a cooperative manner, continuing to circle the Kaaba and working their way towards the outside until they are in sufficient free space to head to their selected exit area. Each agent is randomly assigned an exit according to the probability distribution found in [32].

We have areas defined in the simulation domain for each of the five exits. Once the exit has been randomly selected, we then select a random point in the exit region to serve as the agent's goal point.

To model the cooperative exit behavior exhibited by the pilgrims in the Tawaf, we generate the agent's velocity with a weighted combination of three velocities: a vector from current position towards the exit goal position, a tangential component like that in the CIRCLE state, and an anti-parallel radial component (the opposite of the radial component of the CIRCLE state). The tangential and anti-parallel radial components cause the agent to continue circling the Kaaba while working its way away from the Kaaba.

We blend the exit goal velocity and the circular velocity based on the agent's local density. When the crowd is very dense, the agent continues around the Kaaba. As the local density reduces, the weight between goal and circular velocities changes linearly until an acceptable minimum density is achieved and the agent can move directly towards its end goal.

8.6   Results

We've run several simulations with our system. Our first goal is to achieve a result consistent with observed crowd movement during the Tawaf. To that end, we created a population of 35,000 agents with the following composition: 25% each of young male and female and 25% each of old male and female. Young males had a mean preferred speed of 1.0 m/s and a standard deviation of 0.2 m/s. Similarly old males had a mean preferred speed of 0.85 m/s with a standard deviation of 0.2 m/s. Young and old females had mean preferred speeds of 0.95 and 0.8 m/s, respectively. Both had a standard deviation of 0.15 m/s.

Our approach exploits the efficiency of the underlying pedestrian model. Our simulation used a time step of 0.1 s and was able to generate frames at 26 Hz on an Intel i7 running at 2.67 GHz. The evaluation of the FSM and pedestrian model were parallelized over the set of agents through the use of OpenMP. In essence, our simulator runs faster than real-time. For 35,000 agents, it produces 2.6 s of simulated results for each second of computation.

Figures 8.11 and 8.12 show a single moment from our simulation results. In this image, approximately 25,900 agents are actively circling the Kaaba. The other 9,100 agents are entering, exiting or queueing to touch the Black Stone. The average walking speed of the circumambulating agents is approximately 0.73 m/s. The average completion time for the full Tawaf is 28.1 min. If we assume that the 25,900 circumambulating agents are representative of the portion of the population

Fig. 8.11 The density of the crowd of pilgrims performing the Tawaf in our simulation. The *dark region* in the *center* is the Kaaba. Our simulation reaches a maximum density of 7.3 agents/m². The density field is computed as in [12]

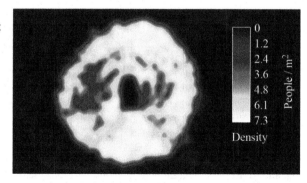

Fig. 8.12 The speed of the individual agents performing the Tawaf in our simulation

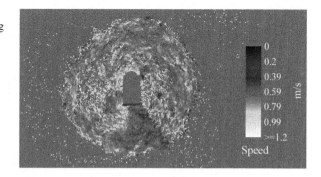

of 35,000 agents that are circling the Kaaba at any time, then this simulation implies a capacity of 55,300 participants per hour.

In 2008, Koshak and Fouda [14] tracked subjects performing the Tawaf with GPS devices. They partitioned the Mataf area into regions and computed the average speed for each region. The results of this analysis are shown in Fig. 8.13. We computed average speed for similar regions in our simulation. The simulated results can be seen in Fig. 8.14. The analysis shows that the simulation compares well with the real data in some ways and diverges in others.

1. Similarities

(a) Region 1, the region immediately preceding the start area, is the slowest region.
(b) Regions 5–7 exhibit higher speeds than regions 1–4.
(c) The top speed of the simulated crowd matches the top speed of the measured crowd.

2. Differences

(a) Simulated data exhibits a much narrower range of speeds.

The disparity observed in the range of speeds can be attributed to two causes. First, when Koshak and Fouda performed their experiments, there was a line on the Mataf floor indicating the starting point. The line has since been removed.

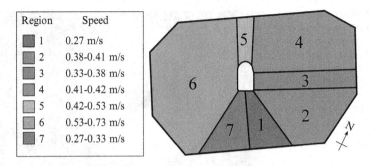

Fig. 8.13 The observed speeds of real pilgrims traversing each region during the Tawaf [14]

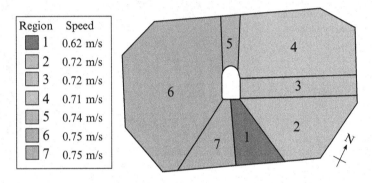

Fig. 8.14 The average speed of simulated agents traversing each region during the Tawaf

At the time, experts felt that as pilgrims approached the line, they would come to a stop while searching for the line. This is considered to be the dominant cause of the extreme slow down in the corresponding region. Our simulation models current behaviors reflecting the removal of the line. Thus, our agents don't come to a stop and the aggregate result is a higher speed through this region.

Secondly, the narrow range of simulated speeds may arise from properties in the pedestrian model. It may be that, as a pedestrian model, RVO is insensitive to density-related effects, such as the fundamental diagram. Further research is required to confirm and correct, if necessary.

Heterogeneity: To explore the impact of the heterogeneous population, we ran two alternative simulations. One consisted of nothing but young males (the fastest pilgrim class). The second simulation consisted solely of old females (the slowest pilgrim class). The simulation consisting only of young males exhibited an average walking speed of 0.82 m/s for 24,900 circumambulating pilgrims and a corresponding Tawaf completion time of 25.5 min. In contrast, the simulation of old females obtained an average walking speed of 0.67 m/s for 26,500 circumambulating pilgrims with a Tawaf completion time of 30.2 min. The implied capacity is 58,600 pilgrims per hour for the young males and 52,700 pilgrims per hour for the old females.

The capacity indicated by the heterogeneous crowd is close to the average capacity of the two homogeneous crowds (although the heterogeneous crowd's capacity is slightly lower). The full impact of heterogeneity is still unclear. It may be that populating the entire crowd with instances of a single, statistically average pedestrian may prove to be sufficient. This requires more study and requires better data concerning the demographics of the pilgrims performing the Tawaf and more flow data of the actual performance.

8.6.1 Limitations

While the results are promising, there are still aspects of the Tawaf it does not capture. In addition to the unknown impact of heterogeneity, these simulations haven't modeled groups. We currently treat the agents as individuals. To more fully capture the dynamics of the Tawaf, we would require a group model such as in [17]. There is, in particular, one instance of group behavior that has often been noted by observers. At times, a group of participants will force their way orthogonally across the crowd flow to get closer to the Kaaba. This behavior, its rate of incidence, and its characteristics are not well understood and, as such, is not included in our model.

More generally, the simulated speeds need to be validated. Although the maximum speed matches that observed by Koshak and Fouda [14], it may prove that at many of the densities observed, the speed of the pedestrians should be lower. Furthermore, since the time of Koshak's and Fouda's experiments, the Mataf area has been changed to improve the flow. We need to validate against more current data collected from coordinated GPS devices and cameras.

8.6.2 Conclusion

The unique nature of the Tawaf exhibits behaviors which are not well modeled by many existing crowd simulation systems. We have presented a framework for simulating many of the complex behaviors and relationships exhibited by pilgrims performing the Tawaf. By coupling a high-level finite-state machine with a low-level pedestrian model, we have been able to model a range of behaviors such as: circumambulating the Kaaba, queuing to touch the Black Stone, entering and exiting the Mataf floor, and pausing to perform Istilam. We've shown how to extend a velocity-obstacle-based pedestrian model to capture asymmetric inter-agent responses and have shown that the model is well behaved even at the extreme densities observed in our simulation. In many important respects, the results of the simulation match those observed in real people performing the Tawaf.

There are still multiple avenues to pursue for future work. The first is to confirm the validity of the pedestrian model with respect to density-dependent effects. In addition, we plan to extend the current set of behaviors to capture the important behaviors currently missing from our simulation, with particular focus on the impact of exiting pilgrims and groups. In addition, we intend to investigate the possibility of using video of the Tawaf to refine the behavior system, both the parameters and structure of the FSM as well as the local collision avoidance parameters.

Acknowledgements This research is supported in part by ARO Contract W911NF-10-1-0506, NSF awards 0917040, 0904990 and 1000579.

References

1. Al-Haboubi, M., Selim, S.: A design to minimize congestion around the ka'aba. Comput. Ind. Eng. **32**(2), 419–428 (1997)
2. Algadhi, S., Mahmassani, H.: Modelling crowd behavior and movement: application to makkah pilgrimage. Transp. Traffic Theory **1990**, 59–78 (1990)
3. Bandini, S., Federici, M., Manzoni, S., Vizzari, G.: Towards a methodology for situated cellular agent based crowd simulations. Engineering societies in the agents world VI, pp. 203–220. Springer, Berlin/Heidelberg (2006)
4. Chraibi, M., Seyfried, A., Schadschneider, A.: Generalized centrifugal-force model for pedestrian dynamics. Phys. Rev. E **82**(4), 046,111 (2010)
5. Crystals project. http://www.csai.disco.unimib.it/CSAI/CRYSTALS/
6. Curtis, S., Snape, J., Manocha, D.: Way portals: Efficient multi-agent navigation with line-segment goals. In: Proceedings of the Symposium on Interactive 3D Graphics and Games (I3D), Costa Mesa, CA, USA (2012)
7. Durupinar, F., Pelechano, N., Allbeck, J., Gudukbay, U., Badler, N.: How the ocean personality model on the perception of crowds. Comput. Graph. Appl. IEEE **31**(3), 22–31 (2010)
8. Fiorini, P., Shiller, Z.: Motion planning in dynamic environments using velocity obstacles. Int. J. Robot. Res. **17**(7), 760–762 (1998)
9. Funge, J., Tu, X., Terzopoulos, D.: Cognitive modeling: knowledge, reasoning and planning for intelligent characters. In: SIGGRAPH, pp. 29–38. ACM, Los Angeles, CA, USA (1999)
10. Guy, S.J., Chhugani, J., Curtis, S., Lin, M.C., Dubey, P., Manocha, D.: Pledestrians: A least-effort approach to crowd simulation. In: Symposium on Computer Animation. ACM, Madrid, Spain (2010)
11. Helbing, D., Molnar, P.: Social force model for pedestrian dynamics. Phys. Rev. E **51**(5), 4282–4286 (1995)
12. Johansson, A., Helbing, D., Al-Abideen, H., Al-Bosta, S.: From crowd dynamics to crowd safety: A video-based analysis. Advances in Complex Systems **11**(04), 497–527 (2008)
13. Ju, E., Choi, M.G., Park, M., Lee, J., Lee, K.H., Takahashi, S.: Morphable crowds. ACM Trans. Graph. **29**(6), 140 (2010)
14. Koshak, N., Fouda, A.: Analyzing pedestrian movement in mataf using gps and gis to support space redesign. In: The 9th International Conference on Design and Decision Support Systems in Architecture and Urban Planning, The Netherlands/Holland (2008)
15. Lee, K.H., Choi, M.G., Hong, Q., Lee, J.: Group behavior from video: a data-driven approach to crowd simulation. In: Symposium on Computer Animation, San Diego, CA, USA pp. 109–118 (2007)
16. Mehran, R., Oyama, A., Shah, M.: Abnormal crowd behavior detection using social force model. CVPR, Miami Beach, FL, USA (2009)

17. Moussaïd, M., Perozo, N., Garnier, S., Helbing, D., Theraulaz, G.: The walking behaviour of pedestrian social groups and its impact on crowd dynamics. PLoS ONE 5(4), e10,047 (2010). doi:10.1371/journal.pone.0010047. http://dx.doi.org/10.1371%2Fjournal.pone.0010047
18. Mulyana, W., Gunawan, T.: Hajj crowd simulation based on intelligent agent. In: 2010 International Conference on Computer and Communication Engineering (ICCCE), pp. 1–4. IEEE, Kuala Lumpur, Malyasia (2010)
19. Narain, R., Golas, A., Curtis, S., Lin, M.C.: Aggregate dynamics for dense crowd simulation. ACM Trans. Graph. 28, 122:1–122:8 (2009). doi:http://doi.acm.org/10.1145/1618452.1618468. http://doi.acm.org/10.1145/1618452.1618468
20. Ondřej, J., Pettré, J., Olivier, A.H., Donikian, S.: A synthetic-vision based steering approach for crowd simulation. In: Proceedings of the SIGGRAPH, Los Angeles, CA pp. 123:1–123:9 (2010)
21. Patil, S., van den Berg, J., Curtis, S., Lin, M., Manocha, D.: Directing crowd simulations using navigation fields. IEEE TVCG, pp. 244–254 (2010)
22. Pelechano, N., Allbeck, J., Badler, N.: Controlling individual agents in high-density crowd simulation. In: SCA07, San Diego, CA, USA (2007)
23. Reynolds, C.: Flocks, herds and schools: A distributed behavioral model. In: SIGGRAPH, Anaheim, CA, USA (1987)
24. Sarmady, S., Haron, F., Talib, A.: A cellular automata model for circular movements of pedestrians during tawaf. Simul. Model. Pract. Theory 19(3), 969–985 (2010)
25. Schadschneider, A.: Cellular automaton approach to pedestrian dynamics – theory. Pedestr. Evacuation Dyn. (2001)
26. Treuille, A., Cooper, S., Popović, Z.: Continuum crowds. In: ACM SIGGRAPH 2006, pp. 1160–1168. ACM, Boston, MA, USA (2006)
27. Ulicny, B., Thalmann, D.: Towards interactive real-time crowd behavior simulation. In: Computer Graphics Forum, vol. 21, pp. 767–775. Wiley Online Library (2002)
28. van den Berg, J., Guy, S.J., Lin, M., Manocha, D.: Reciprocal n-body collision avoidance. In: International Symposium on Robotics Research, Lucerne, Switzerland (2009)
29. Yeh, H., Curtis, S., Patil, S., van den Berg, J., Manocha, D., Lin, M.: Composite agents. Proceedings of SCA, Dublin, Ireland pp. 39–47 (2008)
30. Yersin, B., Maim, J., Pettré, J., Thalmann, D.: Crowd patches: populating large-scale virtual environments for real-time applications. In: I3D09, pp. 207–214. ACM, Boston, MA, USA (2009)
31. Yu, Q., Terzopoulos, D.: A decision network framework for the behavioral animation of virtual humans. In: Symposium on Computer Animation, San Diego, CA, USA pp. 119–128 (2007)
32. Zafar, B.: Analysis of the Mataf – Ramadan 1432 AH. Tech. rep., Hajj Research Institute, Umm al-Qura University, Saudi Arabia (2011)
33. Zainuddin, Z., Thinakaran, K., Abu-Sulyman, I.: Simulating the circumambulation of the ka'aba using simwalk. Eur. J. Sci. Res. 38(3), 454–464 (2009)

Part II
Visual Analysis of Crowds

Chapter 9
Crowd Flow Segmentation Using Lagrangian Particle Dynamics

Saad Ali and Mubarak Shah

Abstract A crowd of people is composed of *groupings* that arise due to interdependence among its members. Advanced visual surveillance and monitoring capabilities for crowded scenes can make use of this inherent group-based composition of human crowds to understand its global motion dynamics and to compartmentalize it into sub-parts for detailed analysis. In this chapter we propose an algorithm that uses motion information to locate such distinct crowd groupings in terms of flow segments in videos of *large dense crowds*. The flow segments are located using a particle-based representation of the motion in the video. This representation enables detection of boundaries between dynamically distinct crowd groupings.

9.1 Introduction

A crowd of people is composed of *groupings* that arise due to interdependence among its members [2, 14]. This interdependence could be a result of a social relationship (e.g. members of the same family or close circle of friends), a common purpose (e.g. to walking towards the same exit door) or an act of participating in a collective activity (e.g. running in a marathon). Advanced visual surveillance and monitoring capabilities for crowded scenes can make use of this inherent group-based composition of human crowds to understand its global motion dynamics and to compartmentalize it into sub-parts for detailed analysis. In this chapter we

S. Ali (✉)
Center for Vision Technologies, SRI International, 201 Washington Road, Princeton, NJ, USA
e-mail: saad.ali@sri.com

M. Shah
Center for Research in Computer Vision, University
of Central Florida, 4000 Central Florida Blvd., Orlando, FL 32816, USA
e-mail: shah@crcv.ucf.edu

S. Ali et al. (eds.), *Modeling, Simulation and Visual Analysis of Crowds*, The International   213
Series in Video Computing 11, DOI 10.1007/978-1-4614-8483-7_9,
© Springer Science+Business Media New York 2013

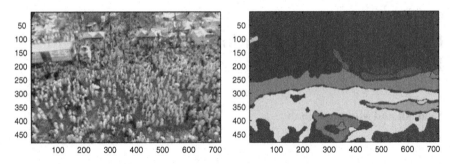

Fig. 9.1 *Left*: A frame depicting groups of people walking in multiple directions. *Right*: The crowd groupings (segments) located by the proposed algorithm

propose an algorithm that uses motion information to locate (or segment) such distinct crowd groupings in terms of flow segments in videos of *large dense crowds*. Figure 9.1 show an instance of a crowded scene where the proposed algorithm discovered various flow segments that belong to distinct crowd groupings in the scene.

Flow segment-based, and in turn group-based, visual analysis of crowded scenes provides multiple benefits: (i) enables a more elaborate and clutter free visualization of various moving groups of people in the scene; (ii) overcomes shortcomings of traditional 'detection and tracking' surveillance approaches that rely on accurate detection of each individual in the scene; (iii) mitigates influence of number of pixels on an individual person and is able to provide reasonable insight into motion of large crowds even at low resolutions.

Lagrangian Particle-based Representation: For segmenting crowd flows, the key idea developed in this chapter is a particle-based representation of the motion in the video. This representation enables detection of boundaries between various dynamically distinct crowd groupings. These boundaries, which are otherwise invisible or imperceivable to human eye, naturally emerge when people walk in different directions or at different speeds.

The proposed particle-based representation consists of particle trajectories that are obtained by examining a cloud of particles (usually in the form of a regular grid) as it mixes and gets transported over time under the action of optical flow generated by the crowd motion. The process of particle propagation using optical field (or motion field in general) is called 'advection'. If we assume that this optical flow field is generated by a certain underlying dynamical system (whose exact form and description is unknown) then one can use these trajectories to reveal representative characteristics of the phase space of this dynamical system where phase space is defined as a space of variables using which all possible states of a dynamical system are represented. The characteristics can include locations of the barriers, mixing properties, location of sources, and sinks in the phase space. Under our assumption the phase space is directly related to the flow field of the crowd, these characteristics can be mapped directly to physical properties of the crowded scene. For example, a barrier in the phase space maps either to a physical obstacle in the scene or to a boundary between crowd groups moving in different directions.

Formally if we assume that the underlying dynamical system is a non-autonomous dynamical system then the barriers are the invariant manifolds of the phase space and are often called *Coherent Structures (CS)* [4]. Generally speaking, Coherent Structures (CS) are separatrices/material lines (i.e. a boundary having two different types of flows on opposite sides) that influence the kinematics of the particle cloud over finite time intervals, and they divide the flow, and in turn the phase space, into dynamically distinct regions where all the particles within the same region have a similar fate or, in other words, coherent behavior. Intuitively speaking, coherent structure is to optical flow data what "edge" is to image data. Note that when coherent structures are studied in terms of quantities derived from particle trajectories, they are named as *Lagrangian Coherent Structures* (LCS).

Note that there are two approaches by which the field of motion can be described: (i) Lagrangian, and (ii) Eulerian. In the Lagrangian approach, properties of the flow are gathered along the path taken by a particle, while in the Eulerian approach properties of the flow are observed at a fixed spatial location. Since in our case particles are allowed to move under the influence of the optical flow, we call our representation a 'Lagrangian particle-based representation'.

LCS Detection: In order to develop an algorithm for detection of LCS (or boundaries between distinct crowd groupings) we make use of several advances in the areas of nonlinear dynamical systems [5, 12], fluid dynamics, [4, 6, 17] and turbulence theory [7, 11]. In these disciplines several approaches have been proposed to compute LCS based on whether the underlying dynamical system is periodic [15], aperiodic [3], or quasi-periodic. The crowd movements are generally aperiodic (i.e. time dependent) in a generic setting as there are no or little prior constraints on its speed and direction over longer durations of time. In this chapter we employ the *Lyapunov Exponent* (LE) approach to locate LCS of the phase space. The LE measures the exponential rate of convergence or divergence between two particle trajectories. For a given crowd video, we use a grid that covers the optical flow field of the video and compute the finite-time estimate of Lyapunov Exponents (LEs) for trajectories starting at each point of the grid. This process returns a finite-time scalar Lyapunov Exponent (FTLE) field over the phase space. We use the result by Haller [4] that show coherent structures appearing as ridges in the FTLE field. In turn these ridges can be used as the boundary between various dynamically distinct crowd groupings for segmentation purposes.

We compute two types of LCS: (1) attracting LCS and (2) repelling LCS. The attracting LCS, represented by a forward FTLE field, are computed by advecting the particle grid forward in time, while the repelling LCS, represented by a backward FTLE field, are computed by advecting the particle cloud grid in time. The two FTLE fields are combined to generate a single scalar field that is segmented using an image segmentation algorithm (e.g., a watershed segmentation algorithm in this case). The steps involved crowd flow segmentation are summarized in the block diagram in Fig. 9.3.

Assumptions: Motion of crowds can exhibit a wide range of behaviors and can be captured using a variety of camera setups (e.g. pole mounted or a ground-based camera). Therefore, it is pertinent to layout the assumptions and constraints on the type of motion and scenes that an processed using the proposed algorithm. Some of these are listed next:

- Crowded scene is viewed from a distance by a camera installed over a tall structure. This constraint results from the abstraction of crowd (or people) as particles. If a scene is viewed from a closer distance, then the algorithm requires a top-down view where only heads of individuals are visible, thereby minimizing artifacts resulting from in-dependent movement of other body parts. Side views of the scene are least preferable within the particle based framework.
- The density of the crowd varies from 3 person per meter square to 7 meter per second square.
- The crowd is formally structured and focused on some collective activity. This constraint results from the fact that LCS detection algorithm exploits in some sense the 'common fate' principle (i.e. trajectories belonging to the same group have the same destination) to localize boundaries between trajectories moving in different directions. If the crowd motion is random or haphazard this may no longer be true.
- Each spatial location in the scene supports one dominant motion. That is, for a any fixed spatial location the distribution of direction and speed of optical flow vectors cannot be multi-modal. This is necessary as algorithm assumes analysis is done only at one time scale and during that time only one type of dominant motion is expected at a location.
- It should be noted that crowd behavior is dynamic in nature and can change drastically. Therefore, in order to perform any video based analysis of crowd motion a sliding window based approach should be adopted. The temporal extent of the window can be kept constant or can be dynamically adopted based on level of activity in the scene. Approaches summarized in this chapter adhere to this principal and performs flow segmentation within of a sliding temporal window.

Chapter Organization The remaining portion of the chapter is organized as follows: Sect. 9.2 provides a overview of the background material and formal definition of various concepts. Section 9.3 discusses the crowd segmentation algorithms and walks the reader through various intermediate steps. Section 9.4 describes experimental setup and presents qualitative results.

9.2 Background, Definitions and Notations

Key background concepts, mathematical notations, and formal definitions are provided in this section. The nomenclature of Shadden et al. [18] is used for this purpose.

Let a compact set $D \subset \mathbb{R}^2$ be the domain of the phase space under study. This domain corresponds to the 2D-spatial extent of the video depicting crowd motion.

Next, define a time-dependent optical flow field $\mathbf{v}(\mathbf{x},t)$ on D that satisfies C^0 and C^2 continuity in time and space, respectively. The C^0 and C^2 assumptions are required to keep the optical flow field smooth. Here, t corresponds to the t-th frame of the video. Then a particle trajectory $\mathbf{x}(t:t_0,\mathbf{x}_0)$, starting at point $\mathbf{x}_0$ at time t_0 can be defined as a solution of

$$\dot{\mathbf{x}}(t;t_0,\mathbf{x}_0) = \mathbf{v}(\mathbf{x}(t;t_0,\mathbf{x}_0),t), \tag{9.1}$$

$$\mathbf{x}(t_0;t_0,\mathbf{x}_0) = \mathbf{x}_0, \tag{9.2}$$

where $\dot{\mathbf{x}}$ is the time derivative. It can also be observed that a trajectory, $\mathbf{x}(t:t_0,\mathbf{x}_0)$, of a particle depends on the initial position $\mathbf{x}_0$ and the initial time t_0. From the above mentioned continuity constraints of optical flow, $\mathbf{v}(\mathbf{x},t)$, it follows that the particle trajectory, $\mathbf{x}(t:t_0,\mathbf{x}_0)$, will be C^1 in time and C^3 in space.

As the goal is to analyze the transport properties (using particle trajectories) of the phase space and, in turn, the underlying crowd, the solution of Eq. (9.1) can be viewed as a transport device or map that takes particles from their initial position $\mathbf{x}_0$ at time t_0 to their position at time t. Formally, this solution is referred as a "flow map," denoted by $\phi_{t_0}^t$, and that satisfies:

$$\phi_{t_0}^t : D \to D : \mathbf{x}_0 \mapsto \phi_{t_0}^t(\mathbf{x}_0) = \mathbf{x}(t;t_0,\mathbf{x}_0). \tag{9.3}$$

In addition, the flow map $\phi_{t_0}^t$ satisfies the following properties:

$$\phi_{t_0}^{t_0}(\mathbf{x}) = \mathbf{x}, \tag{9.4}$$

$$\phi_{t_0}^{t+s}(\mathbf{x}) = \phi_{s}^{t+s}(\phi_{t_0}^{s}(\mathbf{x})) = \phi_{t}^{t+s}(\phi_{t_0}^{t}(\mathbf{x})). \tag{9.5}$$

These properties follow directly from the existence and uniqueness theorem that allows one to conclude that there exists only one solution to a first-order differential equation that satisfies the given initial condition. Next we describes the key concept of FTLE field and discuss the steps involved in its computation from the flow map ϕ.

9.2.1   Finite Time Lyapunov Exponent Field

As mentioned earlier crowd segments/groupings are located using LCS, and the localization of LCS in turn requires computation of the FTLE field. The Lyapunov exponent is an asymptotic quantity that measures the extent to which an infinitely-close pair of particles separate in an infinite amount of time. In the theory of dynamical systems, it is used as a tool for measuring the chaoticity of the system under consideration by measuring the rate of exponential divergence between the neighboring trajectories in the state/phase space. Traditionally, for any given dynamical system, $\dot{x} = f(x)$, the maximum Lyapunov characteristic exponent is defined as $\gamma = lim_{t \to \infty} \chi(t)$, with

$$\chi(t) = \frac{1}{t} ln \frac{|\xi(t)|}{|\xi(0)|}, \tag{9.6}$$

where $\xi(t)$ is the current state of the system, while $\xi(0)$ is the initial state of the given system. These states are usually obtained by solving the differential equation controlling the evolution of the system.

When the Lyapunov exponent analysis is performed over a grid of particles over finite times, it generates a FTLE field. In our formulation, the state of the system is defined as the maximum possible separation between a particle and its neighbors. Essentially, this means that the Lyapunov exponent now can be defined as a ratio of the initial separation to the maximum possible separation between the particle and its neighbors. Using this definition of the Lyapunov exponent, FTLE field $\sigma_T(\mathbf{x}_0, t_0)$ can be computed using the flow map $\phi_{t_0}^{t_0+T}$, which contains the final locations of the particles at the end of particle advection. The flow map, as mentioned earlier, quantifies the transport properties of the phase space by taking a particle from the initial position, $\mathbf{x}_0$, at time t_0 to its later position at time $t_0 + T$.

One important point to note is that the FTLE does not capture the instantaneous separation rate, but rather measures the average, or integrated, separation rate between trajectories. This distinction is important because, in time-dependent complex crowd flows, the instantaneous optical flow is not very informative. However, by accounting for the integrated effect of the crowd-flow using particle trajectories in the FTLE field, we hope to extract information that is more indicative of the actual transport behavior.

The formal derivation of the expression of FTLE proceeds as follows [7, 18]. Consider a particle $\mathbf{x} \in D$ at initial time t_0 (Fig. 9.2). Following advection, the position of the particle after a time interval T is $\mathbf{x} \mapsto \phi_T^{t_0+T}(\mathbf{x})$. Now, when advected through the flow, any arbitrary particle that is infinitesimally close to $\mathbf{x}$ at time t_0 will behave in a manner similar to $\mathbf{x}$ locally in time. However, as the advection time increases the distance between these neighboring particles will change. Now, if we represent the neighboring particle by $\mathbf{y} = \mathbf{x} + \delta\mathbf{x}(0)$ (Fig. 9.2), where $\delta\mathbf{x}(0)$ is an arbitrarily-oriented unit vector, then after a time interval T, the distance between them becomes:

$$\delta\mathbf{x}(t_0 + T) = \phi_{t_0}^{t_0+T}(\mathbf{y}) - \phi_{t_0}^{t_0+T}(\mathbf{x}) \tag{9.7}$$

$$= \frac{d\phi_{t_0}^{t_0+T}(\mathbf{x})}{d\mathbf{x}} \delta\mathbf{x}(0) + O(\|\delta\mathbf{x}(0)\|^2). \tag{9.8}$$

Since the distance $\delta\mathbf{x}(0)$ is infinitesimally small, we can drop the higher order terms in the Taylor series expansion of the flow map around the location $\mathbf{x}$. The magnitude, $\|\delta\mathbf{x}(t_0 + T)\|$, of the final separation can be computed by taking the standard L_2 norm

Fig. 9.2 Computation of FTLE. The initial separation between particle **x** and $\mathbf{y} = \mathbf{x} + \delta\mathbf{x}(0)$ is $\delta\mathbf{x}(0)$. In order to compute the FTLE between them, we need to find out the magnitude of the final separation after a time interval T

$$\|\delta\mathbf{x}(t_0 + T)\|_2 = \left\|\frac{d\phi_{t_0}^{t_0+T}(\mathbf{x})}{d\mathbf{x}}\delta\mathbf{x}(0)\right\|_2. \tag{9.9}$$

We are interested in finding out the maximum possible separation between the particle, **x**, and all its neighbors, which, in other words, means that we seek to maximize $\|\delta\mathbf{x}(t_0 + T)\|_2$ over all possible choices of $\delta\mathbf{x}(0)$:

$$\|\delta\mathbf{x}(t_0 + T)\|_2 = \max_{|\delta\mathbf{x}(0)|=1}\left\|\frac{d\phi_{t_0}^{t_0+T}(\mathbf{x})}{d\mathbf{x}}\delta\mathbf{x}(0)\right\|_2. \tag{9.10}$$

Using the operator norm, the above equation can be written as:

$$\|\delta\mathbf{x}(t_0 + T)\|_2 = \max_{|\delta\mathbf{x}(0)|=1}\left\|\frac{d\phi_{t_0}^{t_0+T}(\mathbf{x})}{d\mathbf{x}}\delta\mathbf{x}(0)\right\|_2 = \left\|\frac{d\phi_{t_0}^{t_0+T}(\mathbf{x})}{d\mathbf{x}}\right\|_2. \tag{9.11}$$

The right-hand side of the above equation is the matrix L_2 norm that can be computed simply by using the standard property that states that, for any matrix A, the matrix L_2 norm is the square root of the maximum eigenvalue of the positive definite symmetric matrix $A^T A$. If we consider $A = \frac{d\phi_{t_0}^{t_0+T}(\mathbf{x})}{d\mathbf{x}}$, then $A^T A$ is

$$\Delta = A^T A = \frac{d\phi_{t_0}^{t_0+T}(\mathbf{x})}{d\mathbf{x}}^* \cdot \frac{d\phi_{t_0}^{t_0+T}(\mathbf{x})}{d\mathbf{x}}, \tag{9.12}$$

where superscript '*' refers to the transpose operator. It is interesting to note that Δ is also known as the finite time version of the Cauchy-Green deformation tensor. The quantity $\frac{d\phi_{t_0}^{t_0+T}(\mathbf{x})}{d\mathbf{x}}$ is the spatial gradient tensor of the flow map. The maximum eigenvalue of Δ is represented by $\lambda_{max}(\Delta)$.

Now, knowing the magnitude of the maximum possible separation, $\lambda_{max}(\Delta)$, and the initial separation, $\delta\mathbf{x}(0)$, between the particle and its neighbors, we can compute the FTLE field, σ, with a finite integration time T corresponding to point $\mathbf{x} \in D$ at time t_0 as:

$$\sigma_{t_0}^T = \frac{1}{T}\ln\sqrt{\lambda_{max}(\Delta)}. \tag{9.13}$$

Since, $\delta\mathbf{x}(0)$ is a unit vector, we eliminated it from the above equation. The above quantity is computed for each $\mathbf{x} \in D$ to obtain the entire FTLE field at time t_0.

9.2.2 Lagrangian Coherent Structures

The LCS corresponds to the boundaries between the crowd flows of distinct dynamics. They appear as ridges in the FTLE field of the video. The relationship between ridges in the FTLE field and the LCS can be explained in the following way. If two regions of a phase space have qualitatively different dynamics, then we expect a coherent motion of particles within each region, and, therefore, the eigenvalues of Δ will be close to 1, an indication that the fate of nearby particles is similar inside the region. At the boundary of the two regions, particles will move in incoherent fashion, and, therefore, will create much higher eigenvalues. These higher values will make the ridge prominent in the FTLE field and point to the locations of the LCS.

We compute two types of LCS, namely "Attracting Lagrangian Coherent Structures" (ALCS) and "Repelling Lagrangian Coherent Structures" (RLCS). The former will emphasize those boundaries between the crowds from which, in a given time interval (*forward in time*), all nearby particle trajectories separate; the later will emphasize those boundaries between the crowds from which in a given time interval (*backward in time*), all nearby particle trajectories separate. For the computation of ALCS, the particle grid is initialized at the first optical flow field and advected forward in time, followed by the computation of forward FTLE field. For the computation of RLCS, the particle grid is initialized at the last optical flow field and advected backward in time, followed by the computation of backward FTLE field.

9.3   Crowd Segmentation: The Algorithm

In this section, we bring together all the concepts explained so far and describe the algorithmic steps that involved in carrying out the segmentation of crowd into dynamically distinct groupings. A block diagram in Fig. 9.3 provides a higher-level view of the algorithmic steps.

9.3.1   Optical Flow Computation

Given a video sequence, the first task is to compute the optical flow between the consecutive frames of the video. We employ two different schemes for this purpose. The first scheme consists of a block-based correlation in the Fourier domain. The process starts by selecting a square patch centered at the same pixel location of two consecutive frames F_1 and F_2, of the given video. The pixel values in both blocks are mean normalized, and a correlation surface is constructed by performing cross correlation in the frequency domain. The peaks are located in the correlation

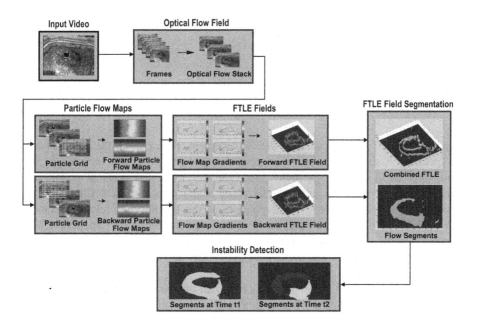

Fig. 9.3 Block diagram of the crowd-flow segmentation algorithm. (1) The input is a video of a crowded scene. (2) Computation of optical flow from the frames of the video. (3) Forward and backward advection of particle grid resulting in forward and backward particle flow maps. (4) Computation of respective FTLE fields from the forward and backward particle flow maps. (5) Fusion of forward and backward FTLE fields and label assignment using the watershed segmentation algorithm. (6) Detection of abnormal events (or crowd-flow instabilities)

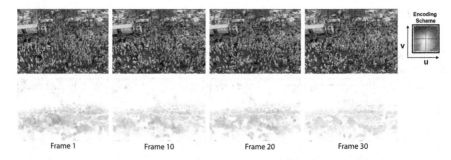

Fig. 9.4 (Color online) Examples of optical flow fields computed by using the algorithm of [1]. *Top Row*: Frames of the video. *Bottom Row*: Color-coded optical flow for the corresponding frames

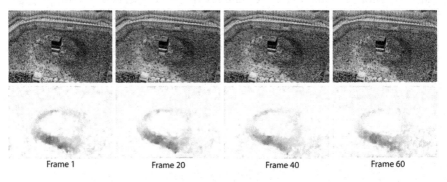

Fig. 9.5 (Color online) Examples of optical flow fields computed by using the algorithm of [1]. *Top Row*: Frames of the video. *Bottom Row*: Color-coded optical flow for the corresponding frames

surface and are used to calculate the displacement. Note that all the pixels inside a block are assigned the same displacement value. The process is repeated for all possible blocks in the given frame. Local outliers in the displacement vectors are replaced in a post-processing step, by using adaptive local median filtering. The removed vectors are filled by interpolation of the neighboring velocity vectors. A typical size of the block employed in our experiments is 16×16 pixels. The second scheme that we used is proposed in [1] where grey value constancy, gradient constancy, smoothness, and multi-scale constraints were used to estimate a high-accuracy optical flow.

To analyze the crowd-flow in a given interval of T frames, we pool the optical flow fields, $\mathbf{v}(1), \mathbf{v}(2), \ldots, \mathbf{v}(T)$, to generate a 3D volume of optical flows. To simplify the notation, we have removed the dependence of $\mathbf{v}$ on location $\mathbf{x}$. This 3D volume of optical flow is used to advect the particles, where parameter T is used as the integration time. we use the symbol B_t^{t+T} to represent a the 3D volume of optical flow fields $\mathbf{v}(t), \mathbf{v}(t+1), \ldots, \mathbf{v}(t+T)$. Figures 9.4–9.7 show color-coded optical flows computed from different sequences in our data set.

Fig. 9.6 (Color online) Examples of optical flow fields computed by using the block-based correlation algorithm. *Top Row*: Frames of the video. *Bottom Row*: Color-coded optical flow for the corresponding frames

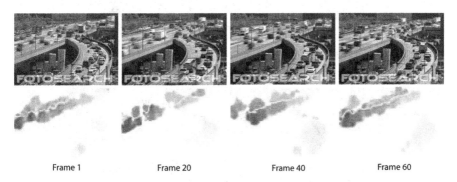

Fig. 9.7 (Color online) Examples of optical flow fields computed by using the block-based correlation algorithm. *Top Row*: Frames of the video. *Bottom Row*: Color-coded optical flow for the corresponding frames

9.3.2 Particle Advection

The next step is to advect a grid of particles through the 3D volume of flow fields, B_t^{t+T}, that corresponds to the time interval t to $t+T$. we start by launching a grid of particles over the first optical flow field, $\mathbf{v}(t)$, in B_t^{t+T}. Ideally, the resolution of the grid should be the same as the number of pixels in each frame of the video. An example of this Cartesian mesh of particles placed over the flow field of a crowd video and the trajectories of particles are provided in Fig. 9.8.

Next, the Lagrangian trajectory $[x(t+T;t,x_0,y_0),y(t+T;t,x_0,y_0)]$ corresponding to a particle at grid location (x_0,y_0) is computed by solving the ordinary differential equations numerically:

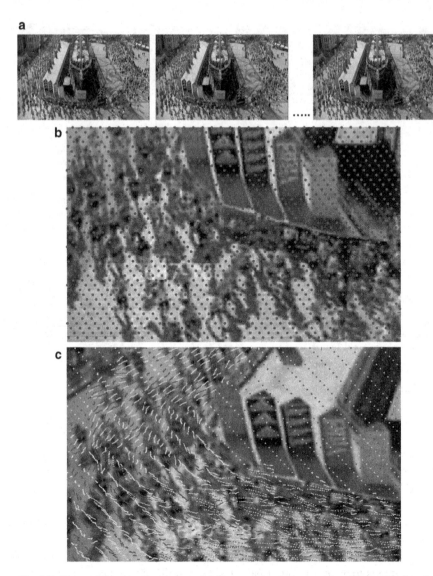

Fig. 9.8 The particle advection process. (**a**) Frames from the input video. (**b**) A grid of particles is overlaid on the flow field of the input sequence. (**c**) Trajectories of the particles are obtained by advecting them through the flow field

$$\frac{dx}{dt} = u(x,y,t), \frac{dy}{dt} = v(x,y,t), \tag{9.14}$$

subject to the initial conditions $[x(0),y(0)] = (x_0,y_0)$. $t+T$ represents the time up-till which we want to compute the trajectory. we use the fourth order Runge-Kutta-Fehlberg algorithm along with cubic interpolation [13] of the velocity field

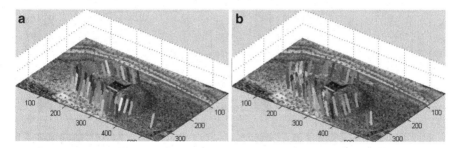

Fig. 9.9 (**a**) The Lagrangian trajectories obtained by forward integration. (**b**) The Lagrangian trajectories obtained by backward integration

to solve this system. The backward particle advection is carried out by initializing the grid of particles over the last optical flow field $\mathbf{v}(t + T)$ in the 3D volume of optical flow fields B_t^{t+T}. The direction of the optical flow vectors is reversed for the backward integration. Figure 9.9a provides a visualization of the Lagrangian trajectories obtained by forward integration, while Fig. 9.9b provides the visualization of the Lagrangian trajectories obtained by the backward integration. The length of integration, $T = 50$, was used for this purpose.

Note that, in our case the domain D is not closed and trajectories can leave the domain. The particles that leave the domain are not advected anymore, and their last available positions are kept in the flow map. That is, we do not perform any re-seeding of the particles if they leave the domain.

9.3.3   Particle Flow Maps and FTLE Field

During forward and backward integration, a separate pair of flow maps, namely ϕ_x and ϕ_y, is maintained for the grid of particles. These flow maps are used to relate the initial position of each particle to its later position obtained after the advection process. This way, the particle flow maps integrate the motion over longer durations of time, which is lacking in the instantaneous optical flow. Here, the first map, ϕ_x, keeps track of how the x coordinate of particles is changing, and, similarly, ϕ_y keeps track of the y coordinate of particles. we use notation ϕ_x^f and ϕ_y^f to refer explicitly to forward flow maps, and ϕ_x^b and ϕ_y^b to refer explicitly to backward flow maps. When the explicit references are not important, we omit the superscripts.

At the start, these maps are populated with the initial positions of the particles, which are the pixel locations at which the particle is placed. The particles are then advected under the influence of B_t^{t+T} using the method described in Sect. 9.3.2. The positions of the particles are updated until the end of the integration time length T.

The computation of the FTLE field from the particle flow maps requires computation of the spatial gradients of the particle flow maps, i.e., $\frac{d\phi_x}{dx}$, $\frac{d\phi_x}{dy}$, $\frac{d\phi_y}{dx}$, and $\frac{d\phi_y}{dy}$. This step is accomplished by using a finite differencing approach for taking

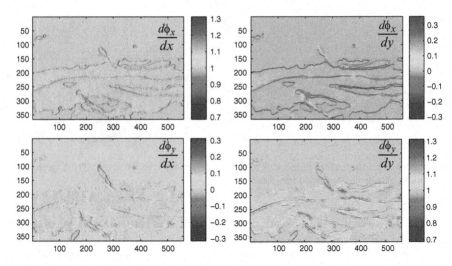

Fig. 9.10 The spatial gradients of the particle flow maps for the sequence shown in Fig. 9.4

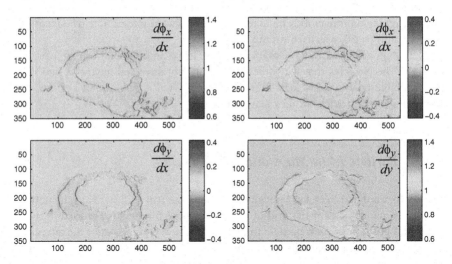

Fig. 9.11 The spatial gradients of the particle flow maps for the sequence shown in Fig. 9.5

derivatives. Figures 9.10 and 9.11 show spatial gradients of particle flow maps for two different sequences in the data set. It can be observed that a high gradient is present where the neighboring particles are behaving differently over the length of the integration. The Cauchy-Green deformation tensor is computed by substituting the spatial gradients of the particle flow maps in Eq. (9.12). Finally, the FTLE field is computed by finding the maximum eigenvalue of the Cauchy-Green deformation tensor and plugging it in Eq. (9.13). Figures 9.12–9.15 show a number of FTLE fields corresponding to different crowd sequences in our data set. In these examples, the combined FTLE field is obtained by adding the forward and backward FTLE

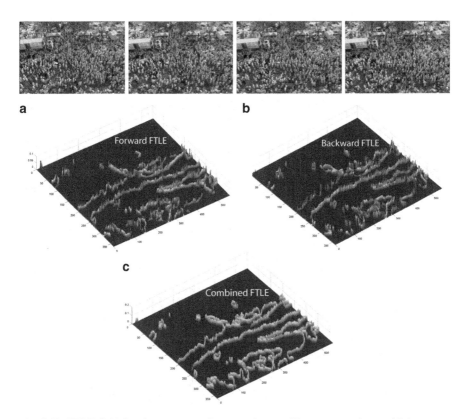

Fig. 9.12 FTLE field for the sequence shown at the *top*. The sequence has multiple groups of people intermingling with each other. The ridges are prominent at the locations where the neighboring crowd groups have dynamically distinct behavior. (**a**) The forward FTLE field obtained by the forward integration of particles. (**b**) The backward FTLE field obtained by the backward integration of particles. (**c**) The combined FTLE field

fields. It can be observed that ridges in these fields (Figs. 9.12–9.15), which point to the location of LCS, are very prominent, and, therefore, can be used to separate regions of the crowd-flow that are dynamically distinct from each other.

The utility of computing forward and backward FTLE fields becomes obvious from the analysis of the FTLE fields shown in Fig. 9.13. In this video sequence traffic from the ramp is merging onto the main highway. When the particles are advected forward in time, no LCS appear at the intersection of the ramp and the main highway (Fig. 9.13a). The reason is that the particles at the intersection move forward coherently in time as the destinations of the underlying traffic flow on the ramp and the main highway are the same. But when these particles are advected backward in time, the LCS appear at the intersection (Fig. 9.13b) since the particles at the intersection do not have the same destination backward in time because the underlying traffic is originating from different locations. In other words, by backward integration, we am able to take into account the origin of the flow

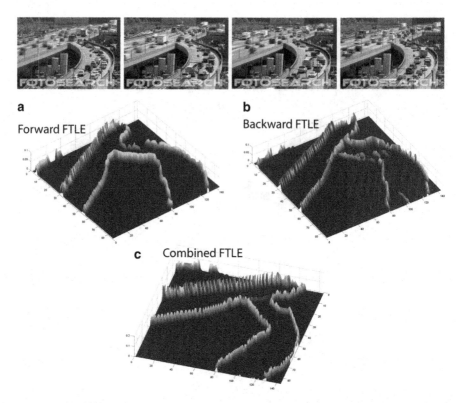

Fig. 9.13 FTLE field for the sequence shown at the *top*. The sequence has multiple lanes of traffic, and the traffic from the ramp is merging onto the main highway. (**a**) The forward FTLE field obtained by the forward integration of particles. Note that no LCS are present at the intersection of the ramp and the highway. (**b**) The backward FTLE field obtained by the backward integration of particles. Note that LCS have now appeared at the intersection of the ramp and the highway. (**c**) The combined FTLE field

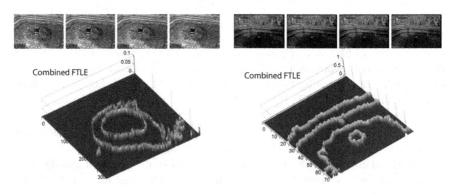

Fig. 9.14 The combined FTLE fields for the sequences shown at the *top*

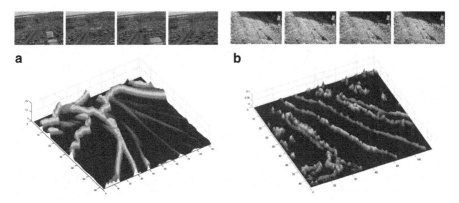

Fig. 9.15 The combined FTLE fields for the sequences shown at the *top*

in addition to its destination. This capability is important to completely resolve different crowd-flow segments present in the scene. This point will become clearer when we present the segmentation results in a later section.

9.3.4 FTLE Field Segmentation

The LCS in the FTLE field can be treated as the watershed lines dividing individual catchment basins. Each catchment basin represents the distinct crowd grouping that is present in the scene. The catchment basins are homogeneous in the sense that all the particles belonging to the same catchment basin have the same origin and destination. To generate a distinct labeling for each catchment basin, we employ the watershed segmentation algorithm [16]. The final segmentation map is created by removing those segments where the magnitude of the flow is zero. we call such segments "vacuum segments." Note that, due to the unique strength of the FTLE field based representation, we do not have to pre-specify the number of crowd-flow segments. This way, we are able to overcome the problem of specifying the number of segments or clusters which is common in most of the clustering and segmentation algorithms [19].

9.4 Experiments and Discussion

This section discusses the experimental setup and the data sets used in the experiments. It also presents the segmentation results along with a discussion of the interpretation of the results.

Fig. 9.16 Example of sequences used in our experiments

9.4.1 Datasets and Experimental Setup

We have tested our approach on videos taken from the stock footage web sites (Getty-Images [9], Photo-Search), and Video Google [10] which are now part of UCF Crowd data set [8]. Two types of crowded scenarios are covered in these videos: the first scenario consists of scenes involving the high-density crowds, while the second scenario consists of high-density traffic scenes. Traffic scenes can be treated as a close approximation of the motion of crowds of people and, therefore, provides us with useful data for testing the performance of the proposed algorithm. Another set of videos were taken from the National Geographic documentary, entitled "Inside Mecca," which covers the yearly ritual of Hajj performed by close to two million people. Therefore, this event provides a unique opportunity for capturing data about the behavior of large gatherings of people in a realistic setting. Figure 9.16 shows key frames from some of these sequences.

For each video, the optical flow is computed by using the algorithms previously described in Sect. 9.3.1. The computation of the optical flow is performed at a coarser resolution than the resolution of the image to reduce the computational cost. Next, a grid of particles is placed over the flow field. The resolution of the grid is kept the same as the number of pixels on which the flow field is computed. The forward and backward particle flow maps are generated using the advection algorithm described in Sect. 9.3.2. The corresponding FTLE fields are computed from the spatial gradient tensor of the flow maps using Eq. (9.13). The backward

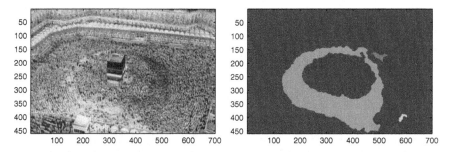

Fig. 9.17 The flow segmentation result on a video taken from the National Geographic documentary "Inside Mecca." *Left*: A frame from the video. *Right*: The crowd-flow segmentation mask

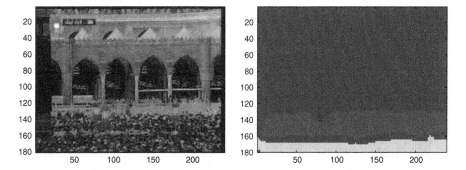

Fig. 9.18 The flow segmentation result on a video from "Video Google." *Left*: A frame from the video. *Right*: The crowd-flow segmentation mask

and forward FTLE fields are fused to generate a combined FTLE field. The fusion is carried out by adding the values of both fields. Finally, the segmentation is performed using the watershed segmentation algorithm.

9.4.2   Segmentation Results

This section presents qualitative analysis of the results obtained on different video sequences. Figures 9.17–9.25 show the segmentation results on all the sequences in the data set.

The first sequence, shown in Fig. 9.17, are extracted from the National Geographic documentary entitled "Inside Mecca". The sequence depicts thousands of people circling the Kabba in a counter-clockwise direction. In this case, the group of people circling in the center is part of the same flow segment because of its common dynamics and desirable goal. The optical flow field of the crowd motion offers a unique challenge as one can observe from the color-coded optical flow shown in Fig. 9.4. The different colors emphasize that the flow vectors along the circular path

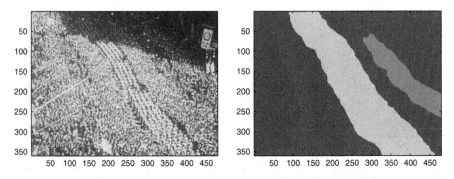

Fig. 9.19 The flow segmentation result on a video taken from the stock footage web site "Getty Images." *Left*: A frame from the video. *Right*: The crowd-flow segmentation mask

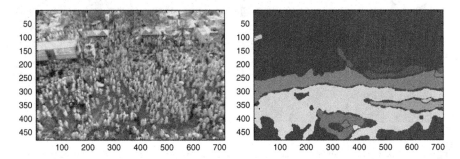

Fig. 9.20 The flow segmentation result on a video taken from the National Geographic documentary "Inside Mecca." *Left*: A frame from the video. *Right*: The crowd-flow segmentation mask

have different directions and magnitudes. This means that a simple clustering of these vectors will not allow us to assign these vectors to the same cluster when, in fact, they all belong to one cluster. The result is shown in Fig. 9.26a, where mean-shift clustering was used to cluster the optical flow vectors extracted from the instantaneous optical flow field. The clustering results are shown for different choices of the band-width parameter. But even with different values of bandwidth, the mean-shift is not able to correctly localize the circular segment. However, using our method where we integrate the motion information over longer durations of time, we are able to correctly segment the complex crowd motions (Fig. 9.17). The LCS structures previously shown in Fig. 9.14a, show that the dynamic behavior of the crowd moving in a circle is preserved by emphasizing the boundaries of the coherent flow regions. Another result of a similar type of motion is presented in Fig. 9.21. In this case, there is an additional group of people that is walking on top of the roof. Our method is able to localize this additional crowd-flow segment as well.

The next result that we would like to discuss is shown in Fig. 9.20. This sequence contains complex motion dynamics as there are several groups of people that are intermingling with each other and moving in various directions. The challenges

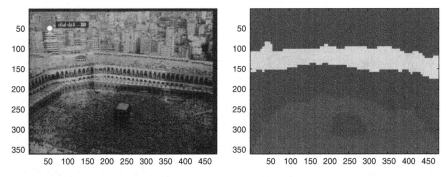

Fig. 9.21 The flow segmentation result on a video from "Video Google." *Left*: A frame from the video. *Right*: The crowd-flow segmentation mask

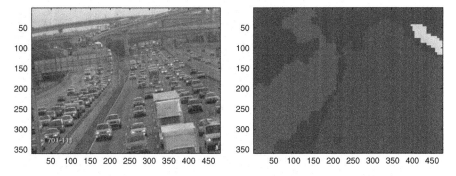

Fig. 9.22 The result of the flow segmentation on a high-density traffic scene. This segmentation was obtained by using both the forward and backward FTLE fields. *Left*: A frame from the video. *Right*: The crowd-flow segmentation mask

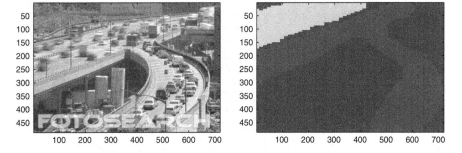

Fig. 9.23 Result of the flow segmentation on a high-density traffic scene. The segments correspond to group of cars that are behaving dynamically different from each other

posed by this sequence are different in that the mixing barriers between various crowd groupings must be correctly located. The segmentation result shown in Fig. 9.20 demonstrate that we am able to localize most of the distinct crowd groupings that were present in the scene. The discovered barriers between the crowd

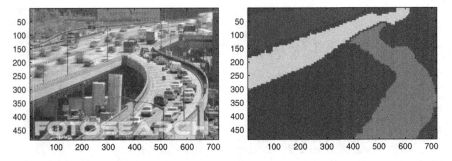

Fig. 9.24 The result of the flow segmentation on a high-density traffic scene. This segmentation was obtained by using only the forward FTLE field. *Left*: A frame from the video. *Right*: The crowd-flow segmentation mask

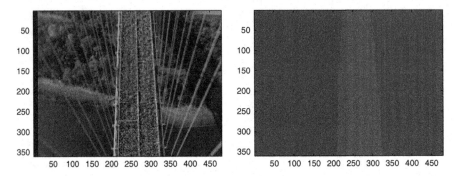

Fig. 9.25 The result of the crowd-flow segmentation on a marathon sequence. *Left*: A frame from the video. *Right*: The crowd-flow segmentation mask

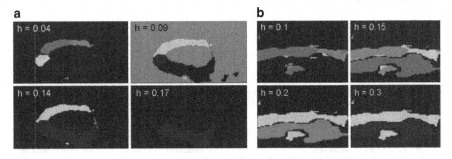

Fig. 9.26 A comparison with respect to the mean shift segmentation. (**a**) The segmentation obtained for the sequence shown in Fig. 9.17. (**b**) The segmentation obtained for the sequence shown in Fig. 9.20

groupings can be observed in the combined FTLE field shown in Fig. 9.12c. The barriers which appear in the form of ridges in the FTLE field, encapsulate each crowd group. A comparison is again performed with the mean-shift clustering approach (Fig. 9.26b), but, again, the mean shift is not able to localize all the crowd-flow segments. This again points to the utility of integrating motion information over longer periods of time, which helps to get a better picture of the crowd motion. Some other example results on sequences involving groups of people are presented in Figs. 9.18, 9.19, and 9.25.

Next, we discuss segmentation results on a high-density traffic sequence (Fig. 9.23). The results on this sequence highlight the utility of using both forward and backward integration of particles through the 3D volume of optical flows. In this sequence, vehicles are moving in two opposite directions on the main highway, while a flow of traffic is merging onto the main highway from the ramp. The challenge in this sequence is to find the right membership of the flow generated by the traffic on the ramp by resolving its origin and destination. If we only use the forward integration, it is obvious that all the particles initialized over the ramp will have the same fate as the particles on the main highway. This means that the traffic on the ramp will become part of the flow generated by the lane on the right-hand side of the highway. Another way to look at the forward integration is from the viewpoint of flow continuity, where out-going flux on the ramp is equal to the additional flux received by the highway at this location. The segmentation result shown in Fig. 9.23 validates the above observation where same labeling is being assigned to the ramp and to the right lane of the main highway. This ambiguity can be resolved by the addition of the backward integration of particles. Since they are considered backwards in time, the particles on the two sections of the road do not share the same origin or, in other words, the outgoing flux is not equal to the flux received by the two sections of the road. The segmentation result shown in Fig. 9.24 demonstrates that by using both forward and backward integration of particles, a flow segmentation that is more refined is obtained. The result on another traffic sequence is shown in Fig. 9.22.

9.5 Summary

This chapter has developed an algorithm for segmenting scenes of crowds of people into 'crowd groupings' that are dynamically distinct. For this purpose, the spatial extent of the video is treated as a phase space of a non-autonomous dynamical system in which transport from one region of the phase space to the other is controlled by the optical flow. Next, a grid of particles is advected forward and backward in time through this phase space and the amount by which the neighboring particles diverged is quantified by using a Cauchy-Green deformation tensor. The maximum eigenvalue of this tensor is used to construct a Finite-Time Lyapunov Exponent (FTLE) field, which revealed the time-dependent invariant manifolds of the phase space called Lagrangian Coherent Structures (LCS). The LCS in turn divided the crowd-flow into regions of different dynamics.

The strength of this approach lies in the fact that it bypasses the need for low-level detection of individual objects altogether, which will be impossible in a high-density crowded scene, and generates a concise representation of the complex mechanics of human crowds using only the global analysis.

References

1. Brox, T., Bruhn, A., Papenberg, N., Weickert, J.: High accuracy optical flow estimation based on a theory for warping. In: ECCV, Prague (2004)
2. Cartwright, D., Zander, A.: Group Dynamics: Research and Theory, 3rd edn. Tavistock Publications, London (1968)
3. Chrisohoides, A., et al.: Experimental visualization of Lagrangian coherent structures in aperiodic flows. Phys. Fluids **15**, L25 (2003)
4. Haller, G.: Distinguished material surfaces and coherent strcutures in three-dimensional fluid flows. Phys. D **149**(4), 248–277 (2001)
5. Haller, G.: Finding finite-time invariant manifolds in two dimensional velocity data. Chaos **10**(1), 99–108 (2000)
6. Haller, G.: Lagrangian strcutures and the rate of strain in partition of two dimensional turbulence. Phys. Fluids **13**(11), p. 3365 (2001)
7. Haller, G., Yuan, G.: Lagrangian coherent structures and mixing in two dimensional turbulence. Physica D **147**(3–4), 352–370 (2000)
8. http://crcv.ucf.edu/data/crowd.php
9. http://www.gettyimages.com
10. http://www.video.google.com
11. Lapeyre, G., et al.: Characterization of finite-time Lyapunov exponents and vectors in two-dimensional turbeulence. Chaos **12**(3), 688–698 (2002)
12. Lekien, F., et al.: Dynamically consistent Lagrangian coherent structures. Am. Inst. Phys.: 8th Exp. Chaos Conf. D **742**, 132–139 (2004)
13. Lekien, F., et al.: Tricubic interpolation in three dimensions. J. Numer. Methods Eng. **63**(3), 455–471 (2005)
14. Lewin, K.: In: Cartwright, D. (ed.) Field Theory in Social Science; Selected Theoretical Papers. Harper & Row, New York (1951)
15. Malhotra, N., et al.: Geometric structures, lobe dynamics and Lagrangian transport in flows with a periodic time dependence, with applications to Rossby wave flow. J. Non-Linear Sci. **8**, 401 (1998)
16. Meyer, F.: Topographic distance and watershed lines. Signal Process. **38**, 113–125 (1994)
17. Poje, A.C., Haller, G.: The geometry and statistics of mixing in aperiodic flows. Phys. Fluids **11**(10), 2963–2968 (1999)
18. Shadden, S.C., et al.: Definition and properties of Lagrangian coherent structures from finite-time Lyapunov exponents in two-dimensional aperiodic flows. Physica D, Nonlinear Phenomena, **212**(3–4), 271–304 (2005)
19. Shi, J., Malik, J.: Normalized cuts and image segmentation. IEEE Trans. Pattern Anal. Mach. Intell. **22**(8), 888–905 (2000)

Chapter 10
Modeling Crowd Flow for Video Analysis of Crowded Scenes

Ko Nishino and Louis Kratz

Abstract In this chapter, we describe a comprehensive framework for modeling and exploiting the crowd flow to analyze videos of densely crowded scenes. Our key insight is to model the characteristic patterns of motion that arise within local space-time regions of the video and then to identify and encode the statistical and temporal variation of those motion patterns to characterize the latent, collective movements of the people in the scene. We show that this statistical crowd flow model can be used to achieve critical analysis tasks for surveillance videos of extremely crowded scenes such as unusual event detection and pedestrian tracking. These results demonstrate the effectiveness of crowd flow modeling in video analysis and point to its use in related fields including simulation and behavioral analysis of people in dense crowds.

10.1 Introduction

Computer vision research, in the past few decades, has made significant strides toward efficient and reliable processing of the ever increasing video data. These advances have mainly been driven by the need for automatic video surveillance that persistently monitors security critical areas from fixed viewpoints. Many methods have been introduced that successfully demonstrate the extraction of meaningful information regarding the scene contents and their dynamics including detecting people, tracking objects and pedestrians, recognizing specific actions by people and scene-wide events, and interactions among people and other scene contents.

K. Nishino (✉) • L. Kratz
Department of Computer Science, Drexel University, 3141 Chestnut Street,
Philadelphia, PA 19104, USA
e-mail: kon@drexel.edu; lak24@drexel.edu

S. Ali et al. (eds.), *Modeling, Simulation and Visual Analysis of Crowds*, The International
Series in Video Computing 11, DOI 10.1007/978-1-4614-8483-7__10,
© Springer Science+Business Media New York 2013

Automated visual analysis of crowded scenes, however, remains a challenging task. As the number of people in a scene increases, nuisances that play against conventional video analysis methods surge. This is particularly true for methods that fundamentally rely on the ability to extract and track individuals. In videos of crowded scenes, the whole body of each person would be hardly visible to the camera, people will occlude each other and other contents in the scene, the notion of foreground and background will start to meld together, and most important the behavior of people will change to accommodate the tightness and clutter in the scene. These are nuisances not only to the computer algorithms for automated analysis but also to human operators that will have to squint through the clutter for hours and days to find a single adverse activity. As such, paradoxically, automated video analysis is most needed where it is actually hardest to do.

The large number of people in a crowd, however, does in turn give rise to invaluable visual cues regarding the scene dynamics. The sheer number of people and their appearance adds texture to the collective movements of the people which we refer to as the crowd flow in this chapter. The crowd flow embodies the latent, coherent motions of individuals which also dynamically varies across the scene and changes as time passes by. If we can model the crowd flow while faithfully encoding its variability both in space and time, we may use it to extract critical contextual information from the dynamic, cluttered scene.

In this chapter, we describe a comprehensive framework for modeling and exploiting crowd flow to analyze videos of densely crowded scenes. Each individual in a crowded scene is not a mere autonomous agent dictated by a set of simple rules, but is an intelligent being that makes judgments on its own movement based on local sensory input with a global perspective in mind. The movements of individuals result in the intricate yet coherent motion that organically evolves in the scene. We will model them as a structured motion field that dynamically changes its form both in space and time. In other words, our approach argues for a scene-centric representation of crowd flow modeling. This is a large departure from conventional object- or people-centric approaches that capture crowd flow as a collection of individuals and their paths.

Our key insight is to exploit the dense local motion patterns created by the large number of people and model their spatio-temporal relationships, representing the underlying intrinsic structure they form in the video. In other words, we model the variations of local spatio-temporal motion patterns to describe common behavior within the scene, and then identify the spatial and temporal relationships between motion patterns to characterize the behavior of the crowd as a whole in the specific scene of interest. We show that modeling the crowd flow can benefit solving critical video analysis tasks that are otherwise challenging to achieve on crowded scenes. Most important, we show that a scene-centric representation of the crowd flow can augment object-centric individual models to track each individual in a highly dense crowd.

We demonstrate the effectiveness of modeling and using the crowd flow for video analysis in two fundamental surveillance tasks: unusual event detection and pedestrian tracking. The experimental results show that exploiting the aggregated movements of people enables robust detection of anomalous behaviors and accurate tracking of individuals. We believe these results have direct implications for human behavior analysis as it enables accurate tracking of individuals in dense crowds, which is essential of longitudinal "in-situ" observations of people in real-world scenes. These results also point to a novel approach of crowd simulation in which the collective movements of people are driven by statistical models learned from observations.

10.2 Related Work

Past work on video analysis has mostly relied on the assumption that the scene content to be analyzed can be reliably extracted in each frame. This is usually achieved by maintaining a background model, subtracting the background from the video frames to extract foreground objects (e.g., pedestrians), and then tracking each of the moving foreground objects. Subsequent analysis then relies on the paths or locations of the tracked foreground objects. Although this paradigm has been largely successful in many video analysis applications, it naturally is limited to videos of relatively sparse scenes where people and other scene contents including static background can be clearly discerned from each other.

Video analysis of crowded scenes has recently attracted interest in the computer vision community, especially to reach beyond such simple scenes and to achieve automated surveillance in more complex, cluttered scenes. Here we review some of the representative approaches to modeling such crowded scenes.

Ali and Shah [1, 2] model the crowd motion by averaging the observed optical flow. Their approach assumes that the crowd does not change over time, and uses the same video clips for learning and applications. Similar work by Mehran et al. [20] use "streaklines," a concept from hydrodynamics, to segment videos of crowds and track pedestrians. Though streaklines encode more temporal information than the average optical flow, they do not encode the temporal relationship between consecutive observations. In contrast, we model the temporal dynamics over local areas, and use our learned crowd model to analyze videos of the same scene recorded at a different time.

Often, the term "motion pattern" is used to describe motion within the scene that are part of the same physical process [11]. Hu and Shah [12] identify motion patterns in crowded scenes by clustering optical flow vectors in similar spatial regions. Such work is applicable to scenes where the motion of the crowd has large, stable patches of heterogeneous flow. In near-view crowded scenes, however, a single physical process (such as the crowd) may be heterogeneous and dynamically varying. Even a single pedestrian may exhibit flow vectors in multiple directions due to their articulated motion. In contrast, we represent the motion in small, space-time areas with a local motion pattern, and capture the dynamically varying heterogeneous crowd with a collection of HMMs.

Andrade et al. [3, 4] also use a collection of hidden Markov models. The observations to the HMM are vectors of pixel locations and optical flow estimates. While these may be viewed as a form of local motion patterns, they do not directly encode the variability in motion that can occur due to poor texture or aperture problems. In contrast, our representation of local motion pattern are directional distributions of optical flow that directly encode the uncertainty in the optical flow estimate as we later demonstrate. In addition, the dimension of their representation increases in dimension with the resolution of the video. Though they scale the frame size, such a lengthy representation still requires more training data to properly capture the covariance of the observations. In contrast, our directional distributions are parameterized by a single 3D flow vector and a concentration parameter, as will show, which reduces the dimensionality of the representation while retaining the variance in the flow.

Other work view frequently occurring motion patterns as an annoyance. Yang et al. [27] argue that high-entropy words, i.e., motions that occur frequently, are not useful for activity recognition since they represent noisy optical flow or areas without motion. Though noise and areas without motion are a factor, they are not the only motion patterns that can occur frequently. In extremely crowded scenes, it is exactly the high frequency local motion patterns that define the characteristic movement of pedestrians within the crowd. In addition, the minor differences between the instances of frequently occurring motion patterns are typically ignored. Hospedales et al. [10], for example, quantize optical flow vectors into one of four primary directions. This disregards the valuable variations of motion patterns that may be used to robustly represent different movements of pedestrians.

Other work that describe the motion in local, space-time volumes assume each cuboid contains motion in a single direction [14, 24]. Often, the optical flow vectors are quantized into a number of discrete directions [10, 25]. These representations disregard the valuable variations of motion patterns that may be used to robustly represent different movements of pedestrians.

Histograms of oriented gradients (HoG) features have been used to describe space-time volumes for human detection [8] and action recognition [16]. The HoG feature is computed on the spatial gradients (the temporal gradient is not used), though they have been extended to the 3D space-time gradient [15]. The orientation of spatial gradients encodes the structure of the pedestrian's appearance, and thus is not suitable when only motion is necessary. Rather than modeling a distribution of the gradients' orientations, we use the relationship between spatio-temporal gradients and optical flow to estimate a directional distribution of optical flow vectors that represent the possible motion within the cuboid.

10.3 Crowd Flow as a Collection of Local Motion Patterns

The key idea underlying our approach is to view the crowd flow as a collection of local spatio-temporal motion patterns in the scene and to model their variation in space and time with a collection of statistical models. In other words, we model the

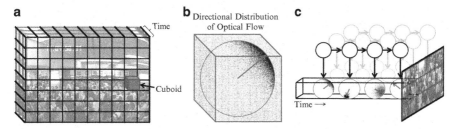

Fig. 10.1 An overview of our crowd flow model. We learn a statistical crowd flow model that encodes the variation of local motion patterns that arise in small space-time volumes (*cuboids*). The temporal variations of these local motion patterns at each local scene region are encoded with a hidden Markov model learned from a training video. The resulting set of HMMs collectively embody the crowd flow in the scene which is then used to detect local unusual events as statistical deviations and track pedestrians by predicting local motion patterns that are used as a prior on each person's movement. (**a**) Training video. (**b**) Local motion pattern. (**c**) Collection of HMMs

crowd flow as a dynamically evolving structure of local motions in the scene and time. This enables us to encode the global and local characteristics of the aggregate movements of people in a scene with a concise analytical expression. Such a model becomes crucial in achieving higher level analysis of the scene contents based on the stationary behavior of the whole.

Figure 10.1 shows an overview of our model. First, as shown in Fig. 10.1a, we divide a training video into spatio-temporal sub-volumes, or "cuboids," defined by a regular grid. Second, as shown in Fig. 10.1b, we model the motion of pedestrians through each cuboid (i.e., the local motion pattern) with a 3D directional distribution of optical flow. Next, as shown in Fig. 10.1c, we train a hidden Markov model (HMM) over the local motion patterns at each grid location. This implies that we assume that the crowd will generate motion patterns that conform to first-order Markov processes at local space-time regions, which may not necessarily be true. Nevertheless, we found that the temporal variation of local crowd motion can be captured well with hidden Markov models which also enables efficient inference of its parameter values. The hidden states of the HMMs encode the multiple possible motions that can occur at each spatial location. The transition probabilities of the HMMs encode the time-varying dynamics of the crowd motion. We represent the crowd motion by the collection of HMMs, encoding the spatially and temporally varying motions of pedestrians that comprise the entire crowd flow.

Our model has three unique characteristics that distinguish it from other methods. First, our model encodes the variability of the crowd flow both in space and time. The collection of HMMs captures the variations of the motion of pedestrians throughout the entire video volume, making it more robust and dynamically adjustable to different crowd behaviors. Second, we model the crowd flow by starting with local motion patterns. This enables the model to scale with the modality of different crowd behaviors, rather than the number of pedestrians. Finally, since our model is a set of statistical models, it may be learned from an example video of the scene and be used to analyze videos of the same scene recorded at a different time.

10.3.1 Modeling Local Motion Patterns

In our method, the video is viewed as a spatio-temporal volume which is subdivided into small volumes that typically span 30 pixels in horizontal and vertical spatial domain as well as 30 frames in the temporal domain. We refer to these small spatio-temporal volumes that collectively form the video as *cuboids*. We first seek to represent the motion in each cuboid in the video volume, i.e., the local motion pattern. The optical flow can be reliably estimated when the cuboid contains motion in a single direction with constant velocity and good texture. The motion in cuboids from real-world crowded scenes, however, may be difficult to estimate reliably. A cuboid may contain complex motion, i.e., motion exhibited by multiple objects moving in multiple directions or a single object that changes direction or speed. In addition, cuboids may contain little or no texture and have indeterminable motion. To handle these different cases, we model each local motion pattern with a distribution of *potential* optical flow vectors whose variance encodes the uncertainty in the optical flow estimate. These potential optical flow vectors can be directly computed from the 3D spatio-temporal gradients observed in the cuboid.

Let $\nabla I(x,y,f)$ be a 3×1 vector representing the 3D spatio-temporal gradient, i.e., gradient of the image intensities computed in the horizontal, vertical, and temporal directions, respectively, at 2D pixel location (x,y) at frame f. The constant brightness constraint [9] dictates the relationship between this 3D spatio-temporal gradient and the 3D optical flow vector $\mathbf{q}$

$$\nabla I(x,y,f)^T \mathbf{q} = 0, \tag{10.1}$$

where $\mathbf{q}$ has two degrees of freedom due to the ambiguity of global scaling and is estimated as a unit vector.

Estimating the optical flow vector $\mathbf{q}$ from a single gradient estimate is ill-posed. For this reason, it is usually assumed that the flow is constant in the space-time area around (x,y,f), and surrounding gradients are used to estimate the optical flow $\mathbf{q}$. Let $\{\nabla I_i | i = 1 \ldots N\}$ be a set of N spatio-temporal gradients (we have dropped x,y,f for notational convenience) computed at the different pixel locations of the cuboid. From the collection of the spatio-temporal gradients in a cuboid, one can estimate the optical flow vector from its Gram matrix (or the structure tensor [26])

$$\mathbf{G}\mathbf{q} = \sum_i^N \nabla I_i \nabla I_i^T \mathbf{q} = \begin{bmatrix} 0 \\ 0 \\ 0 \end{bmatrix}. \tag{10.2}$$

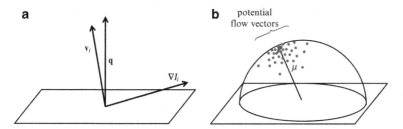

Fig. 10.2 We represent the local motion pattern of a cuboid as a collection of potential optical flow vectors that arise from the spatio-temporal gradients. These potential optical flow vectors are computed from the dominant optical flow and each spatio-temporal gradient and encoded with a directional statistics distribution model. Please see text for details

The optical flow vector $\mathbf{q}$ can easily be computed as the eigenvector of $\mathbf{G}$ with the smallest eigenvalue. Note that this optical flow vector is a unit 3D vector which encodes both the direction and speed.

The single optical flow vector computed from all the spatio-temporal gradients, however, is not a faithful representation of the motion within the cuboid. It represents the dominant motion within the cuboid but assumes that all movements align with that single direction and speed. In reality, the cuboid will contain various motions in different directions and speeds that may be roughly aligned with that single dominant vector but with significant variability. Encoding this variation of motion within each cuboid is critical in arriving at an accurate analytical model of the crowd flow. To capture this variability, we consider the *potential* optical flow vectors that would have arisen from the spatio-temporal gradient vector computed at each pixel in the cuboid.

As illustrated in Fig. 10.2, we consider each spatio-temporal gradient ∇I_i which is not necessarily on the plane defined by the 3D optical flow vector $\mathbf{q}$. Such a point suggests that the actual motion within the cuboid may be in another direction

$$\mathbf{v}_i = \frac{\nabla I_i \times \mathbf{q} \times \nabla I_i}{|\nabla I_i \times \mathbf{q} \times \nabla I_i|}, \tag{10.3}$$

where $\times$ is the cross-product. We call this the potential optical flow vector corresponding to that spatio-temporal gradient. Note that $\mathbf{v}_i$ is orthogonal to ∇I_i, and thus satisfies the optical flow constraint in Eq. (10.1) for ∇I_i.

As shown in Fig. 10.2b, the potential 3D optical flow vectors for each cuboid form a distribution $\{\mathbf{v}_i \,|\, i = 1, \ldots, N\}$ on the unit sphere. We can view this as a probability density function on the unit sphere, which is known as a directional statistics distribution [19]. We choose to model each of these distributions of potential optical flow vectors with the von Mises-Fisher distribution [19]

$$p(\mathbf{x}) = \frac{1}{c(\kappa)} \exp\left\{\kappa \mu^T \mathbf{x}\right\}, \tag{10.4}$$

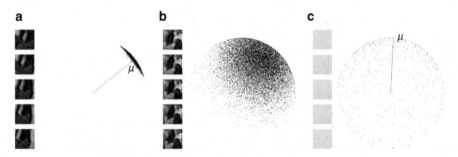

Fig. 10.3 The variance of the potential optical flow distribution encoded by the concentration parameter of the von Mises-Fisher distribution model faithfully characterizes the type of local motion pattern of the corresponding cuboid. (**a**) Uniform motion. (**b**) Complex motion. (**c**) No motion

where μ is the mean direction, $c(\kappa)$ is a normalization constant, and κ is the concentration parameter. We fit a von Mises-Fisher distribution to the potential optical flow vectors $\{\mathbf{v}_i \,|\, i = 1, \ldots, N\}$. Mardia and Jupp [19] show that the sufficient statistic for estimating μ and κ is

$$\mathbf{r} = \frac{1}{N}\sum_i^N \mathbf{v}_i, \tag{10.5}$$

and thus

$$\mu = \frac{1}{|\mathbf{r}|}\mathbf{r}. \tag{10.6}$$

They also show that

$$\kappa = \begin{cases} \dfrac{1}{(1 - |\mathbf{r}|)} & \text{if } |\mathbf{r}| \geq 0.9 \\ A_3^{-1}(|\mathbf{r}|) & \text{otherwise}, \end{cases} \tag{10.7}$$

where

$$A_3(|\mathbf{r}|) = \coth(|\mathbf{r}|) - \frac{1}{|\mathbf{r}|}. \tag{10.8}$$

In our implementation, we use the tabulated data from Mardia and Jupp [19] to compute $A_3^{-1}(|\mathbf{r}|)$.

As illustrated in Fig. 10.3, the concentration parameter κ characterizes the uncertainty in the optical flow estimate within the cuboid. Cuboids containing motion in a single direction have a high concentration parameter, yielding a narrow distribution. Cuboids with complex motion have a wide distribution, indicating motion may occur in different directions. Cuboids with little or no texture have distributions across the entire sphere, indicating that motion may be occurring in any direction. Each local spatio-temporal motion pattern O is defined by a mean 3D optical flow vector μ and a concentration parameter κ that encodes the uncertainty of the estimate.

10.3.2   Modeling the Dynamics of Local Motion Patterns

Now that we have a representation of the local motion patterns, we model their variation in space and time with collection of hidden Markov models (HMMs) to encode the crowd flow in the video.

Let us first briefly review hidden Markov models. Readers are referred to, for instance, [22] for a more detailed account. The latent variables $\{s_t | t = 1, \ldots, T\}$ of an HMM are assumed to follow the Markov property: each latent variable depends upon the previous. We only consider a first degree model, where each latent state is dependent upon the single previous one. Given the latent variables, the observed variables $\{O_t | t = 1, \ldots, T\}$ are conditionally independent. The latent variables are discrete, taking on one of J values. Each HMM is defined by a $J \times 1$ initial state probability vector π, a $J \times J$ state transition matrix $\mathbf{A}$, and the emissions densities $\{p(O_t | s_t = j) | j = 1, \ldots, J\}$. The likelihood of starting in a specific state j is encoded by the initial probability vector

$$\pi(j) = p(s_1 = j). \tag{10.9}$$

The likelihood of transiting from state i to state j is represented by the state transition matrix

$$\mathbf{A}(i, j) = p(s_{t+1} = j | s_t = i). \tag{10.10}$$

A key problem in learning an HMM from data is to compute the likelihood of an observation sequence $\{O_1, \ldots, O_T\}$. This is achieved efficiently using dynamic programming by the Forwards-Backwards algorithm [22]. We review the forwards step here, as we use it extensively in this work. Let $\mathbf{b}_t$ be a $J \times 1$ vector of likelihoods where

$$\mathbf{b}_t(j) = p(O_t | s_t = j). \tag{10.11}$$

In the forwards step, dynamic programming is used to compute the message

$$\alpha_t(j) = p(s_t = j, O_1, \ldots, O_t), \tag{10.12}$$

where $O_1,\ldots,O_t$ are the observations up to time t. After the first observation, the message is initialized

$$\alpha_1(j) = \mathbf{b}_1(j)\pi(j). \tag{10.13}$$

Subsequent messages are computed by the update

$$\alpha_t(j) = \mathbf{b}_t(j)\sum_i^J \alpha_{t-1}(i)\mathbf{A}(i,j). \tag{10.14}$$

Often α_t is scaled by it's magnitude after each update to avoid numerical problems. This yields the posterior

$$\hat{\alpha}_t(j) = \frac{\alpha_t(j)}{|\alpha_t|} = \mathrm{p}(s_t = j|O_1,\ldots,O_t). \tag{10.15}$$

After the backwards step of the Forwards-Backwards algorithm, we may compute the full posterior

$$\gamma_t(j) = \mathrm{p}(s_t = j|O_1,\ldots,O_T) \tag{10.16}$$

which is used during training to update the parameters of the HMM.

An important aspect of the forwards step is that it may be computed online. When each new observation O_t becomes available, the new posterior $\hat{\alpha}_t$ may be computed efficiently by Eq. (10.14). We use this characteristic of HMMs in our applications to achieve online operation.

Next, we turn our attention to the form of the emission density of an HMM $\{\mathrm{p}(O_t|s_t = j)\,|\,j=1,\ldots,J\}$. Each observation $O_t = \{\mu_t, \kappa_t\}$ is a local motion pattern, defined by the 3D mean optical flow vector μ_t and the concentration parameter κ_t. Often complex observations are quantized using a codebook, making the emission densities discrete. This can decrease the training time, but reduces the amount of information represented by each emission density.

Rather than quantizing our local motion patterns, we analytically model the emission densities by imposing priors over μ_t and κ_t. To achieve this, we assume that the mean vector μ_t and concentration parameter κ_t are statistically independent

$$\mathrm{p}(O_t|s_t = j) = \mathrm{p}(\mu_t|s_t = j)\mathrm{p}(\kappa_t|s_t = j). \tag{10.17}$$

We model $\mathrm{p}(\kappa_t|s_t = j)$ as a Gamma distribution defined by a shape parameter a^j and scale parameter θ^j. We model $\mathrm{p}(\mu_t|s_t = j)$ as a von-Mises Fisher distribution (i.e., the conjugate prior on μ_t [18]) defined by a mean direction μ_0^j and a concentration parameter κ_0^j.

A hidden Markov model can be trained using the Baum-Welch [22] algorithm. During training, the posteriors γ_t from Eq. (10.16) are used to update the emission density parameters $\{\mu_0^j, \kappa_0^j, a^j, \theta^j\}$. The mean direction

$$\mathbf{r}_0^j = \frac{\sum_t^T \gamma_t(j)\mu_t}{\sum_t^T \gamma_t(j)} \qquad (10.18)$$

is used to compute

$$\mu_0^j = \frac{1}{|\mathbf{r}_0^j|}\mathbf{r}_0^j \qquad (10.19)$$

and

$$\kappa_0^j = \begin{cases} \dfrac{1}{(1-|\mathbf{r}_0^j|)} & \text{if } |\mathbf{r}_0^j| \geq 0.9 \\[2ex] A_3^{-1}(|\mathbf{r}_0^j|) & \text{otherwise}. \end{cases} \qquad (10.20)$$

There is no closed-form solution to estimating a^j, and thus we use the numerical technique from Choi and Wette [7]. Given an estimate of a^j maximum likelihood is used to estimate the scale

$$\theta^j = \frac{\sum_t^T \gamma_t(j)\kappa_t}{a^j \sum_t^T \gamma_t(j)}. \qquad (10.21)$$

10.4   Using the Crowd Flow Model

The collection of hidden Markov models now encode the spatial and temporal variation of the local motion patterns and capture their dynamics both locally and globally. The crowd flow is encoded in these collection of HMMs as the spatially and temporally stationary behaviors of the local motion patterns. Now we are in a position to exploit this statistical crowd flow model to achieve challenging tasks in highly cluttered scenes. We will demonstrate the power of the model in two important video analysis applications, namely unusual event detection and pedestrian tracking.

10.4.1 Detecting Unusual Local Events

Unusual event detection is a key application in automatic surveillance systems. The sheer number of surveillance cameras deployed produces an abundance of video that is often only viewed after an incident occurs. By automatically detecting disturbances within the scene, the automatic surveillance system can alert security personnel as soon as an incident occurs.

Large-scale unusual events, such as stampedes, incidents of violence, and crowd panic, are rare, even though they are a primary motivation for automatic video surveillance. While these large crowd disturbances are an area of interest, they are not the only disturbance that may need to be detected in crowded scenes. Since crowded scenes may contain any number of moving objects, a key application is the detection of activities by one or few of the scene's constituents that happen in local areas. Detecting such local anomalies is of great interest, especially in very crowded scenes since they can easily go unnoticed or disguised due to the heavy clutter within the scene.

To detect local unusual events, we identify local motion patterns in a query video of the same scene that statistically deviate from the learned model. Specifically, we detect local motion patterns that have low likelihood given the spatio-temporal dynamics of the crowd. We demonstrate with real-world data that the method enables the detection of subtle yet important anomalous activities in high-density crowds, such as individuals moving against the usual flow of traffic or stop in otherwise high motion areas. Such unusual activity may only have a subtle effect on the entire crowd, but still be a disturbance that requires intervention from security personnel.

10.4.1.1 Finding Deviations from the Crowd Flow

The collection of HMMs represent the underlying steady-state motion of the crowd by the spatial and temporal variations of local motion patterns. We seek to identify if a specific local motion pattern contains unusual pedestrian activity. For the purpose of this work, we consider a local motion pattern unusual if it occurs infrequently or is absent from the training video. We derive a probability measure of how much a specific local motion pattern deviates from the crowd motion in order to identify unusual events.

Deviations from the HMM are caused by either an unlikely transition between sequential local motion patterns or a local motion pattern with low emission probabilities. We identify unusual local motion patterns by thresholding the conditional likelihood

$$\mathscr{T}_t = p\left(O_t | O_1, \ldots, O_T\right),$$
(10.22)

where T is the last local motion pattern in the video clip. Exploiting the statistical independence properties of the HMM yields

$$\mathscr{T}_t = \sum_j^J \mathrm{p}(O_t|s_t = j)\gamma_t(j), \qquad (10.23)$$

where γ_t is the posterior $\mathrm{p}(s_t = j|O_1,\ldots,O_T)$ computed from the forwards-backwards algorithm (see Eq. (10.16)). The posterior γ_t encodes the expected latent state given the temporal statistics of the crowd, and decreases $\mathscr{T}_t$ when there is an unlikely transition. The emission likelihoods $\mathrm{p}(O_t|s_t = j)$ are low for all values of j when O_t was absent from the training data.

Computing $\mathscr{T}_t$ requires the entire video clip to be available. In a real-world system, unusual events need to be detected as they occur. To achieve this, we use the predictive distribution to compute an alternative measure

$$\tilde{\mathscr{T}}_t = \sum_j^J \mathrm{p}(O_t|s_t = j)\hat{\alpha}_t(j), \qquad (10.24)$$

where $\hat{\alpha}_t$ is the posterior $\mathrm{p}(s_t = j|O_1,\ldots,O_t)$ computed during the forwards phase of the Forwards-Backwards algorithm (see Eq. (10.15)). The estimate $\tilde{\mathscr{T}}_t$ only requires the local motion patterns up to time t to be available, but does not consider the transition out of the observation at time t. As such, some cuboids are incorrectly classified but anomaly detection can be performed online.

Often, the crowd can display different modalities at different spatial locations of the scene. For example, some areas may regularly contain no motion, while others contain motion in multiple directions. As a result, the ideal threshold value may change with the spatial location. We account for this by dividing our likelihood measure $\tilde{\mathscr{T}}_t$ by the average likelihood of the training data.

10.4.1.2 Experimental Results

After training the crowd flow models on videos of normal activities of target scenes, we detect unusual movements of pedestrian in query videos of the same scene recorded at a different time. For this chapter, we detect anomalies in a concourse and a ticket gate area of a station. The length of training videos varied for each example between 540 and 3,000 frames, depending on the specific example. We use cuboids of size $30 \times 30 \times 20$ for the ticket gate scene and $40 \times 40 \times 20$ for the concourse scene.

Figure 10.4 shows successful detection of unusual movements of pedestrians in local areas. Figure 10.4a, from the ticket gate scene, shows detection of pedestrians reversing directions in the turnstiles. Figure 10.4b shows successful detection of pedestrians in the concourse scene moving from left to right against the regular crowd traffic. The training video used for the examples consists of pedestrians moving in many different directions, but not from the left side of the scene to the right. These examples illustrate the unique ability of our approach to detect irregular local motion patterns within a crowded scene comprised of diverse movements of pedestrians.

a   **b**

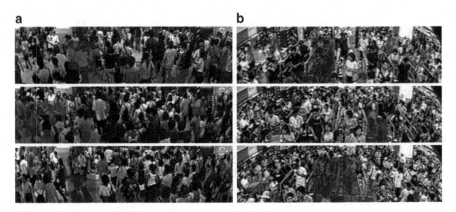

Fig. 10.4 (Color online) Our crowd flow model enables the detection of local unusual activities such as pedestrians moving against the crowd by examining the deviation of observed local motion patterns from the learned crowd flow model. *Blue*: true positive, *Magenta*: false positive, and *Red*: false negative

a   **b**

Fig. 10.5 (Color online) Observing no motion in otherwise high motion areas are also successfully detected as unusual events. *Blue*: true positive, *Magenta*: false positive, and *Red*: false negative

The type of detected events depends entirely on the training data. Figure 10.5 shows detection of pedestrians loitering in otherwise high-traffic areas. Since the training video contains typical crowd motion, the lack of pedestrians (e.g., the empty turnstiles in the ticket gate scene) deviate from the model. This dependency on the training data is not only expected, it is desirable. It allows users of our approach to decide which particular local movements of pedestrian they consider usual by including it in the training video.

It is unreasonable to expect that all possible typical local motion patterns will be contained in the training video. Inevitably some typical local motion patterns will not be captured by the training data, and result in incorrect classifications such as the false positives. These are exasperated by the fact that the events being detected are subtle, local movements of pedestrian. Events that are dramatically different from the training sequence, such as global crowd disturbances, will result in fewer false positives. As shown in Fig. 10.5, the few false negatives in both scenes always occur adjacent to true positives, which suggests they are harmless in practical scenarios.

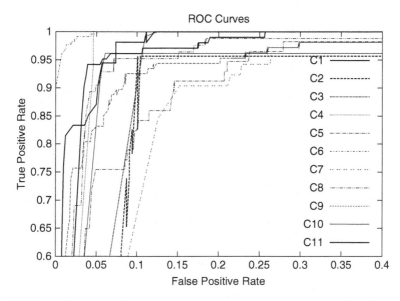

Fig. 10.6 We evaluate the local unusual event detection accuracy on 11 videos (C1 to C11) using manually labeled ground truth data. The receiver operating characteristic curves show the overall accuracy and effectiveness of using the crowd flow model

Figure 10.6 shows the receiver operating characteristic (ROC) curves (generated by varying the likelihood threshold) for all of the clips. Our approach performs with significant accuracy on each of the example videos. In video C5, the upper bodies of the loitering pedestrians move left and right and exhibit motion patterns similar to that of the crowd. This failure indicates that our approach associates similar motion patterns that may be caused by dissimilar movement, a side effect caused by the robustness of the prototypical distributions.

Figure 10.7 shows the detection accuracy using the online likelihood measure (computed from α) compared with the full likelihood measure γ. The online method achieves comparable accuracy to the offline computation.

Figure 10.8 shows the effects of increasing the training data size for video C1. As expected, the performance increases with longer training data, and achieves good performance with 100 observations, or 2,000 frames of video. Using only 50 observations the model achieves significant accuracy with a false positive rate (ratio of false positives to total negatives) of 0.17 and a true positive rate (ratio of true positives to total positives) of 0.88. This strong performance with few observations directly results from the crowd's high density. Since the scene contains a large number of pedestrians, significant variations in local motion patterns occur even in short video clips.

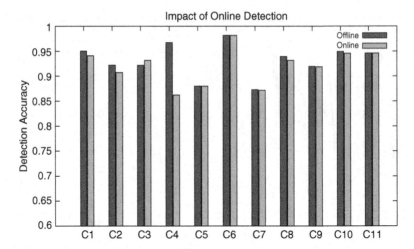

Fig. 10.7 The crowd flow model can also be evaluated to perform local unusual event detection in real time for online operation. As expected the accuracy is lower but still within a small margin from the offline computation results

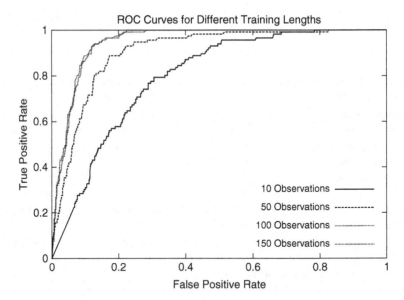

Fig. 10.8 The performance of unusual event detection gracefully increases as more training data is used but stops increasing after sufficient information (local motion patterns and their variations) is provided to faithfully encode the underlying crowd flow

10.4.2 Using the Crowd to Track Individuals

Tracking objects or people is a crucial step in video analysis with a wide range of applications including behavior modeling and surveillance. Conventional tracking methods typically assume a static background or easily discernible moving objects, and as a result are limited to scenes with relatively few constituents. Videos of crowded scenes present significant challenges to tracking due to the large number of pedestrians and the frequent partial occlusions that they produce.

We can leverage the learned crowd flow model to track individual pedestrians in videos of crowded scenes. Specifically, we leverage the crowd flow as a prior in a Bayesian tracking framework. We use the crowd motion to predict local motion patterns in videos containing the pedestrian that we wish to track. Next, we use the predicted local motion pattern as a prior on the state-transition distribution in a particle filter framework to track individuals. We use these predictions as a prior on a particle filter to track individuals. We show that our approach accurately predicts the motion that a target will exhibit during tracking and leads to accurate tracking of individuals which is otherwise extremely challenging.

10.4.2.1 Predicting Motion Patterns

We train the crowd flow model on a video of a crowded scene containing typical crowd behavior. Next, we use it to predict the local motion patterns at each location of a different video of the same scene. Note that, since we create a scene-centric model based on the changing motion in local regions, the prediction is independent of which individual is being tracked. In fact, we predict the local motion pattern at all locations of video volume given only the previous frames of the video.

Given a trained HMM at a specific spatial location and a sequence of observed local motion patterns $O_1, \ldots, O_{t-1}$ from the query video, we seek to predict the next local motion pattern $\tilde{O}_t = \{\tilde{\mu}_t, \tilde{\kappa}_t\}$ that will occur. We achieve this by computing the expected value of the predictive distribution

$$\tilde{O}_t = \mathbb{E}\left[p\left(O_t | O_1, \ldots, O_{t-1}\right)\right], \tag{10.25}$$

by marginalizing the predictive distribution

$$p\left(O_t | O_1, \ldots, O_{t-1}\right) = \sum_j^J p\left(O_t | s_t = j\right) \sum_i^J p\left(s_t = j | s_{t-1} = i\right) p\left(s_{t-1} = i | O_1, \ldots, O_{t-1}\right). \tag{10.26}$$

Note that $p\left(s_{t-1} = i | O_1, \ldots, O_{t-1}\right)$ is the posterior $\hat{\alpha}_{t-1}$ computed during the forwards-backwards algorithm [6] (see Eq. (10.15)), and $p\left(s_t = j | s_{t-1} = i\right)$ is defined by the HMM's state transition matrix $\mathbf{A}$. As such, the second summation in Eq. (10.26) may be represented by

$$\omega_t(j) = \sum_i^J \mathbf{A}(i,j)\hat{\alpha}_{t-1}(i) \tag{10.27}$$

and

$$\tilde{O}_t = \sum_j^J \mathbb{E}\left[\mathrm{p}\left(O_t|s_t=j\right)\right]\omega_t(j). \tag{10.28}$$

Thus $\tilde{O}_t$ is a weighted sum of the expected local motion patterns defined by each emission density.

Recall that each emission density $\mathrm{p}\left(O_t|s_t=j\right)$ is defined by four parameters $\left\{\mu_0^j, \kappa_0^j, a^j, \theta^j\right\}$. Using the means of the Gamma and von Mises-Fisher distributions

$$\mathbb{E}\left[\mathrm{p}\left(O_t|s_t=j\right)\right] = \left\{\mu_0^j, a^j\theta^j\right\}, \tag{10.29}$$

i.e., a local motion pattern with mean direction μ_0^j and concentration parameter $a^j\theta^j$. Thus the predicted local motion pattern $\tilde{O}_t$ is defined by mean direction

$$\tilde{\mu}_t = \frac{1}{\left|\sum_j^J \omega_t(j)\mu_0^j\right|}\sum_j^J \omega_t(j)\mu_0^j \tag{10.30}$$

and concentration parameter

$$\tilde{\kappa}_t = \sum_j^J \omega_t(j)a^j\theta^j. \tag{10.31}$$

During tracking, we use the previous frames of the video to predict the local motion pattern that spans the next M frames (where M is the number of frames in a cuboid). Since the predictive distribution is a function of the HMM's transition probabilities and the hidden states' posteriors, the prediction may be computed on-line and efficiently during the forward phase of the Forwards-Backwards algorithm [22].

10.4.2.2 Crowd Flow Bayesian Tracking

We now use the predicted local motion pattern to track individuals in a Bayesian framework. Specifically, we use the predicted local motion pattern as a prior on the parameters of a particle filter. Our crowd flow model enables these priors to vary in the space-time and dynamically adapt to the changing motions within the crowd.

Tracking can be formulated in a Bayesian framework [13] by maximizing the posterior distribution of the state $\mathbf{x}_f$ of the target at time f given past and current

Fig. 10.9 We use the predicted local motion pattern to impose a prior on the motion of the pedestrian through the space-time volume

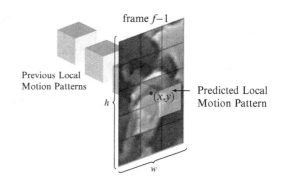

frame $f-1$

Previous Local Motion Patterns

h

(x,y)

Predicted Local Motion Pattern

w

measurements $\mathbf{z}_{1:f} = \{\mathbf{z}_i | i = 1 \dots f\}$. Note that the index of each frame f is different from the temporal index t of the local motion patterns (since the cuboids span many frames). We define state $\mathbf{x}_f$ as a four-dimensional vector $[x, y, w, h]^T$ containing the tracked target's 2D location (in image space), width, and height, respectively. Tracking is performed by maximizing the posterior distribution

$$p(\mathbf{x}_f | \mathbf{z}_{1:f}) \propto p(\mathbf{z}_f | \mathbf{x}_f) \int p(\mathbf{x}_f | \mathbf{x}_{f-1}) \, p(\mathbf{x}_{f-1} | \mathbf{z}_{1:f-1}) \, d\mathbf{x}_{f-1}, \qquad (10.32)$$

where $\mathbf{z}_f$ is the frame at time f, $p(\mathbf{x}_f | \mathbf{x}_{f-1})$ is the transition distribution, $p(\mathbf{z}_f | \mathbf{x}_f)$ is the likelihood, and $p(\mathbf{x}_{f-1} | \mathbf{z}_{1:f-1})$ is the posterior from the previous tracked frame. The transition distribution $p(\mathbf{x}_f | \mathbf{x}_{f-1})$ models the motion of the target between frames $f-1$ and f, and the likelihood distribution $p(\mathbf{z}_f | \mathbf{x}_f)$ represents how well the observed image $\mathbf{z}_f$ matches the state $\mathbf{x}_f$. Often, the distributions are non-Gaussian, and the posterior distribution is estimated using a Markov chain Monte Carlo method such as a particle filter [13] (please refer to [5] for an introduction to particle filters).

As shown in Fig. 10.9, we impose priors on the transition $p(\mathbf{x}_f | \mathbf{x}_{f-1})$ distribution using the predicted local motion pattern at the space-time location defined by $\mathbf{x}_{f-1}$. For computational efficiency, we use the cuboid at the center of the tracked target to define the priors, although the target may span several cuboids across the frame.

Transition Distribution

We use the predicted local motion pattern to hypothesize the motion of the tracked target between frames $f-1$ and f, i.e., the transition distribution $p(\mathbf{x}_f | \mathbf{x}_{f-1})$. Let the state vector $\mathbf{x}_f = \left[\mathbf{k}_f^T, \mathbf{d}_f^T \right]^T$ where $\mathbf{k}_f = [x, y]$ is the target's location (in image coordinates) and $\mathbf{d}_f = [w, h]$ is the size (width and height) of a bounding box around the target. We focus on the target's movement between frames and use a second-degree auto-regressive model [21] for the transition distribution of the size $\mathbf{d}_f$ of the bounding box.

The transition distribution of the target's location $p\left(\mathbf{k}_f|\mathbf{k}_{f-1}\right)$ reflects the 2D motion of the target between frames $f-1$ and f. We model this using the von Mises-Fisher distribution defined by the predicted local motion pattern $\tilde{O}_t = \{\tilde{\mu}_t, \tilde{\kappa}_t\}$ at space-time location $\mathbf{k}_{f-1}$. In the particle filter, a set of N sample locations (i.e., particles) $\{\mathbf{k}^i_{f-1}|i = 1,\dots,N\}$ are drawn from the prior $p(\mathbf{x}_{f-1}|\mathbf{z}_{1:f-1})$. For each sample $\mathbf{k}^i_{f-1}$, we draw a 3D flow vector $\mathbf{v}^i = [v^i_x, v^i_t, v^i_t]$ from the predicted local motion pattern at space-time location $\mathbf{k}^i_{f-1}$. We use these 3D flow vectors to update each particle

$$\mathbf{k}^i_f = \mathbf{k}^i_{f-1} + \begin{bmatrix} v^i_x/v^i_t \\ v^i_y/v^i_t \end{bmatrix}. \tag{10.33}$$

Note that $\tilde{\kappa}_t$ plays a key role in this step: distributions with a large variance will spread the particles over the frame, while those with a small variance (i.e., determinable flow) will keep the particles close together.

Likelihood Distribution

Typical models of the likelihood distribution maintain a template T that represents the target's characteristic appearance in the form of a color histogram [21] or an image [2]. A template T and the region R (the bounding box defined by state $\mathbf{x}_f$) of the observed image $\mathbf{z}_f$ are used to model the likelihood distribution

$$p\left(\mathbf{z}_f|\mathbf{x}_f\right) = \frac{1}{Z}\exp\left[\frac{-d\left(R,T\right)^2}{2\sigma^2}\right], \tag{10.34}$$

where σ is the variance selected empirically, $d(\cdot)$ is a distance measure, and Z is a normalization constant.

Rather than using color histograms or intensity as the defining characteristic of an individual's appearance, we model the template T as an image of the individual's spatio-temporal gradients. This representation is more robust to appearance variations caused by noise or illumination changes. We use a weighted sum of the angles between the spatio-temporal gradient vectors in the observed region and the template to define the distance measure

$$d\left(R,T\right) = \sum_i^M \rho^f_i \arccos\left(\mathbf{t}_i \cdot \mathbf{r}_i\right), \tag{10.35}$$

where M is the number of pixels in the template, $\mathbf{t}_i$ is the normalized spatio-temporal gradient vector in the template, $\mathbf{r}_i$ is the normalized spatio-temporal gradient vector in the region R of the observed image at frame f, and ρ^f_i is the weight of the pixel at location i and frame f,

We model changes in the target's appearance by estimating the weights $\{\rho_i^f | i = 1,\ldots,M\}$ in Eq. (10.35) during tracking. Specifically, pixels that change drastically (due to the pedestrian's body movement or partial occlusions) exhibit a large error between the template and the observed region. We estimate this error E_i^f during tracking to account for a pedestrian's changing appearance. The error at frame f and pixel i is

$$E_i^f = \alpha \arccos\left(\mathbf{t}_i \cdot \mathbf{r}_i\right) + (1 - \alpha) E_i^{f-1}, \tag{10.36}$$

where α is the update rate (set to 0.05) and $\mathbf{t}_i$ and $\mathbf{r}_i$ are again the gradients of the template and observed region, respectively. To reduce the contributions of frequently changing pixels to the distance measure, the weight at frame f and pixel i is inversely proportional to the error

$$\rho_i^f = \frac{1}{Z}\left(\pi - E_i^{f-1}\right), \tag{10.37}$$

where Z is a normalization constant such that $\sum_i \rho_i^f = 1$. To account for changes in appearance, the template is updated each frame by a weighted average.

10.4.2.3   Experimental Results

We evaluated our method on videos of four scenes: the concourse and ticket gate scenes, and the sidewalk and intersection scenes from the UCF dataset [1]. We use a sampling importance re-sampling particle filter as in [13] with 100–800 particles (depending on the subject) to estimate the posterior in Eq. (10.32). We learn a crowd flow model on a video of each scene, and use it to track pedestrians in videos of the same scene recorded at a different time. The training videos for each scene have 300, 350, 300, and 120 frames, respectively. The training videos for the concourse, ticket gate, and sidewalk scenes have a large number of pedestrians moving in a variety of directions. The video for the intersection scene has fewer frames due to the limited length of video available. In addition, the training video of the intersection scene contains only a few motion samples in specific locations, as many of the pedestrians have moved to other areas of the scene in that point in time. Such sparse samples, however, still result in a useful model since most of the pedestrians are only moving in one of two directions (either from the lower left to the upper right, or from the upper right to the lower left).

Due to the perspective projection of many of the scenes, which is a common occurrence in surveillance, the sizes of pedestrians varies immensely. As such, the initial location and size of the targets are selected manually. Many methods exist for automatically detecting pedestrians and their sizes [8] even in crowded scenes [17] and may be used to initialize the tracker.

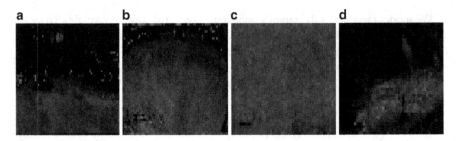

Fig. 10.10 The angular error between the predicted optical flow vector and the observed optical flow vector for all scenes. (**a**) Concourse. (**b**) Ticket gate. (**c**) Sidewalk. (**d**) Intersection

The motion represented by the local motion pattern depends directly on the size of the cuboid. Ideally, we would like to use a cuboid size that best represents the characteristic movements of a single pedestrian. Cuboids the size of a single pedestrian would faithfully represent the pedestrian's local motion and therefore enable the most accurate prediction and tracking. The selection of the cuboid size, however, is entirely scene-dependent, since the relative size of pedestrians within the frame depends on the camera and physical construction of the scene. In addition, a particular view may capture pedestrians of different sizes due to perspective projection. We use a cuboid of size $10 \times 10 \times 10$ on all scenes so that a majority of the cuboids are smaller than the space-time region occupied by a moving pedestrian. By doing so, the cuboids represent the motion of a single pedestrian but still contain enough pixels to accurately estimate a distribution of optical flow vectors. Note that, since the cameras recording the scenes are static, the sizes must be determined only once for each scene prior to training. Therefore, the cuboid sizes may be determined by a semi-supervised approach that approximates the perspective projection of the scene.

We measure the accuracy of the predicted local motion patterns by the angle between the predicted flow $\tilde{\mu}_t$ and the observed optical flow. Figure 10.10 shows the angular error averaged over the entire video for each spatial location in all four scenes. Noisy areas with little motion, such as the concourse's ceiling, result in higher error due to the lack of reliable gradient information. High motion areas, however, have a lower error that indicates a successful prediction of the local motion patterns. The sidewalk scene contains errors in scattered locations due to the occasional visible background in the videos and close-view of pedestrians. There is a larger amount of error in high-motion areas of the intersection scene since a relatively short video was used for training.

Figure 10.11 shows the predicted optical flow, colored by key in the lower left, for four frames from the sidewalk scene. Pedestrians moving from left to right are colored red, those moving right to left are colored green, and those moving from the bottom of the screen to the top are colored blue. As time progresses, our space-time model dynamically adapts to the changing motions of pedestrians within the scene

Fig. 10.11 The crowd flow enables accurate prediction of the optical flow that changes over space and time

Fig. 10.12 The optical flow predicted from the crowd flow model accurately identifies the most likely movement of objects in the specific space time location

as shown by the changing cuboid colors over the frames. Poor predictions appear as noise, and occur in areas of little texture such as the visible areas of the sidewalks or pedestrians with little texture.

Figure 10.12 shows a specific example of the changing predicted optical flow on six frames from the sidewalk scene. In the first two frames the predicted flow is from the left to the right, correctly corresponding to the motion of the pedestrian. In later frames the flow adjusts to the motion of the pedestrian at that point in time. Only by exploiting the temporal structure within the crowd motion are such dynamic predictions possible.

Figure 10.13 shows a visualization of our tracking results on videos from each of the different scenes. Each row shows four frames of our method tracking different targets whose trajectories are shown up to the current frame by the colored curves. The different trajectories in the same spatial locations of the frame demonstrate the ability of our approach to capture the temporal motion variations of the crowd. For example, the green target in row 1 is moving in a completely different direction than the red and pink targets, although they share the spatial location where their trajectories intersect. Similarly, the pink, blue, red, and green targets in row 2 all move in different directions in the center part of the frame, yet our method is able to track each of these individuals. Such dynamic variations that we model using an HMM cannot be captured by a single motion model. Spatial variations are also handled by our approach, as illustrated by the targets concurrently moving

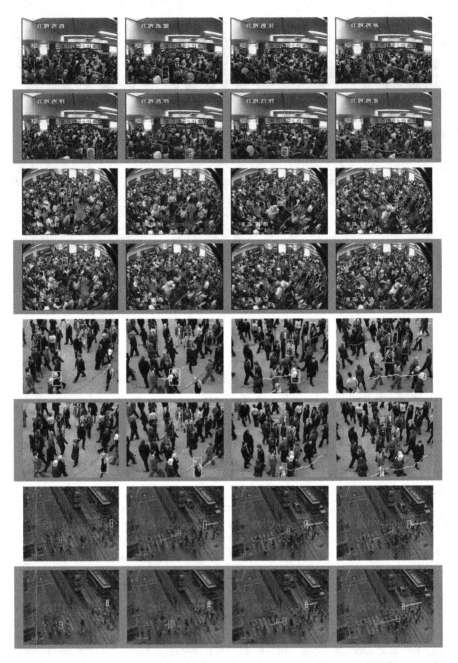

Fig. 10.13 The crowd flow model enables accurate tracking of individual pedestrians even in extremely crowded scenes

Fig. 10.14 Example of a tracking failure due to severe (near full) occlusion

Fig. 10.15 Tracking of pedestrians moving against the crowd

in completely different directions in rows 5 and 6. In addition, our method is robust to partial occlusions as illustrated by the pink target in row 1, and the red targets in rows 3, 5, and 6.

Figure 10.14 shows a failure case due to a severe occlusion. In these instances our method begins tracking the individual that caused the occlusion. This behavior, though not desired, shows the ability of our model to capture multiple motion patterns since the occluding individual is moving in a different direction. Other tracking failures occur due to poor texture. In the sidewalk scene, for example, the occasional viewable background and lack of texture on the pedestrians cause poorly-predicted local motion patterns. On such occasions, a local motion pattern that describes a relatively static structure, such as black clothing or the street, is predicted for a moving target. This produces non-smooth trajectories, such as the pink and red targets in row 5, or the red target in row 6 of Fig. 10.13.

Occasionally, an individual may move in a direction not captured by the training data. For instance, the pedestrian shown on the left of Fig. 10.15 is moving from left to right, a motion not present in the training data. Such cases are difficult to track since the crowd flow model can not predict the pedestrian's motion. On such occasions, the posteriors (given in Eq. (10.27)) are near identical (since the emission probabilities are all close to 0), and thus the predicted optical flow is unreliable. This does not mean the targets can not be tracked, as shown by the correct trajectories in Fig. 10.15, but the tracking depends entirely on the appearance model.

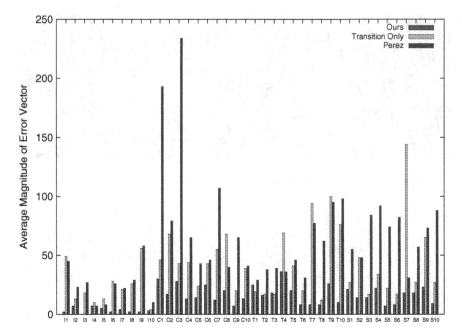

Fig. 10.16 (Color online) Tracking error using our approach compared with a second-degree auto-regressive model and using only our transition distribution with a color-based likelihood

We hand-labeled ground truth tracking results for 40 targets, 10 from each scene, to quantitatively evaluate our approach. Each target is tracked for at least 120 frames. The ground truth includes the target's position and the width and height of a bounding box. The concourse and ticket gate scenes contain many pedestrians whose lower bodies are not visible at all over the duration of the video. On such occasions, the ground truth boxes are set around the visible torso and head of the pedestrian. Given the ground truth state vector $\mathbf{k}_t$, we measure the error of the tracking result $\hat{\mathbf{k}}_t$ as $||\mathbf{k}_t - \hat{\mathbf{k}}_t||_2$.

Figure 10.16 shows the error of our method for each labeled target, averaged over all of the frames in the video, compared to a particle filter using a color-histogram likelihood and second-degree auto-regressive model [21] (labeled as Perez). In addition, we show the results using our predicted state-transition distribution with a color-histogram likelihood (labeled as Transition Only). On many of the targets our state transition distribution is superior to the second-degree autoregressive model, though nine targets have a higher error. Our full approach improves the tracking results dramatically and consistently achieves a lower error than that of Pérez et al. [21].

Figure 10.17 compares our approach with the "floor fields" method by Ali and Shah [2] and the topical model from Rodriguez et al. [23]. Since the other methods do not change the target's template size, we only measured the error in the x, y location of the target. Our approach more accurately tracks the pedestrian's locations in all but a few of the targets. The single motion model by Ali and Shah completely loses many targets that move in directions not represented by their single motion model. The method of Rodriguez et al. [23] models multiple possible movements, but is still limited since it does not include temporal information. Our temporally varying model allows us to track pedestrians in scenes that exhibit dramatic variations in the crowd motion.

Figure 10.18 shows the tracking error over time, averaged over all of the targets, using our approach, that of Ali and Shah [2], and that of Rodriguez et al. [23]. The consistently lower error achieved by our approach indicates that we may track subjects more reliably over a larger number of frames. Our temporally varying model accounts for a larger amount of directional variation exhibited by the targets, and enables accurate tracking over a longer period of time.

10.5   Summary

In this chapter, we introduced a novel, space-time statistical model of the crowd flow in the image space and demonstrated its use in important video analysis applications of crowded scenes. The experimental results show that the model is able to accurately encode the inherent structural patterns of local motions that constitute the crowd flow in concise analytical forms that can then be used to evaluate conformity and predict the motion of local space-time regions in target videos. The results also showed that these information can be successfully used to identify local unusual events and track individuals in videos of high density crowds. We believe the idea of modeling crowd flow from observation has strong implications in applications beyond the two we have demonstrated. In particular, we are hopeful that it will pave the way to finding realistic yet concise models of individual behaviors in crowded scenes that can directly be used in simulating large crowds and validating or even discovering new insights in behavioral studies.

Acknowledgements This work was supported in part by National Science Foundation grants IIS-0746717 and IIS-0803670, and Nippon Telegraph and Telephone Corporation. The authors thank Nippon Telegraph and Telephone Corporation for providing the train station videos.

References

1. Ali, S., Shah, M.: A Lagrangian particle dynamics approach for crowd flow segmentation and stability analysis. In: Proceedings of IEEE International Conference on Computer Vision and Pattern Recognition, pp. 1–6 (2007)
2. Ali, S., Shah, M.: Floor fields for tracking in high density crowd scenes. In: Proceedings of European Conference on Computer Vision (2008)

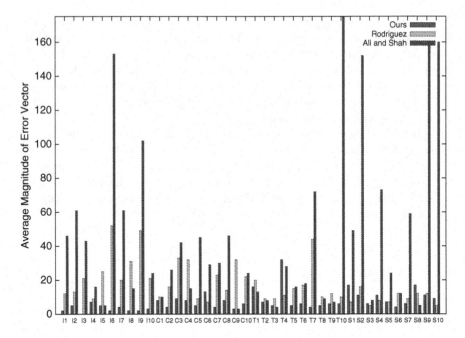

Fig. 10.17 Tracking error of our approach compared with that of Ali and Shah [2] and Rodriguez et al. [23]

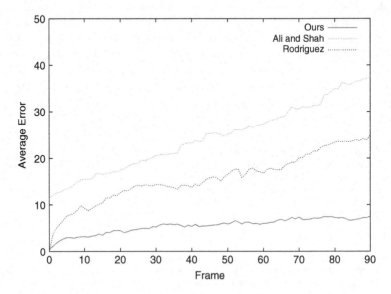

Fig. 10.18 The average error of targets over time using our approach, that of Ali and Shah [2] and Rodriguez et al. [23]

3. Andrade, E., Blunsden, S., Fisher, R.: Modelling crowd scenes for event detection. In: Proceedings of International Conference on Pattern Recognition, pp. 175–178 (2006)
4. Andrade, E.L., Blunsden, S., Fisher, R.B.: Hidden markov models for optical flow analysis in crowds. In: Proceeding of International Conference on Pattern Recognition, pp. 460–463 (2006)
5. Arulampalam, S.M., Maskell, S., Gordon, N.: A tutorial on particle filters for online nonlinear/non-Gaussian Bayesian tracking. IEEE Trans. Signal Process. **50**, 174–188 (2002)
6. Bishop, C.M.: Pattern Recognition and Machine Learning. Springer, New York (2007)
7. Choi, S.C., Wette, R.: Maximum likelihood estimation of the parameters of the gamma distribution and their bias. Technometrics **11**(4), 683–690 (1969)
8. Dalal, N., Triggs, B.: Histograms of oriented gradients for human detection. In: Proceedings of IEEE International Conference on Computer Vision and Pattern Recognition (2005)
9. Horn, B.K.P., Schunck, B.G.: Determining optical flow. Tech. rep., Cambridge, MA (1980)
10. Hospedales, T., Gong, S., Xiang, T.: A Markov clustering topic model for mining behaviour in video. In: Proceedings of IEEE International Conference on Computer Vision (2009)
11. Hu, M., Ali, S., Shah, M.: Detecting global motion patterns in complex videos. In: Proceedings of International Conference on Pattern Recognition (2008)
12. Hu, M., Ali, S., Shah, M.: Learning motion patterns in crowded scenes using motion flow field. In: Proceedings of International Conference on Pattern Recognition, pp. 1–5 (2008)
13. Isard, M., Blake, A.: CONDENSATION-conditional density propagation for visual tracking. Int. J. Comput. Vis. **29**(1), 5–28 (1998)
14. Ke, Y., Sukthankar, R., Hebert, M.: Event detection in crowded videos. In: Proceedings of IEEE International Conference on Computer Vision, pp. 1–8 (2007)
15. Kläser, A., Marszałek, M., Schmid, C.: A spatio-temporal descriptor based on 3D-Gradients. In: Proceedings of British Macine Vision Conference, pp. 995–1004 (2008)
16. Laptev, I., Marszalek, M., Schmid, C., Rozenfeld, B.: Learning realistic human actions from movies. In: Proceedings of IEEE International Conference on Computer Vision and Pattern Recognition (2008)
17. Leibe, B., Seemann, E., Schiele, B.: Pedestrian detection in crowded scenes. In: Proceedings of IEEE International Conference on Computer Vision and Pattern Recognition, pp. 878–885 (2005)
18. Mardia, A., El-Atoum, S.: Bayesian inference for the Von Mises-Fisher distribution miscellanea. Biometrika **63**(1), 203–206 (1976)
19. Mardia, K.V., Jupp, P.: Directional Statistics. Wiley, Chichester (1999)
20. Mehran, R., Moore, B.E., Shah, M.: A streakline representation of flow in crowded scenes. In: European Conference on Computer Vision (ECCV) (2010)
21. Pérez, P., Hue, C., Vermaak, J., Gangnet, M.: Color-based probabilistic tracking. In: Proceedings of European Conference on Computer Vision, pp. 661–675 (2002)
22. Rabiner, L.: A tutorial on hidden Markov models and selected applications in speech recognition. Proc. IEEE **77**(2), 257–286 (1989)
23. Rodriguez, M., Ali, S., Kanade, T.: Tracking in unstructured crowded scenes. In: Proceedings of IEEE International Conference on Computer Vision (2009)
24. Shechtman, E., Irani, M.: Space-time behavior based correlation. In: Proceedings of IEEE Internationl Conference on Computer Vision and Pattern Recognition, pp. 405–412 (2005)
25. Wang, X., Ma, X., Grimson, W.E.L.: Unsupervised activity perception in crowded and complicated scenes using hierarchical Bayesian models. IEEE Trans. Pattern Anal. Mach. Intell. **31**, 539–55 (2009)
26. Wright, J., Pless, R.: Analysis of persistent motion patterns using the 3D structure tensor. In: IEEE Workshop on Motion and Video Computing, pp 14–19 (2005)
27. Yang, Y., Liu, J., Shah, M.: Video scene understanding using multi-scale analysis. In: Proceeding of IEEE International Conference on Computer Vision (2009)

Chapter 11
Pedestrian Interaction in Tracking: The Social Force Model and Global Optimization Methods

Laura Leal-Taixé and Bodo Rosenhahn

Abstract Multiple people tracking consists in detecting the subjects at each frame and matching these detections to obtain full trajectories. In semi-crowded environments, pedestrians often occlude each other, making tracking a challenging task. Tracking methods mostly work with the assumption that each pedestrian moves independently unaware of the objects or the other pedestrians around it. In the real world though, it is clear that when walking in a crowd, pedestrians try to avoid collisions, keep a close distance to a group of friends or avoid static obstacles in the scene. In this chapter, we present an overview of methods that include pedestrian interaction in a tracking framework. This interaction can be expressed in two ways: first, including social and grouping behavior as a physical model within the tracking system, and second, using a global optimization scheme which takes into account all trajectories and all frames to solve the data association problem.

11.1 Introduction

Multiple people tracking is a key problem for many computer vision tasks, such as surveillance, animation or activity recognition. In crowded environments occlusions and false detections are common, and although there have been substantial advances in the last years, tracking is still a challenging task. Tracking is often divided in two steps: detection, finding the objects of interest on every frame, and data association, matching the detections to form complete trajectories in time. Researchers have presented improvements on the object detector [9, 13, 34, 36] as well as on the optimization techniques [18, 23] and even specific algorithms have been developed

L. Leal-Taixé (✉) • B. Rosenhahn
Leibniz University Hannover, Appelstr. 9A, Hannover, Germany
e-mail: leal@tnt.uni-hannover.de; rosenhahn@tnt.uni-hannover.de

S. Ali et al. (eds.), *Modeling, Simulation and Visual Analysis of Crowds*, The International Series in Video Computing 11, DOI 10.1007/978-1-4614-8483-7_11,
© Springer Science+Business Media New York 2013

Fig. 11.1 Terms of the social force model. (**a**) Constant velocity assumption. (**b**) Avoidance forces. (**c**) Group attraction forces

for tracking in crowded scenes [2,32]. Though each object can be tracked separately, recent works have proven that tracking objects jointly and taking into consideration their interaction can give much better results in complex scenes. Current research is mainly focused on two aspects to exploit the interaction between pedestrians: the use of a global optimization strategy [7, 21, 40] and a social motion model [30, 38]. The focus of this chapter is to give a detailed overview of multiple people trackers which include either a global optimization method or social behavior information to improve tracking results in crowded scenarios. Finally, the chapter discusses an approach to marry both concepts and include the social behaviors in a global optimization tracking system (Fig. 11.1).

11.1.1 Related Work

Current research is mainly focused on two aspects to exploit the interaction between pedestrians: the use of a global optimization strategy and a social motion model. In this section, we discuss both research trends.

Global Optimization: The optimization strategy deals with the data association problem, which is usually solved on a frame-by-frame basis or one track at a time. Several methods can be used such as Markov Chain Monte Carlo (MCMC) [19], multi-level Hungarian [20], inference in Bayesian networks [27] or the Nash Equilibrium of game theory [39]. In [6] an efficient approximative Dynamic Programming (DP) scheme is presented, in which trajectories are estimated one after the other. This means that if a trajectory is formed using a certain detection, the other trajectories which are computed later will not be able to use that detection anymore. This obviously does not guarantee a global optimum for all trajectories. Recent works show that global optimization can be more reliable in crowded scenes as it solves the matching problem jointly for all tracks. The multiple object tracking problem is defined as a linear constrained optimization flow problem and Linear Programming (LP) is commonly used to find the global optimum. The idea was first used for people tracking in [16], although this method needs to know a priori the number of targets to track, which limits its application in real tracking situations. In [7], the scene is divided into identical cells, each represented by a node in the constructed graph. Using the information of the Probability Occupancy Map, the problem is formulated either as a max-flow and solved with Simplex, or as a min-cost and solved using k-shortest paths, which is a more efficient solution. Both methods show a far superior performance when compared to the same approach with DP [6]. The authors of [3] also define the problem as a maximum flow on an hexagonal grid, but instead of using matching individual detections, they make use of tracklets. This has the advantage that they can precompute the social forces for each of these tracklets, nonetheless, the fact that the tracklets are chosen locally, means the overall matching is not truly global, and if errors occur during the creation of the tracklets, these cannot be overcome by the global optimization. In [40] the tracking problem is formulated as a Maximum A-Posteriori (MAP) problem, which is mapped to a minimum-cost network flow and then efficiently solved using LP. In this case, each node represents a detection, which means the graph is much smaller compared to [3, 7]. Finally, [37] propose to combine global and local methods to match trajectories across cameras and across time, while a unique global formulation for the multi-view multi-object is presented in [22].

Social Behavior for Tracking: Most tracking systems work with the assumption that the motion model for each target is independent. This simplifying assumption is especially problematic in crowded scenes: imagine the chaos if every pedestrian followed his or her chosen path and completely ignored the other pedestrians in the scene. In order to avoid collisions and reach the chosen destination at the same time, a pedestrian follows a series of social rules or social forces. These have been defined in what is called the Social Force Model (SFM) [15], which has been used for abnormal crowd behavior detection [26], crowd simulation [28] and has only recently been applied to multiple people tracking: in [33], an energy minimization approach is used to predict the future position of each pedestrian considering all the terms of the social force model. In [30] and [24], the social forces are included in the motion model of the Kalman or Extended Kalman filter, while the authors in [4] discuss the type of energy needed to include information about the dynamic model, repulsion, etc. and how to optimize it using the standard conjugate gradient method. In [14] a method is presented to detect small groups of people in a crowd, but it is only recently that grouping behavior has been included in a tracking framework [10, 29, 38]. In [29] groups are included in a graphical model which contains cycles and, therefore, Dual Decomposition [8] is needed to find the solution, which obviously is computationally much more expensive than using Linear Programming. Moreover, the results presented in [29] are only for short time windows. On the other hand, the formulations of [10, 38] are predictive by nature and therefore too local and unable to deal with trajectory changes (e.g. when people meet and stop to talk).

Recently, a new approach [21] includes social and grouping models into a global optimization framework, allowing for a better estimate of the true maximum a-posteriori probability of the trajectories and therefore further improving tracking results, especially in crowded scenes.

11.2 Multiple People Tracking

Tracking is commonly divided in two steps: object detection and data association. First, the objects are detected in each frame of the sequence and second, the detections are matched to form complete trajectories. In this section we define the data association problem and describe how to convert it to a minimum-cost network flow problem, which can be efficiently solved using Linear Programming.

The idea is to build a graph in which the nodes represent the pedestrian detections. These nodes are fully connected to past and future observations by edges, which determine the relation between two observations with a cost. Thereby, the matching problem is equivalent to a minimum-cost network flow problem: finding the optimal set of trajectories is equivalent to sending flow through the graph so as to minimize the cost. This can be efficiently computed using the Simplex algorithm or k-shortest paths [11].

11.2.1 Problem Statement

Let $\mathcal{O} = \{\mathbf{o}_k t\}$ be a set of object detections with $\mathbf{o}_k^t = (\mathbf{p}_k, t)$, where $\mathbf{p}_k = (x, y, z)$ is the 3D position and t is the time stamp. A trajectory is defined as a list of ordered object detections $T_k = \{\mathbf{o}_k^1, \mathbf{o}_k^2, \cdots, \mathbf{o}_k^N\}$, and the goal of multiple object tracking is to find the set of trajectories $\mathcal{T}* = \{T_k\}$ that best explains the detections.

This is equivalent to maximizing the a-posteriori probability of $\mathcal{T}$ given the set of detections $\mathcal{O}$, which is known as *maximum posterior* or *MAP* problem.

$$\mathcal{T}* = \underset{\mathcal{T}}{\mathbf{argmax}}\, P(\mathcal{T}|\mathcal{O}) \tag{11.1}$$

Further assuming that detections are conditionally independent, the objective function is expressed as:

$$\mathcal{T}* = \underset{\mathcal{T}}{\mathbf{argmax}}\, P(\mathcal{O}|\mathcal{T})P(\mathcal{T}) = \underset{\mathcal{T}}{\mathbf{argmax}}\, \prod_k P(\mathbf{o}_k|\mathcal{T})P(\mathcal{T}) \tag{11.2}$$

$P(\mathbf{o}_k|\mathscr{T})$ is the likelihood of the detection. Optimizing Eq. (11.2) directly is intractable since the space of $\mathscr{T}$ is huge, nonetheless we make the assumption that the trajectories cannot overlap (i.e., a detection cannot belong to two trajectories) to obtain:

$$\mathscr{T}* = \underset{\mathscr{T}}{\operatorname{\mathbf{argmax}}} \prod_k P(\mathbf{o}_k|\mathscr{T}) \prod_{T_k \in \mathscr{T}} P(T_k) \qquad (11.3)$$

where the trajectories are represented by a Markov chain:

$$P(\mathscr{T}) = \prod_{T_k \in \mathscr{T}} P_{\text{in}}(\mathbf{o}_k^1) P(\mathbf{o}_k^2|\mathbf{o}_k^1) \ldots P(\mathbf{o}_k^t|\mathbf{o}_k^{t-1}) \ldots P(\mathbf{o}_k^N|\mathbf{o}_k^{N-1}) P_{\text{out}}(\mathbf{o}_k^N) \qquad (11.4)$$

where $P_{\text{in}}(\mathbf{o}_k^t)$ is the probability that a trajectory is initiated with detection $\mathbf{o}_k^t$, $P_{\text{out}}(\mathbf{o}_k^t)$ the probability that the trajectory is terminated at $\mathbf{o}_k^t$ and $P(\mathbf{o}_k^t|\mathbf{o}_k^{t-1})$ is the probability that $\mathbf{o}_k^{t-1}$ is followed by $\mathbf{o}_k^t$ in the trajectory.

11.2.2   Tracking with Linear Programming

In this section, we explain how to convert the MAP problem into a Linear Program, which is a particularly interesting since it can be efficiently solved in polynomial time using any of the available techniques from the optimization community [1].

A linear programming problem consists in minimizing or maximizing a linear function in the presence of linear constraints, which can be both equalities and inequalities.

$$\text{Minimize} \qquad c_1 f_1 + c_2 f_2 + \ldots + c_n f_n \qquad (11.5)$$

$$\text{Subject to} \qquad a_{11} f_1 + a_{12} f_2 + \ldots + a_{1n} f_n \geq b_1 \qquad (11.6)$$
$$a_{21} f_1 + a_{22} f_2 + \ldots + a_{2n} f_n \geq b_2$$

$$\vdots \qquad \vdots \qquad \vdots$$

$$a_{m1} f_1 + a_{m2} f_2 + \ldots + a_{mn} f_n \geq b_m$$

where Eq. (11.5) is the *objective function* and Eq. (11.6) are the *constraints*.

(continued)

(continued)
$c_1, c_2, \ldots, c_n$ denote the known *cost coefficients* and $f_1, f_2, \ldots, f_n$ are the *decision variables* to be determined.

To convert our problem into a linear program, we linearize the objective function by defining a set of flow flags $f_{i,j} = \{0, 1\}$ which indicate if an edge (i, j) is in the path of a trajectory or not.

In a minimum cost network flow problem, the objective is to find the values of the variables that minimize the total cost of the flows over the network. Defining the costs as negative log-likelihoods, and combining Eqs. (11.3) and (11.4), the following objective function is obtained:

$$\mathcal{T}* = \underset{\mathcal{T}}{\text{argmin}} \sum_{T_k \in \mathcal{T}} -\log P(T_k) - \sum_k \log P(\mathbf{o}_k | \mathcal{T})$$

$$= \underset{\mathcal{T}}{\text{argmin}} \sum_i C_{\text{in},i} f_{\text{in},i} + \sum_{i,j} C_{i,j} f_{i,j} + \sum_i C_i f_i + \sum_i C_{i,\text{out}} f_{i,\text{out}} \qquad (11.7)$$

subject to the following constraints:

- Edge capacities: assuming each detection can only correspond to one trajectory, the edge capacities have an upper bound of $u_{ij} \leq 1$ and:

$$f_{\text{in},i} + f_i \leq 1 \qquad f_{i,\text{out}} + f_i \leq 1 \qquad (11.8)$$

- Flow conservation at the nodes:

$$f_{\text{in},i} + f_i = \sum_j f_{i,j} \qquad \sum_j f_{j,i} = f_{i,\text{out}} + f_i \qquad (11.9)$$

- Exclusion property:

$$f_{i,j} = \{0, 1\} \qquad (11.10)$$

The condition in Eq. (11.10) requires us to solve an integer program, which is known to be NP-complete. Nonetheless, we can relax the condition to have the following linear equation:

$$0 \leq f_{i,j} \leq 1. \qquad (11.11)$$

Now the problem is defined and can be solved as a linear program. If certain conditions are fulfilled, the solution $\mathcal{T}*$ will still be integer, and therefore will also be the optimal solution to the initial integer program. We discuss the integrality of the solution in more detail in Sect. 11.4.

To map this formulation into a cost-flow network, we define $G = (N, E)$ to be a directed network with a cost $C_{i,j}$ and a capacity u_{ij} associated with every edge

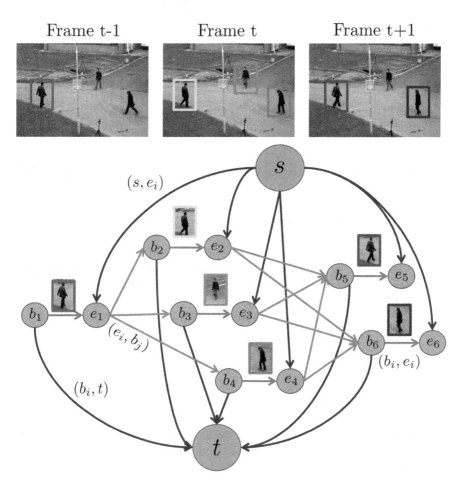

Fig. 11.2 Example of a graph with the special source s and sink t nodes, six detections which are represented by two nodes each: the beginning b_i and the end e_i

$(i, j) \in E$. An example of such a network is shown in Fig. 11.2; it contains two special nodes, the source s and the sink t; all flow that goes through the graph starts at the s node and ends at the t node. Thereby, each flow represents a trajectory T_k and the path that each flow follows indicates which observations belong to each of the trajectories. Each observation $\mathbf{o}_i$ is represented with two nodes, the beginning node $b_i \in N$ and the end node $e_i \in N$ (see Fig. 11.2). A detection edge connects b_i and e_i.

Below we detail the three types of edges present in the graphical model and the cost for each type:

Link Edges: The edges (e_i, b_j) connect the end nodes e_i with the beginning nodes b_j in following frames, with cost $C_{i,j}$ and flow $f_{i,j}$, defined as:

$$f_{i,j} = \begin{cases} 1, & \mathbf{o}_i \text{ and } \mathbf{o}_j \text{ belong to } T_k \text{ and } \Delta f \leq F_{\max} \\ 0, & \text{otherwise} \end{cases} \tag{11.12}$$

where Δf is the frame number difference between nodes j and i and $F_{\max}$ is the maximum allowed frame gap.

The costs of the link edges represent the spatial relation between different subjects. Assuming that a subject cannot move a lot from one frame to the next, we define the costs to be a decreasing function of the distance between detections in successive frames. The time gap between observations is also taken into account in order to be able to work at any frame rate, therefore velocity measures are used instead of distances. The velocities are mapped to probabilities with a Gauss error function as shown in Eq. (11.13), assuming the pedestrians cannot exceed a maximum velocity $V_{\max}$. The effect of parameter $V_{\max}$ is detailed in Sect. 11.5.1.

$$E(V_t, V_{\max}) = \frac{1}{2} + \frac{1}{2} \text{erf} \left(\frac{-V_t + \frac{V_{\max}}{2}}{\frac{V_{\max}}{4}} \right) \tag{11.13}$$

As we can see in Fig. 11.3, the advantage of using Eq. (11.13) over a linear function is that the probability of lower velocities decreases more slowly, while the probability for higher velocities decreases more rapidly. This is consistent with the probability distribution of speed learned from training data.

Therefore, the cost of a link edge is defined as:

$$C_{i,j} = -\log\left(P(\mathbf{o}_j | \mathbf{o}_i)\right) + C(\Delta f) \tag{11.14}$$

$$= -\log E \left(\frac{\|\mathbf{p}_j - \mathbf{p}_i\|}{\Delta t}, V_{\max} \right) + C(\Delta f)$$

where $C(\Delta f) = -\log\left(B_j^{\Delta f - 1}\right)$ is the cost depending on the frame difference between detections.

Detection Edges: The edges (b_i, e_i) connect the beginning node b_i and end node e_i, with cost C_i and flow f_i, defined as:

$$f_i = \begin{cases} 1, & \mathbf{o}_i \text{ belongs to } T_k \\ 0, & \text{otherwise} \end{cases} \tag{11.15}$$

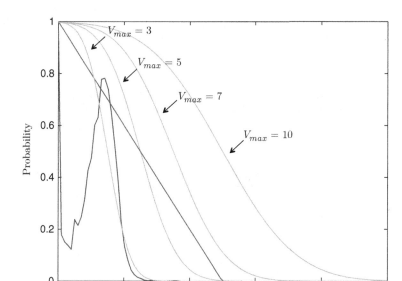

Fig. 11.3 (Color online) *Blue* = normalized histogram of speeds learned from training data. *Red* = probability distribution if cost depends linearly on the velocity. *Green* = probability distribution if the relation of cost and velocities is expressed by Eq. (11.13). An $V_{\max} = 7$ m/s is used in the experiments

If all the costs of the edges are positive, the solution to the minimum-cost problem is the trivial null flow. Consequently, we represent each observation with two nodes and a detection edge with negative cost:

$$C_i = \log\left(1 - P_{det}(\mathbf{o}_i)\right) + \log\left(\frac{\mathrm{BB}_{\min}}{\|\mathbf{p}_{\mathrm{BB}} - \mathbf{p}_i)\|}\right). \qquad (11.16)$$

The higher the likelihood of a detection $P_{det}(\mathbf{o}_i)$ the more negative the cost of the detection edge, hence, confident detections are likely to be in the path of the flow in order to minimize the total cost. If a map of the scene is available, we can also include this information in the detection cost. If a detection is far away from a possible entry/exit point, we add an extra negative cost to the detection edge, in order to favor that observation to be matched. The added cost depends on the distance to the closest entry/exit point $\mathbf{p}_{\mathrm{BB}}$, and is only computed for distances higher than $\mathrm{BB}_{\min} = 1.5m$. This is a probabilistic simple way of including other information present in the scene, such as obstacles or attraction points (shops, doors, etc.).

Entrance and Exit Edges: The edges (s, e_i) connect the source s with all the end nodes e_i, with cost $C_{\mathrm{in},i}$ and flow $f_{\mathrm{in},i}$. Similarly, (b_i, t) connects the end node b_i with sink t, with cost $C_{i,\mathrm{out}}$ and flow $f_{i,\mathrm{out}}$. The flows are defined as:

$$f_{\mathrm{in},i} \text{ (or } f_{i,\mathrm{out}}) = \begin{cases} 1, & T_k \text{ starts (or ends) at } \mathbf{o}_i \\ 0, & \text{otherwise} \end{cases} \qquad (11.17)$$

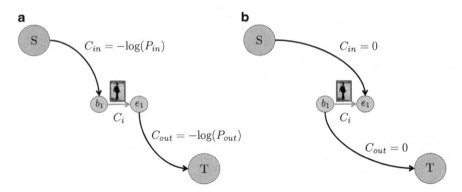

Fig. 11.4 (**a**) Graph structure as used in [40], which requires the computation of P_{in} and P_{out} in an Expectation-Maximization step during optimization. (**b**) Graph structure as used in [21] which does not require the computation of these two parameters; the trajectories are found only with the information of the link and detection edges

This connection, as shown in Fig. 11.4b, was proposed in [21] so that when a track starts (or ends) it does not benefit from the negative cost of the detection edge. Setting $C_{\text{in}} = C_{\text{out}} = 0$ and taking into account the flow constraints of Eqs. (11.8) and (11.9), the trajectories are only created with the information of the link edges.

In contrast, the authors in [40] propose to create the opposite edges (s, b_i) and (e_i, t), which means tracks entering and leaving the scene go through the detection node and therefore benefiting from its negative cost (see Fig. 11.4a). If the costs C_{in} and C_{out} are then set to zero, a track will be started at each detection of each frame, because it will be cheaper to use the entrance and exit edges than the link edges. On the other hand, if C_{in} and C_{out} are very high, it will be hard for the graph to create any trajectory. Therefore, the choice of these two costs is extremely important. In [40], the costs are set according to the entrance and exit probabilities P_{in} and P_{out}, which are data dependent terms that need to be calculated during optimization.

11.3 Modeling Social Behavior

If a pedestrian does not encounter any obstacles, the natural path to follow is a straight line. But what happens when the space gets more and more crowded and the pedestrian can no longer follow the straight path? Social interaction between pedestrians is especially important when the environment is crowded. In this section we consider how to include the social behavior [15], which we divide into the Social Force Model (SFM) and the Group behavior (GR), into the minimum-cost network flow problem.

11.3.1   New MAP and Linear Programming Formulation

The original social force model [15] describes a physical system that predicts the position of a pedestrian in a continuous way, which has been successfully used for crowd simulation [28]. Nonetheless, we use the social information within another paradigm: in our Linear Programming system, we have a set of hypothetical pedestrian positions (in the form of nodes) and we apply the social forces to find out the probability of a certain match (i.e. a certain trajectory being followed by a pedestrian).

When including social and grouping information in the Linear Programming formulation, we can no longer assume that the motion of each subject is independent, which means we have to deal with a much larger search space of $\mathscr{T}$.

We extend this space by including the following dependencies for each trajectory T_k:

- Constant velocity assumption: the observation $\mathbf{o}_k^t \in T_k$ depends on past observations $[\mathbf{o}_k^{t-1}, \mathbf{o}_k^{t-2}]$
- Grouping behavior: If T_k belongs to a group, the set of members of the group $\mathscr{T}_{k,\mathrm{GR}}$ has an influence on T_k
- Avoidance term: T_k is affected by the set of trajectories $\mathscr{T}_{k,\mathrm{SFM}}$ which are close to T_k at some point in time and do not belong to the same group as T_k

The first and third dependencies are grouped into the SFM term. The sets $\mathscr{T}_{k,\mathrm{SFM}}$ and $\mathscr{T}_{k,\mathrm{GR}}$ are disjoint, i.e., for a certain pedestrian k, the set of pedestrians that have an attractive effect (the group to which pedestrian k belongs to), is different from the pedestrians that have a repulsive effect on k. Therefore, we can assume that these two terms are independent and decompose $P(\mathscr{T})$ as:

$$P(\mathscr{T}) = \prod_{T_k \in \mathscr{T}} P(T_k \cap \mathscr{T}_{k,\mathrm{SFM}} \cap \mathscr{T}_{k,\mathrm{GR}}) \qquad (11.18)$$

$$= \prod_{T_k \in \mathscr{T}} P(\mathscr{T}_{k,\mathrm{SFM}}|T_k) P(\mathscr{T}_{k,\mathrm{GR}}|T_k) P(T_k)$$

Let us assume that we are analyzing observation $\mathbf{o}_k^t$. In Fig. 11.5 we summarize which observations influence the matching of $\mathbf{o}_k^t$. Typical approaches [40] only take into account distance (DIST) information, that is, the observation in the previous frame $\mathbf{o}_k^{t-1}$. We introduce the social dependencies (SFM) given by the constant velocity assumption (green nodes) and the avoidance term (yellow nodes). In this case, two observations, $\mathbf{o}_q^t$ and $\mathbf{o}_r^t$ that do not belong to the same group as $\mathbf{o}_k^t$, will be considered to create a repulsion effect on $\mathbf{o}_k^t$. On the other hand, the orange nodes which depict the grouping term (GR), are two other observations $\mathbf{o}_m^t$ and $\mathbf{o}_n^t$ which

Fig. 11.5 Diagram of the
dependencies for each
observation $\mathbf{o}_k^t$

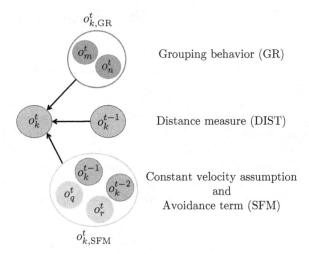

Grouping behavior (GR)

Distance measure (DIST)

Constant velocity assumption
and
Avoidance term (SFM)

do belong to the same group as $\mathbf{o}_k^t$ and therefore have an attraction effect on $\mathbf{o}_k^t$. Note that all these dependencies can only be modeled by high order terms, which means that either we use complex solvers [29] to find a solution in graphs with cycles, or we keep the linearity of the problem by using an iterative approach as we explain later on.

The objective function is accordingly updated:

$$\mathscr{T}* = \operatorname*{\mathbf{argmin}}_{\mathscr{T}} \sum_{T_k \in \mathscr{T}} -\log P(T_k) - \log P(\mathscr{T}_{\text{SFM}}|T_k) \qquad (11.19)$$

$$-\log P(\mathscr{T}_{\text{GR}}|T_k) + \sum_k -\log P(\mathbf{o}_k|\mathscr{T})$$

$$= \operatorname*{\mathbf{argmin}}_{\mathscr{T}} \sum_i C_{\text{in},i} f_{\text{in},i} + \sum_i C_{i,\text{out}} f_{i,\text{out}}$$

$$+ \sum_{i,j} (C_{i,j} + C_{\text{SFM},i,j} + C_{\text{GR},i,j}) f_{i,j} + \sum_i C_i f_i$$

11.3.2 Social Force Model

The social force model states that the motion of a pedestrian can be described as if they were subject to "social forces". There are three main terms that need to be considered: the desire of a pedestrian to maintain a certain speed, the desire to keep a comfortable distance from other pedestrians and the desire to reach a destination. Since we cannot know a priori the destination of the pedestrian in a real tracking system, we focus on the first two terms.

Constant Velocity Assumption: The pedestrian tries to keep a certain speed and direction, therefore we assume that in $t + \Delta t$ we have the same speed as in t and predict the pedestrian's position in $t + \Delta t$ accordingly.

$$\tilde{\mathbf{p}}_i^{t+\Delta t} = \mathbf{p}_i^t + \mathbf{v}_i^t \Delta t$$

Avoidance Term: The pedestrian also tries to avoid collisions and keep a comfortable distance from other pedestrians. This term is modeled as a repulsion field with an exponential distance-decay function with value α learned from training data.

$$\mathbf{a}_i^{t+\Delta t} = \sum_{g_m \neq g_i} \exp\left(-\frac{\|\tilde{\mathbf{p}}_i^{t+\Delta t} - \tilde{\mathbf{p}}_m^{t+\Delta t}\|}{\alpha \Delta t} \right) \qquad (11.20)$$

To compute the cost of edge (i, j), the constant velocity assumption is used to predict the position of $\mathbf{o}_i$ and $\mathbf{o}_j$ as well as the rest of pedestrians $\tilde{\mathbf{p}}_m^{t+\Delta t}$, and the repulsion acceleration each pedestrian has on i is also taken into account. The only pedestrians that have this repulsion effect on subject i are the ones which do not belong to the same group as i and $\|\tilde{\mathbf{p}}_i^{t+\Delta t} - \tilde{\mathbf{p}}_m^{t+\Delta t}\| \leq 1m$. The different avoidance terms are combined linearly.

Now the prediction of the pedestrian's next position is also influenced by the avoidance term (acceleration) from all pedestrians:

$$\tilde{\mathbf{p}}_i^{t+\Delta t} = \mathbf{p}_i^t + (\mathbf{v}_i^t + \mathbf{a}_i^{t+\Delta t} \Delta t)\Delta t \qquad (11.21)$$

The distance between prediction and real measurements is used to compute the cost:

$$C_{\text{SFM},i,j} = -\log E\left(\frac{\|\tilde{\mathbf{p}}_i^{t+\Delta t} - \mathbf{p}_j^{t+\Delta t}\|}{\Delta t}, V_{\max} \right) \qquad (11.22)$$

where the function E is detailed in Eq. (11.13).

In Fig. 11.6 we plot the probability distribution computed using different terms. Note, this is just for visualization purposes, since we do not compute the probability for each point on the scene, but only for the positions where the detector has fired. There are four pedestrians in the scene, the purple one and three green ones walking in a group. As shown in Fig. 11.6b, if we only use the predicted positions (yellow heads) given the previous speeds, there is a collision between the purple pedestrian and the green marked with a 1 collide. The avoidance term shifts the probability mode to a more plausible position.

11.3.3   Group Model

The social behavior [15] also includes an attraction force which occurs when a pedestrian is attracted to a friend, shop, etc. In this section, we show how to model

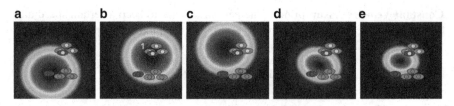

Fig. 11.6 (Color online) Three green pedestrians walk in a group, the predicted positions in the next frame are marked by *yellow heads*. The purple pedestrian's linearly predicted position (*yellow head*) clearly interferes with the trajectory of the group. Representation of the probability (blue is 0 red is 1) distribution for the purple's next position using: (**a**) only distances, (**b**) only SFM (constant velocity assumption and avoidance term), (**c**) only GR (considering the purple pedestrian belongs to the group), (**d**) distances+SFM and (**e**) distances+SFM+GR

the attraction between members of a group. Before modeling group behavior we need to determine which tracks form each group and at which frame the group begins and ends (to deal with splitting and formation of groups). The idea is that if two pedestrians are close to each other over a reasonable period of time, they are likely to belong to the same group. From the training sequence in [30], the distance and speed probability distributions of the members of a group P_g vs. individual pedestrians P_i is learned. If m and n are two trajectories which appear on the scene at $t = [0,N]$, we compute the flag $G_{m,n}$ that indicates if m and n belong to the same group.

$$G_{m,n} = \begin{cases} 1, & \sum_{t=0}^{N} P_g(m,n) > \sum_{t=0}^{N} P_i(m,n) \\ 0, & \text{otherwise} \end{cases} \qquad (11.23)$$

For every observation $\mathbf{o}_i$, we define a group label g_i which indicates to which group the observation belongs to, if any. If several pedestrians form a group, they tend to keep a similar speed, therefore, if i belongs to a group, we can use the mean speed of all the other members of the group to predict the next position for i:

$$\tilde{\mathbf{p}}_i^{t+\Delta t} = \mathbf{p}_i^t + \sum_{g_m=g_i} \mathbf{v}_m^t \Delta t \qquad (11.24)$$

The distance between this predicted position and the real measurements is used in (11.13) to obtain the cost for the grouping term.

An example is shown in Fig. 11.6c, where we can see that the maximum probability provided by the group term keeps the group configuration. In Fig. 11.6d we show the combined probability of the distance and SFM information, which narrows the space of probable positions. Finally, Fig. 11.6e represents the combined probability of DIST, SFM and GR. As we can see, the space of possible locations for the purple pedestrian is considerably reduced as we add the social and grouping behaviors, which means we have less ambiguities for data association. This is specially useful to decrease identity switches as we present in Sect. 11.5.

11.4 Optimization

To compute the SFM and grouping costs, we need to have information about the velocities of the pedestrians, which can only be obtained if we already have the trajectories. This chicken-and-egg problem is solved iteratively as shown in Algorithm 3; on the first iteration, the trajectories are estimated only with the information defined in Sect. 11.2.2, for the rest of iterations, the SFM and GR is also used. The algorithm stops when the trajectories do not change or when a maximum number of iterations M_i is reached.

Algorithm 1 Iterative optimization

---

  while $\mathcal{T}_i \neq \mathcal{T}_{i-1}$ and $i \leq M_i$ **do**

    if $i == 1$ **then**

        1.1. Create the graph using only DIST information

    else

        1.2. Create the graph using DIST, SFM and GR information

    end if

    2. Solve the graph to find $\mathcal{T}_i$

    3. Compute velocities and groups given $\mathcal{T}_i$

  end while

11.4.1 Linear Programming Solvers

The minimum cost solution is found using the Simplex algorithm [11], with the implementation given in [25]. Though Simplex has an exponential worst-case complexity, most sequences can be tracked in just a few seconds; this is because each node represents one detection, and therefore the dimension of the graph is quite small. For larger graphs [7] or more crowded environments, we can use the k-shortest paths solver [7, 31] which has a worst case complexity of $O(k(m + n \cdot \log(n)))$. For more details on network flows and Simplex we refer the reader to [1], and to [35] for more information on the k-shortest path algorithm.

11.4.2 Integrality of the Solution

When defining the program to be solved, we saw that Eq. (11.10) defined an integer program, which is known to be NP-complete. The condition is relaxed into Eq. (11.11) in order to use efficient Linear Programming solvers to find the optimum solution to our problem. If the solution to the relaxed version of the program is integer, then we know it is an optimal solution of the original problem [1]. The question is, can we guarantee that the solution will be always integer?

Let us assume the conditions of the Linear Program are expressed as: $Ax = b$. If all entries of A and b are integer, as it is our case, we can determine that $Ax = b$ has an integer solution by Cramer's rule:

$$Ax = b \quad \Leftrightarrow \quad x = A^{-1}b \quad \Leftrightarrow \quad \forall i : x_i = \frac{\det(A^i)}{\det(A)} \qquad (11.25)$$

where A^i is equal to A except on the i-th column where it is equal to b. From here, we can determine that x will be integer when $\det(A)$ is equal to $+1$ or -1. A matrix $A \in Z^{m \times n}$ is *totally unimodular* if the determinant of all the subsquare matrices of A is either 0, $+1$ or -1.

Theorem 11.1. *If A is totally unimodular, every vertex solution of $Ax \leq b$ is integer.*

A well-known case of totally unimodular matrices are the node arc incidence matrices N of a directed network. Therefore, our defined constraint matrix is totally unimodular, and the solutions we will obtain will always be integer.

11.4.3 Computationally Reduction

To reduce the computational cost, the graph can be pruned using the physical constraints represented by the edge costs. If any of the costs C_{ij}, $C_{SFM,i,j}$ or $C_{GR,i,j}$ is infinite, the two detections i and j are either two far away to belong to the same trajectory or they do not match according to social and grouping rules, therefore the edge (i, j) is erased from the graphical model. For long sequences, the video can be divided into several batches and optimize for each batch. For temporal consistency, the batches have an overlap of $F_{max} = 10$ frames. The runtime of [21] for a sequence of 800 frames (114 s), 4,837 detections, batches of 100 frames and 6 iterations is 30 s on a 3 GHz machine.

11.5 Experimental Results

In this section we show the tracking results of several state-of-the-art methods on three publicly available datasets and compare them using the CLEAR metrics [17], which split the measuring scores into *accuracy* and *precision*:

- Detection Accuracy (DA): measures how many detections where correctly found and therefore is based on the count of missed detections m_t and false alarms f_t for each frame t.

$$DA = 1 - \frac{\sum_{t=1}^{N_f} m_t + f_t}{\sum_{t=1}^{N_f} N_G^t}$$

where N_f is the number of frames of the sequence and N_G^t is the number of ground truth detections in frame t. A detection is considered to be correct when it is found within 50 pixels from the ground truth and the bounding boxes of both ground truth and detection have some overlap.

- Tracking Accuracy (TA): similar to DA but also including the identity switches i_t. In this case, the measure does not penalize identity switches as much as a missing detection or a false alarm as we use a $\log_{10}$ weight.

$$DA = 1 - \frac{\sum_{t=1}^{N_f} m_t + f_t + \log_{10}(1 + i_t)}{\sum_{t=1}^{N_f} N_G^t}$$

- Detection Precision (DP): precision measurements represent how well the bounding box detections match the ground truth. For this, an overlap measure between bounding boxes is used:

$$Ov^t = \sum_{i=1}^{N_{\text{mapped}}^t} \frac{|G_i^t \cap D_i^t|}{|G_i^t \cup D_i^t|}$$

where N_{mapped}^t is the number of mapped objects in frame t, i.e., the number of detections that are matched to some ground truth object. G_i^t is the ith ground truth object of frame t and D_i^t the detected object matched to G_i^t. The DP measure is then expressed as:

$$DP = \frac{\sum_{t=1}^{N_f} \frac{Ov^t}{N_{\text{mapped}}^t}}{N_f}$$

- Tracking Precision (TP): measures the spatiotemporal overlap between ground truth trajectories and detected ones, taking into account also split and merged trajectories.

$$TP = \frac{\sum_{i=1}^{N_{\text{mapped}}^t} \sum_{t=1}^{N_f} \frac{|G_i^t \cap D_i^t|}{|G_i^t \cup D_i^t|}}{\sum_{t=1}^{N_f} N_{\text{mapped}}^t}$$

11.5.1 Analysis of the Effect of the Parameters

All parameters defined in previous sections are learned from training data using one sequence of the publicly available dataset [30]. In this section we study the effect of the few parameters needed in [21], and show the method works well for a wide range of these parameters and therefore no parameter tuning is needed to obtain a good performance. The analysis is done on two publicly available datasets: a crowded town center [5] and the well-known PETS2009 dataset [12], to see the different effects of each parameters on each dataset.

11.5.2 Number of Iterations

The first parameter we analyze is the number of iterations M_i allowed. This determines how many times the loop between computing social forces and computing trajectories is performed as explained in Algorithm 1. Looking at the results on the PETS 2009 dataset in Fig. 11.7b, we can see that after just two iterations the results remain very stable. Actually, the algorithm reports no changes in the trajectories after three iterations, and therefore stops even though the maximum number of iterations allowed is higher. The result with one and two iterations is also not very different, which means the social and grouping behavior do not significantly improve the results for this particular dataset. This is due to the fact that this dataset is very challenging from a social behavior point of view, with subjects often changing direction and groups forming and splitting frequently. More details and comments on these results can be found in Sect. 11.5.6.2. On the other hand, we observe a different effect on the TownCenter dataset, shown in Fig. 11.7a. In this case, there is a clear improvement when using social and grouping behavior (i.e. the result improves when we use more than one iteration. We also observe a pattern on how the Tracking Accuracy of the dataset evolves: there is a cycle of three iterations for which the accuracy increases and decreases in a similar pattern. This means that the algorithm is jumping between two solutions and will not converge to neither one of them. This happens when pedestrians are close together for a long period of time but are not forming a group, which means that even with social forces, it is hard to say which paths they will follow.

11.5.3 Maximum Speed

This is the parameter that determines the maximum speed of the pedestrians that we are observing. In this case, we can see in Fig. 11.7c, d a clear trend in which the results are very bad when we force the pedestrians to walk more slowly that they actually do, since we are artificially splitting trajectories. The results converge when

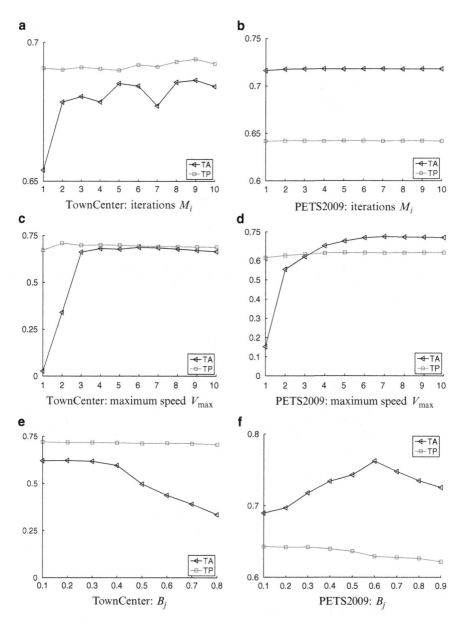

Fig. 11.7 (Color online) Tracking accuracy (*black*) and precision (*magenta*) obtained for the Town Center dataset (*left column*) and the PETS 2009 dataset (*right column*) given varying parameter values. (**a**) TownCenter: iterations M_i. (**b**) PETS2009: iterations M_i. (**c**) TownCenter: maximum speed V_{max}. (**d**) PETS2009: maximum speed V_{max}. (**e**) TownCenter: B_j V_{max}. (**f**) PETS2009: B_j

the maximum speed allowed is around 3–5 m/s, which is the reported mean speed of pedestrians in a normal situation. More interestingly, we observe that the results are kept constant when using higher maximum speed values. This is a positive effect of the global optimization framework, since we can use a much higher speed limit and this will still give us good results and will allow us to track a person running through the scene, a case of panic when people start running, etc.

11.5.4 Cost for the Frame Difference

The last parameter, B_j, appears in Eq. (11.15) and represents the penalization term that we apply when the frame difference between two detections that we want to match is larger than 1. This term is used in order to give preference to matches that are close in time. Here we can again see different effects on the two datasets. In Fig. 11.7e, we see that the results are stable until a value of 0.4. The lower the value, the higher is the penalization cost for the frame difference, which means it is more difficult to match those detections which are more than 1 frame apart. When the value of B_j is higher than 0.4, there are more ambiguities in the data association process because it is easier to match detections which are many frames apart. In the TownCenter dataset, there is no occluding object in the scene, which means missing detections are sporadic within a given trajectory. In this scenario, a lower value for B_j is better, since small gaps can be filled and there are less ambiguities. Nonetheless, we see different results in the PETS 2009 dataset in Fig. 11.7f, since here there is a clear occluding object in the middle of the scene (see Fig. 11.8) which occludes the pedestrians for longer periods of time. In this case, a higher value of B_j allows to overcome these large gaps of missing data, and that is why the best value for this dataset is around 0.6.

11.5.5 Evaluation with Missing Data, Noise and Outliers

We evaluate the impact of every component of the approach in [21] with one of the sequences of the dataset [30], which contains images from a crowded public place, with several groups as well as walking and standing pedestrians. The sequence is 11,601 frames long and contains more than 300 trajectories. First of all, the group detection method is evaluated on the whole sequence with ground truth detections: 61% are correctly detected, 26% are only partially detected, 13% are not found and an extra 7% groups are detected wrongly. All experiments are performed with 6 iterations, a batch of 100 frames, $V_{max} = 7$ m/s, $F_{max} = 10$, $\alpha = 0.5$ and $B_j = 0.3$.

Fig. 11.8 Four frames of the PETS2009 sequence (separation of nine frames), showing several occlusions, both created by the obstacle on the scene and between pedestrians. All the occlusions can be recovered with the proposed method

Using the ground truth (GT) pedestrian positions as the baseline for our experiments, we perform three types of tests, missing data, outliers and noise, and compare the results obtained with:

- DIST: proposed network model with distances
- SFM: adding the Social Force Model (Sect. 11.3.2)
- SFM+GR: adding SFM and grouping behavior (Sect. 11.3.3)

Missing Data: This experiment shows the robustness of our approach given missed detections. This is evaluated by randomly erasing a certain percentage of detections from the GT set. The percentages evaluated are $[0, 4, 8, 12, 16, 20]$ from the total number of detections over the whole sequence. As we can see in Fig. 11.9, both SFM and SFM+GR increase the tracking accuracy when compared to DIST.

Outliers: With an initial set of detections of GT with 2% missing data, tests are performed with $[0, 10, 20, 30, 40, 50]$ percentage of outliers added in random positions over the ground plane. In Fig. 11.9, the results show that the SFM is especially important when the tracker is dealing with outliers. With 50% of outliers, the identity switches with SFM+GR are reduced 70% w.r.t the DIST results.

Noise: This test is used to determine the performance of our approach given noisy detections, which are very common mainly due to small errors in the 2D-3D mapping. From the GT set with 2% missing data, random noise is added to every detection. The variances of the noise tested are $[0, 0.002, 0.004, 0.006, 0.008, 0.01]$

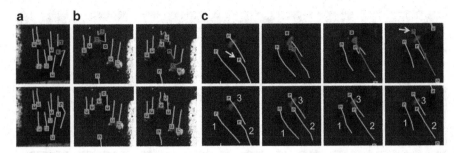

Fig. 11.9 (Color online) Experiments are repeated 50 times and average result, maximum and minimum are plotted. *Blue star* = results with DIST, *Green diamond* = results with SFM, *Red square* = results with SFM+GR. *From left to right*: Experiment with simulated missing data, with outliers, and with random noise

of the size of the scene observed. As expected, group information is the most robust to noise; if the position of pedestrian A is not correctly estimated, other pedestrians in the group will contribute to the estimation of the true trajectory of A.

These results corroborate that having good behavioral models becomes more important as the observations deteriorate. In Fig. 11.10 we plot the tracking results of a sequence with 12% simulated missing data. Only using distance information can see identity switches as shown in Fig. 11.10a. In Fig. 11.10b we can see how missing data affects the matching results. The matches are shifted, this chain reaction is due to the global optimization. In both cases, the use of SFM allows the tracker to interpolate the necessary detections and find the correct trajectories. Finally, in Fig. 11.10c we plot the wrong result which occurs because track 3 has two consecutive missing detections. Even with SFM, track 2 is switched for 3, since the switch does not create extreme changes in velocity. In this case, the grouping information is key to obtaining good tracking results. More results are shown in Fig. 11.13, first row.

11.5.6 Tracking Results

In this section, we compare results of several state-of-the-art methods on two publicly available datasets: a crowded town center [5] and the well-known PETS2009 dataset [12]. We compare results obtained with:

- Benfold et al. [5]: using the results provided by the authors for full pedestrian detections. The HOG detections are also given by the authors and used as input for all experiments.
- Zhan et al. [40]: globally optimum tracking based on network flow linear programming.
- Pellegrini et al. [30]: tracker based on Kalman Filter which includes social behavior.

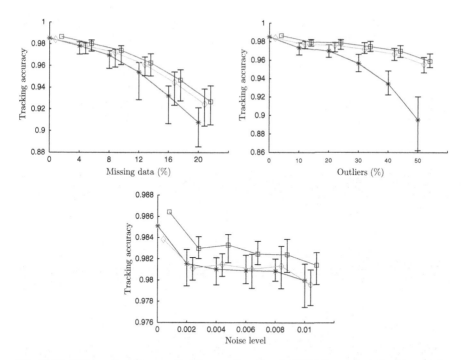

Fig. 11.10 (Color online) *Top row*: Tracking results with only DIST. *Bottom row*: Tracking results with SFM+GR. *Green* = correct trajectories, *Blue* = observation missing from the set, *Red* = wrong match. (**a**) Wrong match with DIST, corrected with SFM. (**b**) Missing detections cause the matches to shift due the global optimization; correct result with SFM. (**c**) Missed detection for subject 3 on two consecutive frames. With SFM, subject 2 in the first frame (*yellow arrow*) is matched to subject 3 in the last frame (*yellow arrow*), creating an identity switch; correct result with grouping information

- Yamaguchi et al. [38]: tracker based on Kalman Filter which includes social and grouping behavior.
- Leal-Taixé et al. [21]: globally optimum tracking based on network flow linear programming and including social and grouping behavior.

For a fair comparison, we do not use appearance information for any method. The methods [5,30,38] are online, while [21,40] processes the video in batches. For these last two methods, all experiments are performed with six iterations, a batch of 100 frames, $V_{max} = 7$ m/s, $F_{max} = 10$, $\alpha = 0.5$ and $B_j = 0.3$.

11.5.6.1 Town Center Dataset

We perform tracking experiments on a video of a crowded town center [5], using one of every ten frames (simulating 2.5 fps). We show detection accuracy (DA), tracking

Table 11.1 Town Center sequence

	DA	TA	DP	TP	IDsw
HOG detections	63.1	–	71.9	–	–
Benfold et al. [5]	64.9	64.8	**80.5**	**80.4**	259
Zhang et al. [40]	66.1	65.7	71.5	71.5	114
Pellegrini et al. [30]	64.1	63.4	70.8	70.7	183
Yamaguchi et al. [38]	64.0	63.3	71.1	70.9	196
Leal-Taixé et al. [21]	**67.6**	**67.3**	71.6	71.5	**86**

Fig. 11.11 Predictive approaches [30, 38] (*first row*) vs. Proposed method (*second row*)

accuracy (TA), detection precision (DP) and tracking precision (TP) measures as well as the number of identity switches (IDsw).

Note, the precision reported in [5] is about 9% higher than the input detections precision; this is because the authors use the motion estimation obtained with a KLT feature tracker to improve the exact position of the detections, while we use the raw detections. Still, our algorithm reports 64% less ID switches. As shown in Table 11.1, [21] algorithm outperforms [30, 38], both of which include social behavior information, by almost 4% in accuracy and with 50% less ID switches. In Fig. 11.11 we can see an example where [30, 38] fail. The errors are created in the greedy phase of predictive approaches, where people fight for detections. The red false detection in the first frame takes the detection in the second frame that should belong to the green trajectory (which ends in the first frame). In the third frame, the red trajectory overtakes the yellow trajectory and a new blue trajectory starts where the green should have been. None of the resulting trajectories violate the SFM

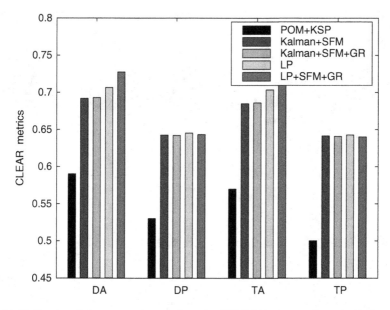

Fig. 11.12 Results of the proposed method on the PETS2009 dataset views 1. (a) Detection accuracy, *DA*. (b) Detection precision, *DP*. (c) Tracking accuracy, *TA*. (d) Tracking precision, *TP*

and GR conditions. On the other hand, a global optimization framework takes full advantage of the SFM and GR information and correctly recovers all the trajectories. More results of the proposed algorithm can be seen in Fig. 11.13, last row.

11.5.6.2   Results on the PETS2009 Dataset

In addition, we present results of monocular tracking on the PETS2009 sequence L1, View 1 with the detections obtained using the Mixture of Gaussians (MOG) background subtraction method. We compare the results with the previously described methods plus the monocular result of View 1 presented in [7], where the detections are obtained using the Probabilistic Occupancy Map (POM) and the tracking is done using k-shortest paths (Fig. 11.12).

The first observation that we make is that the linear programming methods (LP and LP+SFM+GR) clearly outperform predictive approaches in accuracy. This is because this dataset is very challenging from a social behavior point of view, because the subjects often change direction and groups form and split frequently. Approaches based on a probabilistic framework [21, 40] are better suited for unexpected behavior changes (like destination changes), where other predictive approaches fail [30,38]. We can also see that the LP+SFM+GR method has a higher accuracy than the LP method, which does not take into account social and grouping behavior. The grouping term is specially useful to avoid identity switches between

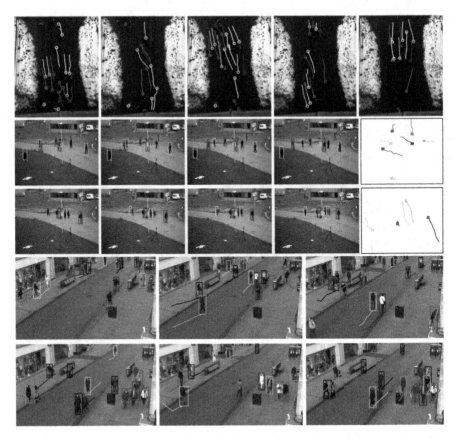

Fig. 11.13 *First row*: Results on the BIWI dataset (Sect. 11.5.5). The scene is heavily crowded, social and grouping behavior are key to obtaining good tracking results. *Second* and *third rows*: Results on the PETS2009 dataset (Sect. 11.5.6.2). *Last two rows*: Results on the Town Center dataset (Sect. 11.5.6.1)

member of a group (see an example in Fig. 11.13, third row, the cyan and green pedestrian who walk together). Precision is similar for all methods since the same detections have been used for all the experiments and we do not apply smoothing or correction of the bounding boxes.

11.6 Conclusions

In this chapter, we presented an overview of methods that integrate pedestrian interaction into a tracking framework in two ways: using a globally optimum solver or improving the dynamic model with social forces. Furthermore, we presented how to combine the strength of both approaches by finding the MAP estimate of the trajectories total posterior including social and grouping models using a minimum-cost network flow with an improved novel graph structure that outperforms existing approaches. People interaction is persistent rather than transient, hence the probabilistic formulation fully exploits the power of behavioral models as opposed to standard predictive and recursive approaches such as Kalman filtering. Experiments on three public datasets reveal the importance of using social interaction models for tracking in difficult conditions such as in crowded scenes with the presence of missed detections, false alarms and noise.

Acknowledgements This work was partially funded by the German Research Foundation, DFG projects RO 2497/7-1 and RO 2524/2-1.

References

1. Ahuja, R., Magnanti, T., Orlin, J.: Network Flows: Theory, Algorithms and Applications. Prentice Hall, Englewood Cliffs (1993)
2. Ali, S., Shah, M.: Floor fields for tracking in high density crowded scenes. In: ECCV, Marseille, France (2008)
3. Andriyenko, A., Schindler, K.: Globally optimal multi-target tracking on an hexagonal lattice. In: ECCV, Crete, Greece (2010)
4. Andriyenko, A., Schindler, K.: Multi-target tracking by continuous energy minimization. In: CVPR, Colorado Springs, USA (2011)
5. Benfold, B., Reid, I.: Stable multi-target tracking in real-time surveillance video. In: CVPR, Colorado Springs, USA (2011)
6. Berclaz, J., Fleuret, F., Fua, P.: Robust people tracking with global trajectory optimization. In: CVPR, New York, USA (2006)
7. Berclaz, J., Fleuret, F., Türetken, E., Fua, P.: Multiple object tracking using k-shortest paths optimization. In: TPAMI (2011)
8. Bertsekas, D.: Nonlinear Programming. Athena Scientific, Belmont (1999)
9. Breitenstein, M., Reichlin, F., Leibe, B., Koller-Meier, E., van Gool, L.: Robust tracking-by-detection using a detector confidence particle filter. In: ICCV, Kyoto, Japan (2009)
10. Choi, W., Savarese, S.: Multiple target tracking in world coordinate with single, minimally calibrated camera. In: ECCV (2010)
11. Dantzig, G.: Linear Programming and Extensions. Princeton University Press, Princeton (1963)
12. Ferryman, J.: Pets 2009 dataset: Proc. 11th IEEE Int'l Workshop performance and evaluation of tracking and surveillance, http:/pets2009.net (2011)
13. Gall, J., Yao, A., Razavi, N., van Gool, L., Lempitsky, V.: Hough forests for object detection, tracking, and action recognition. In: TPAMI (2011)
14. Ge, W., Collins, R., Ruback, B.: Automatically detecting the small group structure of a crowd. In: WACV, Snowbird, Utah (2009)
15. Helbing, D., Molnár, P.: Social force model for pedestrian dynamics. Phys. Rev. E **51**, 4282 (1995)
16. Jiang, H., Fels, S., Little, J.: A linear programming approach for multiple object tracking. In: CVPR, Minneapolis, Minnesota (2007)
17. Kasturi, R., Goldgof, D., Soundararajan, P., Manohar, V., Garofolo, J., Boonstra, M., Korzhova, V., Zhang, J.: Framework for performance evaluation for face, text and vehicle detection and tracking in video: data, metrics, and protocol. TPAMI (2009)

18. Kaucic, R., Perera, A., Brooksby, G., Kaufhold, J., Hoogs, A.: A unified framework for tracking through occlusions and across sensor gaps. In: CVPR, San Diego, USA (2005)
19. Khan, Z., Balch, T., Dellaert, F.: Mcmc-based particle filtering for tracking a variable number of interacting targets. In: TPAMI (2005)
20. Leal-Taixé, L., Heydt, M., Rosenhahn, A., Rosenhahn, B.: Automatic tracking of swimming microorganisms in 4d digital in-line holography data. In: IEEE Workshop on Motion and Video Computing (WMVC), Snowbird, Utah (2009)
21. Leal-Taixé, L., Pons-Moll, G., Rosenhahn, B.: Everybody needs somebody: modeling social and grouping behavior on a linear programming multiple people tracker. In: ICCV Workshops. 1st Workshop on Modeling, Simulation and Visual Analysis of Large Crowds, Barcelona, Spain (2011)
22. Leal-Taixé, L., Pons-Moll, G., Rosenhahn, B.: Branch-and-price global optimization for multi-view multi-object tracking. In: CVPR, Providence, USA (2012)
23. Leibe, B., Schindler, K., Cornelis, N., van Gool, L.: Coupled detection and tracking from static cameras and moving vehicles. TPAMI (2008)
24. Luber, M., Stork, J., Tipaldi, G., Arras, K.: People tracking with human motion predictions from social forces. In: ICRA, Anchorage, USA (2010)
25. Makhorin, A.: Gnu linear programming kit (glpk). http://www.gnu.org/software/glpk/ (2010)
26. Mehran, R., Oyama, A., Shah, M.: Abnormal crowd behavior detection using social force model. In: CVPR, Miami, USA (2009)
27. Nillius, P., Sullivan, J., Carlsson, S.: Multi-target tracking – linking identities using bayesian network inference. In: CVPR (2006)
28. Pelechano, N., Allbeck, J., Badler, N.: Controlling individual agents in high-density crowd simulation. In: Eurographics/ACM SIGGRAPH Symposium on Computer Animation, Prague, Czech Republic (2007)
29. Pellegrini, S., Ess, A., van Gool, L.: Improving data association by joint modeling of pedestrian trajectories and groupings. In: ECCV (2010)
30. Pellegrini, S., Ess, A., Schindler, K., van Gool, L.: You'll never walk alone: modeling social behavior for multi-target tracking. In: ICCV (2009)
31. Pirsiavash, H., Ramanan, D., Fowlkes, C.: Globally-optimal greedy algorithms for tracking a variable number of objects. In: CVPR (2011)
32. Rodriguez, M., Sivic, J., Laptev, I., Audibert, J.: Data-driven crowd analysis in videos. In: ICCV (2011)
33. Scovanner, P., Tappen, M.: Learning pedestrian dynamics from the real world. In: ICCV (2009)
34. Shu, G., Dehghan, A., Oreifej, O., Hand, E., Shah, M.: Part-based multiple-person tracking with partial occlusion handling. In: CVPR (2012)
35. Suurballe, J.: Disjoint paths in a network. Networks **4**, 125–145 (1974)
36. Wu, B., Nevatia, R.: Detection and tracking of multiple, partially occluded humans by bayesian combination of edgelet part detectors. In: IJCV **75**(2) (2007)
37. Wu, Z., Kunz, T., Betke, M.: Efficient track linking methods for track graphs using network-flow and set-cover techniques. In: CVPR (2011)
38. Yamaguchi, K., Berg, A., Ortiz, L., Berg, T.: Who are you with and where are you going? In: CVPR (2011)
39. Yang, M., Yu, T., Wu, Y.: Game-theoretic multiple target tracking. In: ICCV (2007)
40. Zhang, L., Li, Y., Nevatia, R.: Global data association for multi-object tracking using network flows. In: CVPR (2008)

Chapter 12
Surveillance of Crowded Environments: Modeling the Crowd by Its Global Properties

Antoni B. Chan and Nuno Vasconcelos

Abstract In this chapter, we consider aspects of the crowd that can be modeled holistically, by analyzing global properties. We first discuss the dynamic texture model for representing holistic motion flow, which treats the video as a sample from a linear dynamical system. By defining appropriate distances and kernels between dynamic textures, crowd motion can be recognized with standard classification algorithms. Besides motion flow, crowd size, i.e., the number of objects within a crowd can also be modeled holistically. From a suitable set of low-level features, crowd counts can be estimated with a regression function that directly maps features into the number of objects within the crowd. In both cases, the surveillance task is solvable by analyzing global scene properties, and there is no need to detect or track individual objects. In result, the solutions tend to be robust even when the crowd is large, there are substantial occlusions, complex object interactions, or the objects are small.

12.1 Introduction

There is currently a great interest in vision technology for monitoring all types of environments. This could have many goals, e.g., security, resource management, or advertising. From the technological standpoint, computer vision solutions typically focus on detecting, tracking, and analyzing individuals in the scene. However, there are many problems in environment monitoring that can be solved without

A.B. Chan (✉)
Department of Computer Science, City University of Hong Kong, Hong Kong, China
e-mail: abchan@cityu.edu.hk

N. Vasconcelos
Department of Electrical and Computer Engineering, University of California,
San Diego, CA, USA
e-mail: nuno@ece.ucsd.edu

S. Ali et al. (eds.), *Modeling, Simulation and Visual Analysis of Crowds*, The International   295
Series in Video Computing 11, DOI 10.1007/978-1-4614-8483-7__12,
© Springer Science+Business Media New York 2013

explicit tracking of individuals (e.g., people, cars, etc.). These are problems where all the information required to perform the task can be gathered by analyzing the environment *globally*: e.g., monitoring of traffic flows, detection of disturbances in public spaces, detection of speeding on highways, or estimation of the size of moving crowds. By definition, these tasks are based on either properties of (1) the "crowd" as a whole, or (2) an individual's "deviation" from the crowd. In both cases, to accomplish the task it should suffice to build good *models for the patterns of crowd behavior*. Events could then be detected as *variations in these patterns*, and abnormal individual actions could be detected as *outliers* with respect to the crowd behavior.

One property of the crowd (or, in general, groups of similar objects) that can be modeled holistically is the motion that it induces. Traditional motion representations, based on optical flow, are inherently local and have significant difficulties when faced with aperture problems and noise. The classical solution to this problem is to regularize the optical flow field [4, 42, 43, 56], but this introduces undesirable smoothing across motion edges or regions where the motion is, by definition, not smooth (e.g., vegetation in outdoors scenes). Recently, there has been more success in modeling complex scenes as *dynamic textures* or, more precisely, samples from stochastic processes defined over space and time [34]. This work has demonstrated that the dynamic texture has a surprising ability to abstract a wide variety of complex global patterns of motion and appearance into a *simple* spatio-temporal model. Since most of the information required for the classification of crowd events is contained in the interaction between the many motions that it contains, the dynamic texture can be used to capture the variability of the global motion, through a holistic representation of the video pixels, without the need for segmenting or tracking individual components. By defining an appropriate distance or kernel function between dynamic textures, crowd motion or events can be recognized using standard classification algorithms, such as nearest neighbors [73], support vector machines [11, 83], or boosting [80].

In addition to motion, a crowd property of interest for surveillance is the crowd size, e.g., the number of people it contains. Traditional computer vision solutions typically focus on detecting and tracking individuals in the crowd [52]. However, it is also possible to accurately count crowds from *global low-level features*, without the need for detection and tracking individual objects. One possibility [18] is to segment the crowd into regions of interest (e.g., groups of people moving in different directions), extract features (e.g., area, edge, and texture features) from each segment, and map them into estimates of the crowd count per segment. The mapping can be implemented with sophisticated statistical inference methods, e.g., Gaussian process regression, that directly map features into counts. This work has shown that accurate crowd counts are possible without people detection, even when the crowd is sizable and inhomogeneous. Since these solutions avoid the detection of individual objects, they tend to be robust even when the crowd is large, there are complex occlusions or object interactions, or the objects are small. The remainder of the chapter is organized as follows. In Sect. 12.2, we present the dynamic texture model used to represent global motion in video. In Sect. 12.3, we discuss distance functions between dynamic textures and present a framework for motion classification using the dynamic texture representation. Finally, in Sect. 12.4, we discuss methods for crowd counting using non-linear regression and global low-level features.

12.2   Modeling Global Motion Flow with Dynamic Textures

Figure 12.1 presents a sample from a large collection of visual processes that have proven remarkably challenging for traditional motion representations, based on modeling of the individual trajectory of pixels [43, 56], particles [46], or objects in a scene. Since most of the information required for the perception of these processes is contained in the interaction between the many motions that compose them, a holistic representation of the associated motion field should be capable of capturing its variability without the need for segmentation or tracking of individual components. In this regard, one promising holistic representation is to model the motion field as a collection of layers [84]. Another promising representation, which is the focus of this chapter, is to model these processes as *dynamic textures*, i.e., realizations of an auto-regressive stochastic process with both a spatial and temporal component [34, 77]. Like many other recent advances in vision, the success of these methods derives from the adoption of representations based on generative probabilistic models that can be learned from collections of training examples.

Various representations of a video sequence as a spatio-temporal texture have been proposed in the vision literature over the last decade. Earlier efforts were aimed at the extraction of features that capture both the spatial appearance of a texture and the associated motion flow field. For example, [67] represents temporal textures by the first and second order statistics of the normal flow of the video. Subsequently, various authors proposed to model a temporal texture as a generative process, resulting in representations that can be used for both synthesis and recognition.

Fig. 12.1 Examples of visual processes that are challenging for traditional spatio-temporal representations: fire, smoke, the flow of a river stream, or the motion of an ensemble of objects, e.g., a flock of birds, a bee colony, a school of fish, the traffic on a highway, or the flow of a crowd

For example, [3] uses a multi-resolution analysis tree method, which represents a temporal texture as the hierarchical multi-scale transform associated with a 3D wavelet, while in [77], a spatio-temporal autoregressive (STAR) representation models the interaction of pixels within a local neighborhood over both space and time. By relying on spatio-temporally localized image features these representations are incapable of abstracting the video into a pair of holistic appearance and motion components.

This problem is addressed by the dynamic texture (DT) model of [34], an autoregressive random process (specifically, a linear dynamical system (LDS)) that includes a hidden state variable that captures the motion flow, and an observation variable that determines the appearance component, conditioned on the state variable. Both the hidden state vector and the observation vector are representative of the entire image, enabling a holistic characterization of the motion for the entire sequence.

12.2.1 Dynamic Texture Model

A dynamic texture [34] (DT) is a generative model for both the appearance and the dynamics of video sequences. It can be thought of as an extension of the hidden Markov models commonly used in speech recognition, and is the model that underlies the Kalman filter frequently employed in control systems. The model consists of a random process containing an *observation variable* y_t, which encodes the appearance component (vectorized video frame at time t), and a *hidden state variable* x_t, which encodes the dynamics (evolution of the video over time). The appearance component is drawn at each time instant, conditionally on the current hidden state. The state and observation variables are related through the *linear dynamical system* (LDS) defined by

$$x_t = Ax_{t-1} + v_t, \tag{12.1}$$

$$y_t = Cx_t + w_t + \bar{y}, \tag{12.2}$$

where $x_t \in \mathbb{R}^n$ and $y_t \in \mathbb{R}^m$ are real vectors (typically $n \ll m$). The matrix $A \in \mathbb{R}^{n \times n}$ is a *state transition matrix*, which encodes the dynamics or evolution of the hidden state variable (i.e., the motion of the video), and the matrix $C \in \mathbb{R}^{m \times n}$ is an *observation matrix*, which encodes the appearance component of the video sequence. The vector $\bar{y} \in \mathbb{R}^m$ is the mean of the dynamic texture (i.e., the mean video frame). v_t is a *driving noise process*, which injects randomness into the hidden state, and is distributed as an n-dimensional multivariate Gaussian with zero mean and covariance matrix $Q \in \mathbb{R}^{n \times n}$, i.e., $v_t \sim \mathcal{N}(0, Q)$. w_t is the *observation noise*, which models the noise in the pixel observations, and is distributed as an m-dimensional multivariate Gaussian with zero mean and covariance $R \in \mathbb{R}^{m \times m}$. Typically, it is assumed the observation noise is independent and identically distributed (i.i.d.)

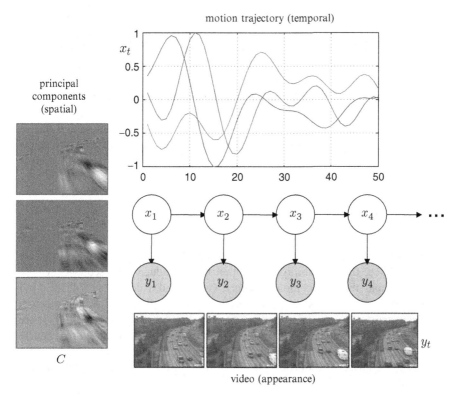

Fig. 12.2 Example of the dynamic texture model: (*bottom*) the video frames from a traffic sequence; (*left*) the first three principal components; (*top*) the hidden-state space trajectory of the corresponding coefficients. (*middle*) The graphical model of the DT. x_t and y_t are the hidden state and observed video frame at time t

between the pixels, and hence $R = rI_m$ is a scaled identity matrix. Finally, the *initial condition* is specified as $x_1 \sim \mathcal{N}(x_0, Q)$, where $x_0 \in \mathbb{R}^n$ is a fixed vector.[1] The dynamic texture is specified by the parameters $\Theta = \{A, Q, C, R, x_0, \bar{y}\}$, and can be represented by the graphical model of Fig. 12.2.

When C is orthonormal, the columns of C can be interpreted as the *principal components* of the sequence of video frames. In this case, the video frame at time t, y_t, is a linear combination of these principal components, and the corresponding weights (i.e., the PCA coefficients) are given by the hidden state vector x_t. These PCA coefficients evolve according to a linear Gauss-Markov process, given by (12.1). Hence, the dynamic texture can be interpreted as a time-varying PCA representation of video. Figure 12.2 shows an example of a traffic sequence, its first three principal components, and the corresponding hidden-state coefficients.

---

[1] Here we focus on the case where the initial state x_0 is fixed. More generally, the initial state could be distributed as a Gaussian, $x_1 \sim \mathcal{N}(\mu, S)$

An alternative interpretation considers a single pixel as it evolves over time. Each dimension of the state vector x_t defines a one-dimensional temporal trajectory over time, as shown in Fig. 12.2 (top). The evolution of a pixel over time is then a weighted sum of these trajectories, according to the weighting coefficients in the corresponding row of C. This is analogous to the discrete Fourier transform, where a 1-D signal is represented as a weighted sum of complex exponentials but, for the DT, the trajectories are not necessarily orthogonal (and in fact random processes). This interpretation illustrates the ability of the DT to model a given motion at different intensity levels (e.g., cars moving from shade into sunlight) by simply scaling the rows of C.

12.2.2 Inference

Since the noise processes are Gaussians and the states and observations are linearly related, the probability distribution of the hidden-state sequence $x_{1:\tau} = (x_1, \ldots, x_\tau)$ and the observation sequence (vectorized video) $y_{1:\tau} = (y_1, \ldots, y_\tau)$ are also multivariate Gaussian [47]. Hence, by the conditional Gaussian theorem [47], any marginal, e.g., $p(x_t)$, or conditional distributions, e.g., $p(x_t|y_{1:t})$, are Gaussian. The conditional distributions can be computed efficiently with recourse to the Kalman and Kalman smoothing filters. The Kalman filter [39, 47] estimates the mean and covariance of the state x_t of an LDS, conditioned on only the past observations,

$$\hat{x}_{t|t-1} = \mathbb{E}(x_t|y_{1:t-1}), \qquad\qquad \hat{V}_{t|t-1} = \mathrm{cov}(x_t|y_{1:t-1}), \qquad (12.3)$$

$$\hat{x}_{t|t} = \mathbb{E}(x_t|y_{1:t}), \qquad\qquad \hat{V}_{t|t} = \mathrm{cov}(x_t|y_{1:t}), \qquad (12.4)$$

where (12.3) is the one-step-ahead prediction using only previous observations $y_{1:t-1}$, and (12.4) the corrected estimate after inclusion of the current observation y_t. This calculation is implemented with a set of forward recursive equations. Likewise, the Kalman smoothing filter [39, 75] estimates the state x_t, conditioned on the *entire* observation sequence $y_{1:\tau}$,

$$\hat{x}_{t|\tau} = \mathbb{E}(x_t|y_{1:\tau}), \quad \hat{V}_{t|\tau} = \mathrm{cov}(x_t|y_{1:\tau}), \quad \hat{V}_{t,t-1|\tau} = \mathrm{cov}(x_t, x_{t-1}|y_{1:\tau}), \quad (12.5)$$

using a set of forward-backward recursive equations. This results in a refinement of the state estimates of the Kalman filter, using all the observed data.

Finally, the data log-likelihood can be computed efficiently from the Kalman filter using the "innovations" form [39, 47, 75],

$$\log p(y_{1:\tau}) = \sum_{t=1}^{\tau} \log p(y_t|y_{1:t-1}) = \sum_{t=1}^{\tau} \log \mathcal{N}(y_t|C\hat{x}_{t|t-1} + \bar{y}, R + C\hat{V}_{t|t-1}C^T)$$

$$(12.6)$$

where $\mathcal{N}(x|\mu,\Sigma)$ is a multivariate Gaussian density function with mean μ and covariance Σ, and $\hat{x}_{t|t-1}$ and $\hat{V}_{t|t-1}$ are calculated with the Kalman filter, as in (12.3).

12.2.3 Parameter Estimation

A number of methods are available to learn the parameters of the dynamic texture from a training video sequence $y_{1:\tau} = (y_1,\ldots,y_\tau)$, including maximum-likelihood methods (e.g., expectation-maximization [75]), non-iterative subspace methods (e.g., N4SID [65], CCA [5, 50]) or a suboptimal, but computationally efficient, "greedy" least-squares procedure [34].

The expectation-maximization (EM) algorithm [31] is an iterative algorithm to determine maximum-likelihood parameter estimates [47],

$$\Theta^* = \underset{\Theta}{\operatorname{argmax}} \log p(y_{1:\tau};\Theta), \tag{12.7}$$

when the model contains hidden-state variables. Each EM iteration alternates between estimating the hidden-states from the current parameters (E-step) and updating the parameters given the estimate of the hidden-states (M-step),

$$\text{E} - \text{Step}: \mathcal{Q}(\Theta;\hat{\Theta}) = \mathbb{E}_{x_{1:\tau}|y_{1:\tau};\hat{\Theta}}[\log p(x_{1:\tau},y_{1:\tau};\Theta)] \tag{12.8}$$

$$\text{M} - \text{Step}: \hat{\Theta}^* = \underset{\Theta}{\operatorname{argmax}} \mathcal{Q}(\Theta;\hat{\Theta}), \tag{12.9}$$

where $\hat{\Theta}$ is the current set of parameter estimates. For a DT, the EM algorithm has the following form [75]: the E-step computes the statistics of the hidden-state variables, conditioned on the training video, using the Kalman smoothing filter; the M-step then updates the parameters from these aggregated statistics.

Subspace methods, such as N4SID [65] and CCA [5,50], attempt to first recover the hidden states (or its state space) directly, by regressing between windows of past and future observation sequences. Under some conditions,[2] the non-iterative CCA subspace method is also a maximum-likelihood estimator [5]. However, this (and several other) sub-space methods require computing the singular value decomposition (SVD) of a $(dm) \times (dm)$ matrix, where d is the length of the sequence window. The large dimensionality of the video frame frequently makes this operation infeasible for DT models learned from video. For CCA, typically $8 \le d \le 15$, and the SVD is infeasible for large m.

---

[2]One of these conditions is that the parameter n must be set to the true state-space dimension! Another condition is that the state noise and observation noise are realized from the same white noise process.

Algorithm 2 Greedy least-squares for dynamic textures [34]

Input: observed sequence $y_{1:\tau}$.

Compute sample mean: $\bar{y} = \frac{1}{\tau}\sum_{t=1}^{\tau} y_t$.

Subtract mean: $\tilde{y}_t = y_t - \bar{y}, \forall t, \qquad \tilde{Y}_{1:\tau} = [\tilde{y}_1 \cdots \tilde{y}_\tau]$.

Compute SVD: $\tilde{Y}_{1:\tau} = USV^T$.

Estimate observation matrix: $C = [u_1 \cdots u_n]$.

Estimate state-space variables: $\hat{X}_{1:\tau} = [\hat{x}_1 \cdots \hat{x}_\tau] = C^T \tilde{Y}_{1:\tau}$.

Estimate remaining parameters:

$$
\begin{aligned}
A &= \hat{X}_{2:\tau}(\hat{X}_{1:\tau-1})^\dagger, & x_0 &= \hat{x}_1, \\
\hat{V}_{1:\tau-1} &= \hat{X}_{2:\tau} - A\hat{X}_{1:\tau-1}, & Q &= \frac{1}{\tau-1}\hat{V}_{1:\tau-1}(\hat{V}_{1:\tau-1})^T, \\
\hat{W}_{1:\tau} &= \tilde{Y}_{1:\tau} - C\hat{X}_{1:\tau}, & R &= \frac{1}{\tau}\hat{W}_{1:\tau}\hat{W}_{1:\tau}^T, \quad r = \frac{1}{m}\mathrm{tr}(R).
\end{aligned}
\tag{12.10}
$$

Output: $\Theta = \{A, Q, C, R, x_0, \bar{y}\}$.

One popular alternative to the EM or sub-space methods is the computationally-efficient method proposed in [34], which is summarized in Algorithm 2. This method learns the spatial and temporal parameters of the model separately, by exploiting the interpretation of C as a principal component matrix. Given a mean-subtracted sequence $\tilde{Y}_{1:\tau} = [\tilde{y}_1 \cdots \tilde{y}_\tau]$ (in matrix form), principal component analysis (PCA) is performed by applying the SVD to $\tilde{Y}$. The columns of C are then estimated as the first n principal components, and the hidden-states $\hat{X}$ by the corresponding PCA coefficients. The transition matrix A is calculated via least-squares regression between neighboring states, and the noise parameters Q and R are calculated from the residual and reconstruction errors. In summary, the procedure uses several least-squares steps successively (e.g. PCA and pseudo-inverse) to greedily obtain the parameter estimates.

Although suboptimal in both the maximum-likelihood and least-squares sense, this greedy least-squares procedure has been shown to produce good estimates of DT parameters in various applications, including video synthesis [34] and recognition [11, 73]. It was shown in [21] that this procedure can be viewed as a single-iteration approximation of the EM algorithm, where C is approximated by the PCA basis of the observations, and the conditional expectations of the E-step are approximated by the PCA coefficients. In practice, the greedy least-squares estimate serves as a good initialization for the EM algorithm, which then typically converges in a few iterations.

12.2.4 Model Extensions and Applications

The original DT model has been extended in various ways in the literature. Some of these extensions aim to improve the video synthesis capabilities of the model, e.g., by modifying the learning algorithm [76], adding a closed-loop to the model [89], using higher-order SVD [27], or adopting a multi-scale auto-regressive process [36]. Other extensions aim to increase the representational power of the DT model, e.g., by introducing non-linear observation functions [12, 55], or phase-based non-parametric models [40]. Doretto and Soatto [33] proposes a dynamic shape and appearance model for image sequences, which is a generalization of the DT. Another line of research aims to model multiple dynamic textures simultaneously, either as a dynamic texture mixture [13], for modeling collections of video samples, or as a layered dynamic texture [15, 16], for modeling a single video as a composition of several distinct DT regions. Finally, [20, 71] propose methods for grouping DTs into similar clusters.

The DT and its extensions have a variety of computer vision applications, including video texture synthesis [34, 40, 89], video clustering [13], image and video registration [38, 70], motion segmentation [13, 15, 26, 35, 41, 79, 81], lip synthesis and classification [8], human activity and gait recognition [6, 22], background subtraction [21, 64, 91], motion saliency [57], anomaly detection [58], and motion classification [11, 12, 20, 40, 71, 73, 80, 83, 86]. The wide variety of applications highlight both the modeling capabilities of the DT, and the robustness of the underlying probabilistic framework.

An alternative holistic representation of motion is based on high-level models of optical flow. One line of work is inspired by fluid dynamics, simulating particles as moving according to the mean optical flow field (average flow over all video frames). Ali and Shah [1] discovers coherent structures in the flow field, using Lagrangian particle dynamics to examine how particle clouds mix and move. Similarly, [62] records the social force between particles to form a force flow image. Local regions in the force flow are then quantized into a bag-of-forces and the descriptor is used for anomaly detection. Finally, [63] proposes "streaklines" that represent all particle trajectories which pass through a particular point. Another line of research is based on direct examination of optical flow patterns. Hu et al.[44] finds clusters in the mean motion flow field using a mean-shift like algorithm, whereas [45] employs hierarchical agglomerative clustering with two distance functions that measure spatial proximity or flow similarity. Yang et al. [88] quantizes the motion flow field in both space and direction, to obtain a bag-of-words representation, where each subword corresponds to the quantized flow in a video patch. Noisy or outlier subwords are removed based on their conditional entropy, and patterns of motion are learned via a diffusion map embedding. Finally, [74] estimates Gaussian mixture models (GMMs) from the optical flow vectors, and motion patterns are found by grouping the GMMs based on thresholding the KL divergence between them.

12.3   Motion Classification with Dynamic Textures

In this section, we present a framework for video and motion classification using the DT representation, which is summarized in Fig. 12.3. After preprocessing, the parameters of a DT model are learned for each video clip. Next, the similarities between DT models are computed with a suitable measure of distance between DTs.

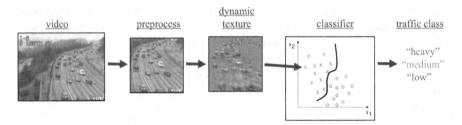

Fig. 12.3 DT classification: video is preprocessed, a DT learned, and classified using a suitable distance function and standard classification techniques

Finally, the distance matrix is used in conjunction with standard classification algorithms (e.g., support vector machines, nearest neighbors, or boosting) to perform motion classification and event detection. The key idea is that, by representing each video by a DT, the high-dimensional video clip is reduced to a compact parametric model, which leads to increased classifier robustness. Several distance functions have been proposed in the literature, including those based on observable subspaces, output trajectories, and probability distributions of the DT, which we present next.

12.3.1 Distances and Kernels Between DT

Because the DT is a generative probabilistic model with hidden-states, various distance functions can be defined by comparing different properties of the model. One family of distances [73] is based on the observable subspace of the DT, i.e., the space of all possible output sequences when ignoring the noise terms. Another distance function [83] focuses on the initial condition, by comparing trajectories of DT outputs. Finally, a probabilistic kernel function [11] can be defined between the probability distributions of the output sequence or the hidden-state sequence of the DT. In the remainder of this section, we present several distance functions between two DT, Θ_1 and Θ_2, with parameters $\{A_1, C_1, Q_1, R_1, x_{01}, \bar{y}_1\}$ and $\{A_2, C_2, Q_2, R_2, x_{02}, \bar{y}_2\}$, respectively.

12.3.1.1 Principal Angle-Based Distances

One method of calculating the distance between two DTs is to use the subspace of all possible noiseless observations. Consider a single DT of parameters $\Theta = \{A, C, Q, R, x_0, \bar{y}\}$. Ignoring the noise terms v_t and w_t, the (vectorized) video sequence generated by the DT is

$$
\begin{bmatrix} y_1 \\ y_2 \\ y_3 \\ \vdots \end{bmatrix} = \begin{bmatrix} Cx_0 \\ CAx_0 \\ CA^2x_0 \\ \vdots \end{bmatrix} = \begin{bmatrix} C \\ CA \\ CA^2 \\ \vdots \end{bmatrix} x_0 = \mathscr{O}x_0, \qquad \text{where } \mathscr{O} = \begin{bmatrix} C \\ CA \\ CA^2 \\ \vdots \end{bmatrix}. \tag{12.11}
$$

Hence, in the noiseless case, the output of the DT belongs to the subspace spanned by the n columns of $\mathscr{O}$. In other words, $\mathscr{O}$, which is known as the extended observability matrix, characterizes all possible noiseless output sequences of the DT (or all possible mean output sequences of the DT in the noisy case). Hence, two DTs, Θ_1 and Θ_2, can be compared using the distance between their observability subspaces, $\mathscr{O}_1$ and $\mathscr{O}_2$. Several distances between subspaces are based on principal angles, which are a greedy selection of principal directions, $\mathscr{O}_1x_i$ and $\mathscr{O}_2y_i$, that minimize the angle between subspaces, while being orthogonal to the previous selected principal directions,

$$
\cos(\theta_i) = \max_{\substack{x_i,y_i\in\mathbb{R}^n, \\ \mathscr{O}_1x_i\perp\mathscr{O}_1x_j, \ \mathscr{O}_2y_i\perp\mathscr{O}_2y_j, \\ j\in\{1,\cdots,i-1\}}} \frac{|x_i^T \mathscr{O}_1^T \mathscr{O}_2 y_i|}{\|\mathscr{O}_1x_i\|\,\|\mathscr{O}_2y_i\|}, \quad i=1,\cdots,n. \tag{12.12}
$$

The principal angles can be efficiently computed by solving a generalized eigenvalue problem involving $2n \times 2n$ matrices [24].

Several distances between DTs can be defined from the principal angles, including the Martin distance [61,73],

$$
d_M(\Theta_1,\Theta_2)^2 = -\log\prod_{i=1}^{n}\cos^2\theta_i, \tag{12.13}
$$

the Finsler distance [8],

$$
d_F(\Theta_1,\Theta_2) = \max_{i\in\{1,\cdots,n\}}\theta_i, \tag{12.14}
$$

and the Frobenius distance [24],

$$
d_f(\Theta_1,\Theta_2) = 2\sum_{i=1}^{n}\sin^2\theta_i. \tag{12.15}
$$

Note that principal-angle-based distances only consider the A and C parameters of the DTs. Hence, these distances do not exploit the stochastic nature or the initial condition of the DT model. This may hinder the effectiveness of the distances when the stochastic variations of the DT are discriminant, or the initial conditions important.

12.3.1.2 Binet-Cauchy Kernel

Another possibility for comparing DTs is through their output sequences. Vishwanathan et al. [83] proposes a family of Binet-Cauchy kernel functions, based on the expectation of the weighted inner product between two DT outputs,

$$k_{BC}(\Theta_1,\Theta_2) = \mathbb{E}_{v,w}\left[\sum_{t=0}^{\infty} e^{-\lambda t} y_t^T W y_t'\right] \tag{12.16}$$

where W is a user-defined weight matrix, $\lambda \geq 0$ a discounting factor, and y_t, y_t' the observation variables of Θ_1, Θ_2, respectively. Several kernels can be derived from (12.16) using different noise assumptions. Under the assumption that the two DTs evolve with the *same noise realization*, the kernel is

$$k_{BC}(\Theta_1,\Theta_2) = x_{01}^T M x_{02} + \frac{1}{1-e^{-\lambda}}\text{tr}[QM + WR], \tag{12.17}$$

where M satisfies the Sylvester equation $M = e^{-\lambda}A_1^T M A_2 + C_1^T W C_2$. When the two DTs evolve with *independent noise realizations*, this kernel simplifies to

$$k_{BC}(\Theta_1,\Theta_2) = x_{01}^T M x_{02}. \tag{12.18}$$

Finally, the kernel can be made independent of the initial conditions by taking the expectation over x_{01} and x_{02}. This leads to

$$k_{BC}(\Theta_1,\Theta_2) = \text{tr}[\Sigma M] + \frac{1}{1-e^{-\lambda}}\text{tr}[QM + WR] \tag{12.19}$$

where $\Sigma = \mathbb{E}_{x_{01},x_{02}'}[x_{01}x_{02}^T]$ is the correlation matrix between the two initial conditions (or covariance matrix when x_{01} and x_{02} are zero mean). It is interesting to contrast the Martin distance and the Binet-Cauchy kernel. On the one hand, the Martin distance compares the subspace of all possible noiseless outputs of the two DT. On the other, the Binet-Cauchy kernel focuses on specific realizations of the output when given particular initial conditions, or distributions thereof.

12.3.1.3 Kullback-Leibler Kernel

Neither the Martin distance nor the Binet-Cauchy kernel utilize the full probabilistic description provided by the DT model. The Martin distance ignores the noise terms altogether, the Binet-Cauchy kernel assumes that the two DTs share the same noise processes. Since the DT is a generative probabilistic model, two DT can also be compared as probability distributions. Chan and Vasconcelos [11] proposes a probabilistic kernel function between DTs based on the Kullback-Leibler (KL) divergence [28] between the associated distributions. The KL divergence rate between two distributions of a time-series $y_{1:\tau}$, $p(y_{1:\tau})$ and $q(y_{1:\tau})$, is

$$D(p(y_{1:\tau})\|q(y_{1:\tau})) = \frac{1}{\tau}\int p(y_{1:\tau})\log\frac{p(y_{1:\tau})}{q(y_{1:\tau})}dy_{1:\tau}. \qquad (12.20)$$

The DT model provides two probability distributions, based on either the image observation variables or the hidden-state variables. This allows the derivation of two probabilistic kernels, which can discriminate either the appearance or the motion flow of the dynamic texture.

The KL divergence between image distributions $p(y_{1:\tau})$ and $q(y_{1:\tau})$ of DTs Θ_1 and Θ_2 can be rewritten as a sum of conditional KL divergence terms

$$D_Y(\Theta_1\|\Theta_2) = \frac{1}{\tau}\sum_{t=1}^{\tau} D_Y(p(y_t|y_{1:t-1})\|q(y_t|y_{1:t-1})), \qquad (12.21)$$

where

$$D_Y(p(y_t|y_{1:t-1})\|q(y_t|y_{1:t-1})) = \int p(y_{1:t-1})\int p(y_t|y_{1:t-1})\log\frac{p(y_t|y_{1:t-1})}{q(y_t|y_{1:t-1})}dy_t dy_{1:t-1}. \qquad (12.22)$$

In (12.22), $p(y_t|y_{1:t-1})$ and $q(y_t|y_{1:t-1})$ are both conditional Gaussians, as in (12.6). Hence, the inner integral yields the standard formula for the KL divergence between Gaussians, whereas the outer integral takes the expectation over $y_{1:t-1}$. This expectation can be computed efficiently via sensitivity analysis of the Kalman filter (see [9] for more details). Since it is based on distributions of image pixels, the KL divergence in image space tends to favor iconic pixel matches. It has best performance when the goal is to differentiate between DTs of different visual appearance (e.g., a flock of birds from a school of fish in Fig. 12.1). Under this measure, two sequences of distinct textures subject to similar motion are identified as distinct.

The KL divergence can also be computed between the distributions of the hidden-state sequences. However, since each DT uses a different hidden-state space (defined by C_1 and C_2), the state KL divergence cannot be computed directly from the state distributions. One solution is to project one state space onto the other by applying a sequence of two transformations: (1) from the original state space into image space, and (2) from image space into the target state space, yielding the new parameters,

$$\hat{A}_1 = FA_1 F^{-1}, \ \hat{Q}_1 = FQ_1 F^T, \ \hat{x}_{01} = Fx_{01}, \ \hat{S}_1 = FS_1 F^T, \qquad (12.23)$$

where $F = C_2^T C_1$ (or $F = (C_2^T C_2)^{-1}C_2^T C_1$ if C_2 is not orthogonal). The KL divergence between state spaces can now be computed with the transformed parameters

$$D_X(\Theta_1 \| \Theta_2) = \frac{1}{2} \log \frac{|Q_2|}{|\hat{Q}_1|} + \frac{1}{2} \mathrm{tr}(Q_2^{-1} \hat{Q}_1) - \frac{n}{2} + \frac{1}{2\tau} \| \hat{x}_{01} - x_{02} \|_{Q_2}^2$$
$$+ \frac{1}{2} \mathrm{tr}\left(\bar{A}^T Q_2^{-1} \bar{A} \frac{1}{\tau} \sum_{t=1}^{\tau-1} P_t \right)$$

(12.24)

where $\bar{A} = \hat{A}_1 - A_2$, and $P_t = \mathbb{E}_{\Theta_1}[x_t x_t^T]$ is the second-moment of the state variable x_t for DT Θ_1. The state KL divergence is useful in discriminating motion flow when the appearances are similar, e.g., differentiating levels of traffic on a highway, or detecting outliers and unusual events (e.g., cars speeding or committing other traffic violations). Finally, given the image or state-space KL divergences, the KL kernel is defined as a radial basis function of the symmetric KL divergence,

$$k_{KL}(\Theta_p, \Theta_q) = e^{-\gamma(D(\Theta_p \| \Theta_q) + D(\Theta_q \| \Theta_p))},$$

(12.25)

where γ is the RBF bandwidth parameter. Kernels derived from image and state-space can then be combined with standard multiple kernel learning techniques [2, 49].

12.3.2 Dynamic Textures Classification

Given an appropriate distance function, DTs can be classified by adopting any of the standard classification algorithms. One example is the simple nearest neighbor classifier with the Martin distance, as originally proposed in [73]. DT kernel functions can also be combined with support vector machines (SVM) [78] to obtain a large-margin discriminative classifier [11, 83]. In these cases, the "kernel trick" embeds the DT parameters into a high-dimensional non-linear feature-space, which is more suitable for comparing DTs than just computing the Euclidean distance between the parameters. When using a probabilistic kernel function [11], the DT classification framework inherits the advantages of both the probabilistic representation (e.g., support for complex statistical inference and regularization) and the discriminative learning framework (e.g., good generalization guarantees of large-margin classifiers). Finally, another type of large-margin classifier, AdaBoost, can be applied to dynamic textures by defining a suitable weak classifier for DT. Vidal and Favaro [80] proposes a DT weak classifier of the form, $h(\Theta) = \mathrm{sign}(\sum_{t=0}^{\infty} \lambda^t h_t^T y_t - \phi)$, where y_t is the observation variable of DT Θ, h_t is the prototype output for the weak classifier, and λ is a discounting factor. The prototype h_t takes the form of a noiseless DT output, $h_t = \tilde{C} \tilde{A}^t \tilde{x}_0 + \bar{y}$, which results in a closed-form solution to the infinite sum similar to [83].

Fig. 12.4 Frames from a highway traffic video (Courtesy of Washington State Department of Transportation)

12.3.3   Examples: Traffic Classification and Crowd Event Detection

We first apply DT classification to the task of highway traffic monitoring in scenes such as those of Fig. 12.4. The holistic motion representation of the DT captures the variability of the entire motion field, without the need for segmenting or tracking the individual cars. Since only the motion is modeled, the framework is inherently invariant to lighting changes. In addition, because the model does not rely on a dense motion field derived from pixel similarities (e.g., correlation or optical flow), it is robust to occlusion, blurring, image resolution, and other image transformations.

The dataset contains videos of light, medium, and heavy traffic, taken from a stationary traffic camera overlooking a highway in Seattle [85] during daytime. A variety of traffic patterns and weather conditions (e.g., overcast, raining, sunny, rain drops on the camera lens) are present. An example video is shown in Fig. 12.4. Each video clip (50 frames at 10 fps) was converted to grayscale, and reduced and clipped to a 48×48 pixel window. In order to reduce the impact of different lighting conditions, each video clip was normalized to have zero image mean and overall pixel variance of one. Finally, a DT was learned for each video clip by estimating the parameters using the sub-optimal least-squares method [34] with $n = 15$.

Because the video was obtained from a fixed camera facing the same stretch of road, the motion is always in the same direction and confined to the same area. Hence, the state KL divergence is an appropriate distance to discriminate between flow patterns. To illustrate the effectiveness of this distance, we retrieved the most similar videos to a query clip, according to the state KL divergence ($\tau = 250$). Figure 12.5a–c shows the retrieval results for several queries involving light, medium, and heavy traffic. Note that the retrieval operation is robust to variable lighting conditions due to overcast or sunny periods. Nighttime sequences outside of the original database were also used as queries, with the retrieval results presented in Fig. 12.5d–f. Even in this extreme scenario for lighting variation, the retrieved traffic scenes have similar motion. In addition, the framework is robust to

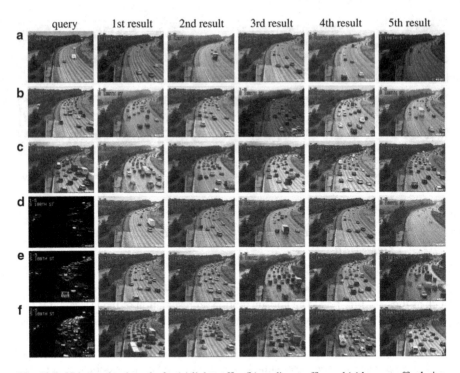

Fig. 12.5 Video retrieval results for (**a**) light traffic, (**b**) medium traffic, and (**c**) heavy traffic during the day. Retrieval using a night sequence outside the original database for (**d**) light, (**e**) medium, and (**f**) heavy traffic shows robustness to lighting conditions

occlusion and blurring due to raindrops on the camera lens, as seen in the third and fifth results of Fig. 12.5e.

An SVM classifier based on the KL kernel was then trained to label video as containing light, medium, or heavy traffic. Figure 12.6 shows several classification examples under different lighting conditions: (a) sunny lighting, including strong shadows; and (b) overcast lighting, including raindrops on the camera lens. The overall classification accuracy was 94.5 % for daytime video. Several night time videos, outside the original database, were also classified. Even though the classifier was trained with video taken during the day, it is still able to correctly label the nighttime sequences, including the event of a traffic jam (heavy traffic) at night. This is particularly interesting because, at night, cars appear as a combination of headlights and a pair of tail lights. These results provide evidence that the DT model is indeed extracting relevant motion information, and that the proposed classification framework is capable of very robust discrimination between classes of motion.

A final experiment was conducted to characterize the variation of highway traffic patterns during the day. The SVM classifier was trained from 61 sequences spanning 4 h of the first day, and tested on 193 sequences spanning 15 h of the following day. The ground truth classification and the outputs of the state-KL SVM are shown in

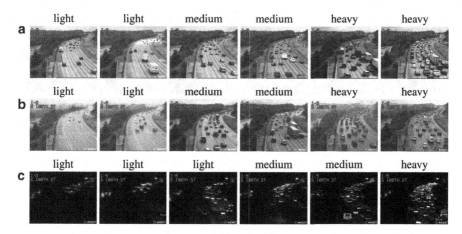

Fig. 12.6 Classification of traffic congestion under variable lighting conditions: (**a**) sunny, (**b**) overcast, and (**c**) nighttime

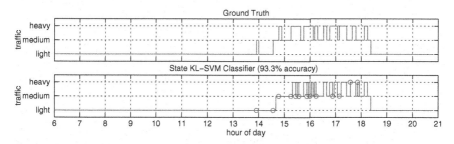

Fig. 12.7 Classification of traffic congestion in sequences spanning 15 h: (*top* to *bottom*) ground truth; classification using state KL SVM. Errors are highlighted with circles

Fig. 12.7. The increased traffic due to the rush hour can be seen between 2:30 PM and 6:30 PM. Further details on these traffic classification experiments can be found in [9, 11].

Next we consider the classification of crowd events from the PETS 2009 dataset. The dataset consists of videos of six classes of crowd behavior (walking, running, merging, splitting, evacuation, and dispersion), taken simultaneously from four viewpoints. DT classifiers based on NN and Martin distance and state KL divergence were learned for each behavior and viewpoint. A probability score for classification was obtained by combining the decisions of the Martin distance and state KL divergence NN classifiers (over the four views) using a voting scheme. Figure 12.8 shows several examples of event detection on video 14-33. In the beginning of the video, people walk towards the center of the frame, and the "walking" and "merging" events are detected. Next, the people form a group in the center of the

Fig. 12.8 (Color online) Examples of event recognition on PETS-2009 (video 14-33). *Light gray text (green in color version)* indicates that the class was detected. Detection probabilities are given in parenthesis

frame, and no events are detected since the people are not moving. Finally, when the people run away from the center, the "running", "evacuation", and "dispersion" events are detected. More details on these experiments can be found in [19].

12.4   Crowd Counting from Global Properties

In this section, we present a formulation of the pedestrian counting problem based on the analysis of global scene properties. Pedestrian counting is a canonical example of problems that vision technology addresses with object-centric approaches. The mainstream approach to this problem involves detecting the people in the scene [51, 54, 82, 87, 90], tracking them over time [7, 52, 68], and counting the number of tracks. However, it is also possible to accurately estimate the size of a crowd from *global low-level features*, without the need for object detection and tracking. Feature-based methods were first applied to subway platform monitoring, through a combination of: (1) background subtraction; (2) extraction of various features of foreground pixels, such as total area [23, 30, 66], edge count [23, 30, 72], or texture [59]; and (3) estimation of crowd density or size with a regression function, e.g., linear [30, 66], piece-wise linear [72], or neural networks [23, 59]. In recent years, feature-based regression has also been applied to outdoor scenes. For example, [48] applies neural networks to the histograms of foreground segment areas and edge orientations, while [32] estimates the number of people in each foreground segment by matching its shape to a database containing the silhouettes of possible people configurations. The approach of [18] first segments the crowd into sub-regions of interest (e.g., groups of people moving in different directions), and then maps features extracted from each segment into a count, using Gaussian process (GP) regression. These works have shown that accurate crowd counts are possible without people detection, even when the crowd is *sizable and inhomogeneous*, e.g., has *sub-components with different dynamics,* as illustrated in Fig. 12.9. In the remainder of the section we discuss this crowd counting framework in more detail.

Fig. 12.9 Examples of a low-resolution scene containing a sizable crowd with inhomogeneous dynamics, due to pedestrian motion in different directions

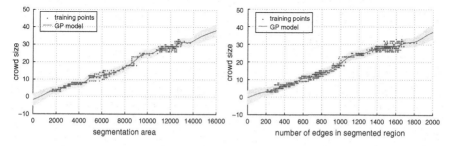

Fig. 12.10 Correspondence between crowd size and two simple features: (*left*) segmentation area, and (*right*) the number of edge pixels in the segmented region. The GP regression function is also plotted with two standard deviations error bars (*gray area*)

12.4.1   Crowd Counting Framework

Figure 12.9 shows examples of a crowded scene on a pedestrian walkway. The goal is to estimate the number of people moving in each direction. Given a segmentation into the two motion sub-components, the key insight is that it is possible to estimate crowd size from low-level features extracted per crowd segment. For example, as shown in Fig. 12.10, features such as the segment area or the number of edge pixels within the segmented region are approximately linear in crowd size, assuming proper normalization for scene perspective. An outline of the crowd counting architecture is given in Fig. 12.11. The video is segmented into crowd regions moving in different directions. For each segment, various features are extracted, under a perspective map that weighs each image location by its approximate size in the real scene. Finally, GP regression is used to estimate the number of people per segment.

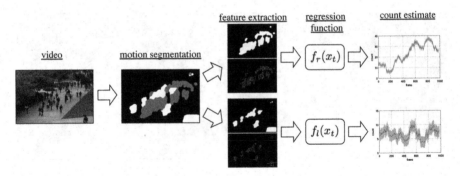

Fig. 12.11 Overview of the crowd counting architecture: the scene is segmented into sub-regions of crowds moving in different directions. Features, normalized to account for perspective, are extracted from each segment. The number of people is finally estimated with a regression function

12.4.1.1 Crowd Segmentation

The first step is to segment the video regions containing the crowd. This can be achieved with either background subtraction, if the goal is to estimate the size of the whole crowd irrespective of direction, or motion segmentation, if the goal is to count how many people move in different directions. In the latter case, a robust motion segmentation algorithm is the *mixture of dynamic textures* [15], which segments motion by clustering spatio-temporal video cubes. This procedure tends to work well, and was used to segment a full hour of video in [13]. The resulting segmentations are illustrated in Fig. 12.14.

12.4.1.2 Perspective Normalization and Feature Extraction

Before extracting features from the video segments, it is important to consider the effects of perspective. Because objects closer to the camera appear larger, any feature extracted from a foreground object will account for a smaller portion of the object than one extracted from an object farther away. This makes it important to normalize the features for perspective. One possibility is to weight each pixel according to a perspective normalization map. The pixel weight is based on the expected depth of the object which generated the pixel, with larger weights given to far objects [18, 48]. Using this weighting counteracts the foreshortening problem, allowing for more consistent features across scene depth.

Ideally, features such as segmentation area or number of edges should vary linearly with the number of people in the scene [30, 66], given proper perspective normalization. Figure 12.10 plots the segmentation area versus the crowd size. While the overall trend is indeed linear, there exist local non-linearities that arise from a variety of factors, including occlusion, segmentation errors, and pedestrian configuration (e.g., spacing within a segment). To model these non-linearities, additional features are extracted from each segment. These features can be divided into three categories: segment features (e.g., area [30, 66], perimeter length [18], perimeter edge orientation histogram [18], perimeter-area ratio [18], the number of connected components [17]); internal edge features (e.g., total edge length [30, 66], edge orientation histogram [48], Minkowski dimension [60]); and texture features (e.g., entropy, homogeneity, and energy of the gray-level co-occurrence matrix [59]). More details on the various features are available in the references given above.

12.4.1.3   Count Regression

Crowd size estimates are obtained directly from the feature space using a counting regression function. There are many possible choices of regression functions, e.g., linear regression [30,66], piece-wise linear regression [72], or neural networks [23,59]. The GP regression [69] of [18] has two advantages. First, because the GP is a Bayesian regression model, the learned mappings are robust to the number of training examples, due to the inherent marginalization over all possible parameters. This reduces the number of hand-annotated training examples needed to learn the counting function. Second, GP regression can implement non-linear counting functions by specification of a non-linear kernel. Non-linear regression becomes increasingly more important as the size and density of the crowd increase, due to more frequent partial occlusions between objects.

The GP defines a distribution over functions, which is "pinned down" at the training points. The classes of functions that the GP can model is dependent on the kernel function used. For example, Bayesian linear regression uses a linear kernel,

$$k(x_p, x_q) = \alpha_1 (x_p^T x_q + 1) + \alpha_2 \delta(x_p, x_q), \qquad (12.26)$$

where $\alpha = \{\alpha_1, \alpha_2\}$ are hyperparameters that weight the linear term (first term) and the observation noise (second term), and $\delta(a, b)$ is the delta function (1 when $a = b$, and 0 otherwise). In the context of pedestrian counting, we have noted that the dominant trend of many of the features (e.g., segment area) is linear, as shown in Fig. 12.10. This is complemented by some local non-linearities due to occlusions and segmentation errors. To capture both the dominant trend and the local non-linearities, we combine a linear and an RBF kernel, i.e.,

$$k(x_p, x_q) = \alpha_1 (x_p^T x_q + 1) + \alpha_2^2 e^{-\frac{\|x_p - x_q\|^2}{2\alpha_3^2}} + \alpha_4^2 \delta(x_p, x_q) \qquad (12.27)$$

with hyperparameters $\alpha = \{\alpha_1, \alpha_2, \alpha_3, \alpha_4\}$. The (first) linear term models the overall trend and the (second) RBF term the local non-linearities (with a small length scale α_3), while the third term models observation noise. The kernel hyperparameters α can be learned from the training data by maximizing the marginal likelihood of the model [69]. Figure 12.10 shows an example of GP regression for the segment area feature. Note that the regression function captures both the dominant trend and

Fig. 12.12 (Color online) Ground-truth pedestrian annotations: *dark gray* and *light gray* tracks (*red* and *green* in color version) are people moving away from, and towards the camera. The ROI used in the experiments is highlighted

the local non-linearities. Finally, while the same feature set is used throughout the crowd counting architecture, a different regressor is learned per direction of crowd motion. This is necessary because appearance changes with traveling direction.

12.4.2 Example Counting Results

The counting system was applied to video of a pedestrian walkway at UCSD. The video contains 2,000 frames (200 s) and is annotated with ground-truth counts for two motion classes, "away" from or "towards" the camera. An example annotation is shown in Fig. 12.12.

The counting architecture was trained with 800 frames of the annotated data, using a set of 29 features (area, segment, and texture features) and GP regression with the linear-RBF kernel in (12.27). Crowd size was estimated on the remaining video, by rounding the GP prediction to the nearest integer. Figure 12.13 shows the crowd size estimates as a function of time, for the two crowd directions. Figure 12.14 shows the original image, segmentation, and crowd estimates for several frames in the test set. The estimates track the ground-truth well in most of the test frames. The overestimate of the size of the "away" crowd in frames 180–300 is caused by two bicyclists traveling quickly through the scene, as shown in the second image of Fig. 12.14. The average per-frame absolute error on the test set was $1.621/0.869$ for the away/towards classes, when using all the features. For comparison, the error was $2.037/1.307$ when using only the area features, $1.894/1.172$ when using just segment features, and $1.767/1.122$ when using segment and edge features. This demonstrates the complementary nature of the different feature subsets: the segment features provide a coarse linear estimate, which is refined by the edge and texture features that account for various non-linearities.

We next compare this global feature regression approach to an implementation of counting with two state-of-the-art people detection algorithms: (1) SVM and histogram-of-gradients [29], which we denote "HOG"; and (2) a deformable parts

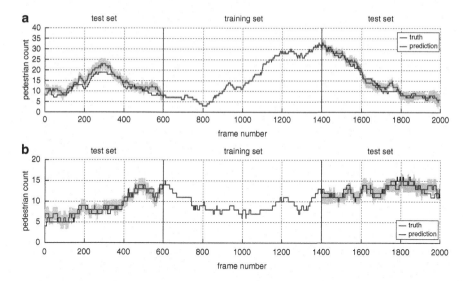

Fig. 12.13 Crowd counting results over both the training and test sets for: (**a**) people moving away and (**b**) people moving towards the camera. The *gray bars* show the two standard-deviations error bars of the GP

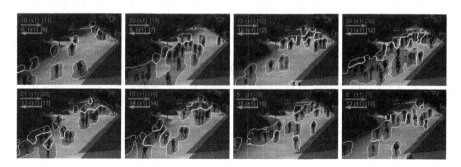

Fig. 12.14 (Color online) Crowd counting results: The *dark gray* and *light* gray segments (*red* and *green* in the color version) are the "away" and "towards" crowds. The estimated crowd count for each segment is in the *top-left*, with the (rounded standard-deviation of the GP) and the (ground-truth). The ROI is also highlighted

model [37], denoted as "DPM". The detectors were provided by the respective authors, and were run on each full-size video frame. A filter was applied to remove detections outside the region of interest or inconsistent with the geometry of the scene. Finally, detection results were filtered by confidence level thresholding. The threshold was selected to minimize the counting error on the training set. Figure 12.15 presents plots of counts for each detection algorithm, as well as the GP regression, when counting all people in the scene (regardless of direction). On the test frames, DPM had a lower average absolute error than HOG (4.02 versus 5.71).

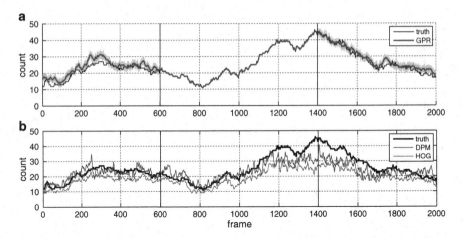

Fig. 12.15 (Color online) Counting results on the whole crowd: (**a**) low-level features and GP; and (**b**) people detection algorithms, HOG [29] and DPM [37]

However, both detectors performed significantly worse than GP regression with low-level features, which had an absolute error rate of 1.95. These results highlight the benefits of regression-based counting for crowded scenes, where significant amounts of occlusion can be quite probematic for people detectors.

Finally, the counting system was tested on the PETS2009 dataset. Figure 12.16 presents results on three regions of interest (R0, R1, and R2) of two videos, 13-57 and 13-59. The average absolute error on video 13-57 was 2.308/1.697/1.072 (for regions R0/R1/R2), and 1.647/0.685/1.282 in video 13-59. Again, the estimated counts track the ground-truth, even when the crowd is large, dense, and contains significant occlusions. More details on the counting architecture and experiments can be found in [17–19].

12.4.3   Extensions

The basic counting via regression framework has been extended in several ways. One extension aims to improve the regression algorithm. While the GP regression has real-valued outputs, the actual counts are restricted to non-negative integers. Chan and Vasconcelos [14] addresses this issue, by developing a Bayesian version of Poisson regression. This yields a GP-like method for regressing to integer counts. Chan and Dong [10] further extends this idea by incorporating a more flexible data likelihood, a Conway-Maxwell Poisson distribution.

Other extensions extract different types of counts from the video. Cong et al. [25] counts the number of people crossing a line-of-interest. After background

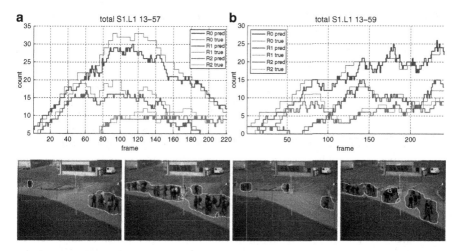

Fig. 12.16 Count results on PETS2009 videos (**a**) `13-57` and (**b**) `13-59`. The total count is predicted on three regions R0 (*large box*), R1 (*medium box*), and R2 (*small box*). The *first row* shows the plots of counts over time. The *second row* presents several example frames with segmentation and count estimates

subtraction, a "flow-mosaic" is generated by adaptively sampling a slice of the video, where the slice thickness is proportional to the flow along the line. Features are extracted from each crowd blob of the mosaic, and the count of each blob crossing the line is estimated with regression. Lempitsky and Zisserman [53] proposes a supervised learning framework for generating a count *density* image, whose integral over a region-of-interest yields its count. In this framework, the linear mapping from local feature vector to the count density of a pixel is learned by minimizing an upper bound on the counting error over all rectangular sub-regions.

12.5   Conclusions and Future Directions

In this chapter, we presented two holistic representations of crowd video. The first representation is based on holistic modeling of motion flow. The video is modeled with dynamic textures, and motion classification is performed by defining appropriate distances and kernels between these models. The effectiveness of this approach to motion classification was demonstrated on a traffic surveillance problem. When compared to previous solutions, motion analysis with these models has several advantages: (1) it does not require segmentation or tracking of objects; (2) it does not require estimation of a dense motion vector field; and (3) it is robust to lighting variation, blurring, occlusion, and low image resolution. The second representation was based on global low-level features and regression models. It does not depend on object detection or feature tracking and was successfully used to estimate the size of inhomogeneous crowds, composed of pedestrians traveling in different directions. Regression-based counting was shown more robust than counting based on people detection, which can be hindered by the complex occlusions and object interactions that occur within large crowds, lighting variations, or low image resolution.

Dynamic texture classification was shown successful for scenes that are texture-like (e.g., moving cars, moving crowds, or video textures). One interesting direction of future research is to apply the DT to problems where the observations are not video textures, but times-series of feature vectors. In this case, a kernelized representation of the DT [12] can be adopted to model non-linear feature trajectories. For example, [22] applies this idea to human action recognition, by extracting histogram-of-optical-flow (HOOF) feature trajectories and learning a kernel DT, based on a histogram kernel. A similar technique could be applied to motion flow. In the area of crowd counting, future research directions include the development of view-invariant features that scale to larger and denser crowds, and better training procedures supporting multiple viewpoints and knowledge transfer. Finally, there is a need for the deployment of crowd counting systems in real world environments, for extended periods of time, from different viewpoints, and analyzing crowd trends over long periods of time. Longer-term modeling of crowds has important applications in outlier event detection and resource management.

Acknowledgements The authors wish to thank the Washington State DOT for the videos of highway traffic [85], Jeffrey Cuenco and Zhang-Sheng John Liang for annotating part of the pedestrian video data, Navneet Dalal and Pedro Felzenszwalb for the people detection algorithms [29, 37], and Piotr Dollar for running these algorithms. This work was supported by NSF CCF-0830535, IIS-0812235, IIS-0534985, NSF IGERT award DGE-0333451, and the Research Grants Council of the Hong Kong Special Administrative Region, China (CityU 110610).

References

1. Ali, S., Shah, M.: A Lagrangian particle dynamics approach for crowd flow segmentation and stability analysis. In: IEEE Conference on Computer Vision and Pattern Recognition, IEEE (2007)
2. Bach, F., Lanckriet, G., Jordan, M.: Multiple kernel learning, conic duality, and the SMO algorithm. In: International Conference on Machine Learning, ACM Press (2004)
3. Bar-Joseph, Z., El-Yaniv, R., Lischinski, D., Werman, M.: Texture mixing and texture movie synthesis using statistical learning. IEEE Trans. Vis. Comput. Graph. **7**(2), 120–135 (2001)
4. Barron, J., Fleet, D., Beauchemin, S.: Performance of optical flow techniques. Int. J. Comput. Vis. **12**, 43–77 (1994)
5. Bauer, D.: Comparing the CCA subspace method to pseudo maximum likelihood methods in the case of no exogenous inputs. J. Time Ser. Anal. **26**, 631–668 (2005)
6. Bissacco, A., Chiuso, A., Ma, Y., Soatto, S.: Recognition of human gaits. In: IEEE Conference on Computer Vision and Pattern Recognition 20, IEEE (2001)
7. Brostow, G.J., Cipolla, R.: Unsupervised Bayesian detection of independent motion in crowds. In: IEEE Conference on Computer Vision and Pattern Recognition, IEEE, vol 1, pp. 594–601 (2006)

8. Cetingul, E., Chaudhry, R., Vidal, R.: A system theoretic approach to synthesis and classification of lip articulation. In: International Workshop on Dynamical Vision, Springer LNCS (2007)
9. Chan, A.B.: Beyond dynamic textures: a family of stochastic dynamical models for video with applications to computer vision. PhD thesis, UCSD (2008)
10. Chan, A.B., Dong, D.: Generalized gaussian process models. In: IEEE Conference on Computer Vision and Pattern Recognition, IEEE (2011)
11. Chan, A.B., Vasconcelos, N.: Probabilistic kernels for the classification of auto-regressive visual processes. In: IEEE Conference on Computer Vision and Pattern Recognition, IEEE, vol. 1, pp. 846–851 (2005)
12. Chan, A.B., Vasconcelos, N.: Classifying video with kernel dynamic textures. In: IEEE Conference on Computer Vision and Pattern Recognition, IEEE (2007)
13. Chan, A.B., Vasconcelos, N.: Modeling, clustering, and segmenting video with mixtures of dynamic textures. IEEE Trans. Pattern Anal. Mach. Intell. 30(5), 909–926 (2008)
14. Chan, A.B., Vasconcelos, N.: Bayesian Poisson regression for crowd counting. In: IEEE International Conference on Computer Vision, IEEE (2009a)
15. Chan, A.B., Vasconcelos, N.: Layered dynamic textures. IEEE Trans. Pattern Anal. Mach. Intell.: Spec. Issue Probab. Graph. Models Comput. Vis. 31(10), 1862–1879 (2009b)
16. Chan, A.B., Vasconcelos, N.: Variational layered dynamic textures. In: IEEE Conference on Computer Vision and Pattern Recognition, IEEE (2009c)
17. Chan, A., Vasconcelos, N.: Counting people with low-level features and Bayesian regression. IEEE Trans. Image Process. 21(4), 2160–2177 (2012)
18. Chan, A.B., Liang, Z.S.J., Vasconcelos, N.: Privacy preserving crowd monitoring: counting people without people models or tracking. In: IEEE Conference on Computer Vision and Pattern Recognition, IEEE (2008)
19. Chan, A., Morrow, M., Vasconcelos, N.: Analysis of crowded scenes using holistic properties. In: 11th IEEE International Workshop on Performance Evaluation of Tracking and Surveillance (PETS'09) (online) (2009)
20. Chan, A.B., Coviello, E., Lanckriet, G.R.G.: Clustering dynamic textures with the hierarchical EM algorithm. In: IEEE Conference on Computer Vision and Pattern Recognition, IEEE (2010a)
21. Chan, A.B., Mahadevan, V., Vasconcelos, N.: Generalized Stauffer-Grimson background subtraction for dynamic scenes. Mach. Vis. Appl. 22(5) 751–766 (2011)
22. Chaudry, R., Ravichandran, A., Hager, G., Vidal, R.: Histograms of oriented optical flow and Binet-Cauchy kernels on nonlinear dynamical systems for the recognition of human actions. In: IEEE International Conference on Computer Vision and Pattern Recognition, IEEE (2009)
23. Cho, S.Y., Chow, T.W.S., Leung, C.T.: A neural-based crowd estimation by hybrid global learning algorithm. IEEE Trans. Syst. Man Cybern. 29, 535–541 (1999)
24. Cock, K.D., Moor, B.D.: Subspace angles between linear stochastic models. In: IEEE Conference on Decision and Control, Proceedings, IEEE, pp. 1561–1566 (2000)
25. Cong, Y., Gong, H., Zhu, S.C., Tang, Y.: Flow mosaicking: real-time pedestrian counting without scene-specific learning. In: IEEE CVPR, IEEE (2009)
26. Cooper, L., Liu, J., Huang, K.: Spatial segmentation of temporal texture using mixture linear models. In: Dynamical Vision Workshop in the IEEE International Conference of Computer Vision, Springer LNCS (2005)
27. Costantini, R., Sbaiz, L., Süsstrunk, S.: Higher order SVD analysis for dynamic texture synthesis. IEEE Trans. Image Process. 17(1), 42–52 (2008)
28. Cover, T., Thomas, J.: Elements of Information Theory. Wiley, New York (1991)
29. Dalal, N., Triggs, B.: Histograms of oriented gradients for human detection. In: IEEE Conference on Computer Vision and Pattern Recognition, IEEE, vol. 2, pp. 886–893 (2005)
30. Davies, A.C., Yin, J.H., Velastin, S.A.: Crowd monitoring using image processing. Electron. Commun. Eng. J. 7, 37–47 (1995)
31. Dempster, A.P., Laird, N.M., Rubin, D.B.: Maximum likelihood from incomplete data via the EM algorithm. J. R. Stat. Soc. B 39, 1–38 (1977)

32. Dong, L., Parameswaran, V., Ramesh, V., Zoghlami, I.: Fast crowd segmentation using shape indexing. In: IEEE International Conference on Computer Vision, IEEE (2007)
33. Doretto, G., Soatto, S.: Dynamic shape and appearance models. IEEE Trans. Pattern Anal. Mach. Intell. **28**(12), 2006–2019 (2006)
34. Doretto, G., Chiuso, A., Wu, Y.N., Soatto, S.: Dynamic textures. Int. J. Comput. Vis. **51**(2), 91–109 (2003a)
35. Doretto, G., Cremers, D., Favaro, P., Soatto, S.: Dynamic texture segmentation. In: IEEE International Conference on Computer Vision, IEEE, vol. 2, pp. 1236–1242 (2003b)
36. Doretto, G., Jones, E., Soatto, S.: Spatially homogeneous dynamic textures. In: ECCV, Springer-Verlag LNCS 3021–3024 (2004)
37. Felzenszwalb, P., McAllester, D., Ramanan, D.: A discriminatively trained, multiscale, deformable part model. In: IEEE Conference on Computer Vision and Pattern Recognition, IEEE (2008)
38. Fitzgibbon, A.W.: Stochastic rigidity: image registration for nowhere-static scenes. In: IEEE International Conference on Computer Vision, IEEE, vol. 1, pp. 662–670 (2001)
39. Gelb, A.: Applied Optimal Estimation. MIT, Cambridge (1974)
40. Ghanem, B., Ahuja, N.: Phase based modelling of dynamic textures. In: IEEE Internationl Conference on Computer Vision, IEEE (2007)
41. Ghoreyshi, A., Vidal, R.: Segmenting dynamic textures with Ising descriptors, ARX models and level sets. In: Dynamical Vision Workshop in the European Conference on Computer Vision, Springer LNCS (2006)
42. Horn, B.K.P.: Robot Vision. McGraw-Hill, New York (1986)
43. Horn, B., Schunk, B.: Determining optical flow. Artif. Intell. **17**, 185–204 (1981)
44. Hu, M., Ali, S., Shah, M.: Detecting global motion patterns in complex videos. In: IEEE International Conference on Pattern Recognition, IEEE (2008a)
45. Hu, M., Ali, S., Shah, M.: Learning motion patterns in crowded scenes using motion flow field. In: IEEE International Conference on Pattern Recognition, IEEE (2008b)
46. Isard, M., Blake, A.: Condensation – conditional density propagation for visual tracking. Int. J. Comput. Vis. **29**(1), 5–28 (1998)
47. Kay, S.M.: Fundamentals of Statistical Signal Processing: Estimation Theory. Prentice-Hall, Upper Saddle River (1993)
48. Kong, D., Gray, D., Tao, H.: Counting pedestrians in crowds using viewpoint invariant training. In: British Machine Vision Conference, BMVA (2005)
49. Lanckriet, G., Cristianini, N., Bartlett, P., Ghaoui, L.E., Jordan, M.: Learning the kernel matrix with semidefinite programming. J. Mach. Learn. Res. **5**, 27–72 (2004)
50. Larimore, W.E.: Canonical variate analysis in identification, filtering, and adaptive control. In: IEEE Conference on Decision and Control, IEEE, vol. 2, pp. 596–604 (1990)
51. Leibe, B., Seemann, E., Schiele, B.: Pedestrian detection in crowded scenes. In: IEEE Conference on Computer Vision and Pattern Recognition, IEEE, vol. 1, pp. 875–885 (2005)
52. Leibe, B., Schindler, K., Van Gool, L.: Coupled detection and trajectory estimation for multi-object tracking. In: IEEE International Conference on Computer Vision, IEEE (2007)
53. Lempitsky, V., Zisserman, A.: Learning to count objects in images. In: Advances in Neural Information Processing Systems, NIPS (2010)
54. Lin, S.F., Chen, J.Y., Chao, H.X.: Estimation of number of people in crowded scenes using perspective transformation. IEEE Trans. Syst. Man Cybern. **31**(6), 645–654 (2001)
55. Liu, C.B., Lin, R.S., Ahuja, N., Yang, M.H.: Dynamic texture synthesis as nonlinear manifold learning and traversing. In: British Machine Vision Conference, vol. 2, pp. 859–868. BMVA (2006)
56. Lucas, B., Kanade, T.: An iterative image registration technique with an application to stereo vision. In: Proceeding on DARPA Image Understanding Workshop, pp. 121–130. Morgan Kaufmann Publishers, (1981)
57. Mahadevan, V., Vasconcelos, N.: Spatiotemporal saliency in highly dynamic scenes. IEEE Trans. Pattern Anal. Mach. Intell. **32**(1), 171–177 (2010)

58. Mahadevan, V., Li, W., Bhalodia, V., Vasconcelos, N.: Anomaly detection in crowded scenes. In: IEEE Conference on Computer Vision and Pattern Recognition (CVPR), IEEE (2010)
59. Marana, A.N., Costa, L.F., Lotufo, R.A., Velastin, S.A.: On the efficacy of texture analysis for crowd monitoring. In: IEEE Proceedings of Computer Graphics, Image Processing, and Vision, IEEE, pp. 354–361 (1998)
60. Marana, A.N., Costa, L.F., Lotufo, R.A., Velastin, S.A.: Estimating crowd density with minkoski fractal dimension. In: IEEE Proceedings of International Conference Acoustics, Speech, Signal Processing, IEEE, vol. 6, pp. 3521–3524 (1999)
61. Martin, R.J.: A metric for ARMA processes. IEEE Trans. Signal Process. **48**(4), 1164–1170 (2000)
62. Mehran, R., Oyama, A., Shah, M.: Abnormal crowd behavior detection using social force model. In: IEEE Conference on Computer Vision and Pattern Recognition, IEEE (2009)
63. Mehran, R., Moore, B., Shah, M.: A streakline representation of flow in crowded scenes. In: European Conference on Computer Vision, LNCS (2010)
64. Monnet, A., Mittal, A., Paragios, N., Ramesh, V.: Background modeling and subtraction of dynamic scenes. In: CVPR, IEEE (2003)
65. Overschee, P.V., Moor, B.D.: N4SID: subspace algorithms for the identification of combined deterministic-stochastic systems. Automatica **30**, 75–93 (1994)
66. Paragios, N., Ramesh, V.: A MRF-based approach for real-time subway monitoring. In: IEEE Conference on Computer Vision and Pattern Recognition, IEEE, vol. 1, pp. 1034–1040 (2001)
67. Polana, R., Nelson, R.C.: Recognition of motion from temporal texture. In: IEEE Conference on Computer Vision and Pattern Recognition, IEEE, pp. 129–134 (1992)
68. Rabaud, V., Belongie, S.J.: Counting crowded moving objects. In: IEEE Conference on Computer Vision and Pattern Recognition, IEEE (2006)
69. Rasmussen, C.E., Williams, C.K.I.: Gaussian Processes for Machine Learning. MIT, Cambridge (2006)
70. Ravichandran, A., Vidal, R.: Video registration using dynamic textures. IEEE Trans. Pattern Anal. Mach. Intell. **33**(1), pp. 158–171 (2011)
71. Ravichandran, A., Chaudhry, R., Vidal, R.: View-invariant dynamic texture recognition using a bag of dynamical systems. Video Registration using Dynamic Textures. In: IEEE International Conference on Computer Vision and Pattern Recognition, IEEE **33**(1) 158–171 (2011)
72. Regazzoni, C.S., Tesei, A.: Distributed data fusion for real-time crowding estimation. Signal Process. **53**, 47–63 (1996)
73. Saisan, P., Doretto, G., Wu, Y., Soatto, S.: Dynamic texture recognition. In: IEEE Conference on Computer Vision and Pattern Recognition, IEEE, vol. 2, pp. 58–63 (2001)
74. Saleemi, I., Hartung, L., Shah, M.: Scene understanding by statistical modeling of motion patterns. In: IEEE Conference on Computer Vision and Pattern Recognition, IEEE (2010)
75. Shumway, R.H., Stoffer, D.S.: An approach to time series smoothing and forecasting using the EM algorithm. J. Time Ser. Anal. **3**(4), 253–264 (1982)
76. Siddiqi, S.M., Boots, B., Gordon, G.J.: A constraint generation approach to learning stable linear dynamical systems. In: Advances in Neural Information Processing Systems, NIPS (2007)
77. Szummer, M., Picard, R.: Temporal texture modeling. In: IEEE Conference on Image Processing, IEEE, vol. 3, pp. 823–826 (1996)
78. Vapnik, V.N.: The nature of statistical learning theory. Springer, New York (1995)
79. Vidal, R.: Online clustering of moving hyperplanes. In: Neural Information and Processing Systems, NIPS (2006)
80. Vidal, R., Favaro, P.: Dynamicboost: boosting time series generated by dynamical systems. In: IEEE International Conference on Computer Vision, IEEE
81. Vidal, R., Ravichandran, A.: Optical flow estimation & segmentation of multiple moving dynamic textures. In: IEEE Conference on Computer Vision and Pattern Recognition, vol. 2, pp. 516–521 (2005)
82. Viola, P., Jones, M., Snow, D.: Detecting pedestrians using patterns of motion and appearance. Int. J. Comput. Vis. **63**(2), 153–161 (2005)

83. Vishwanathan, S.V.N., Smola, A.J., Vidal, R.: Binet-cauchy kernels on dynamical systems and its application to the analysis of dynamic scenes. Int. J. Comput. Vis. **73**(1), 95–119 (2007)
84. Wang, J., Adelson, E.: Representing moving images with layers. IEEE Trans. Image Proc. **3**(5), 625–638 (1994)
85. Washington State Department of Transportation. http://www.wsdot.wa.gov (2005)
86. Woolfe, F., Fitzgibbon, A.: Shift-invariant dynamic texture recognition. In: ECCV, Springer LNCS (2006)
87. Wu, B., Nevatia, R.: Detection of multiple, partially occluded humans in a single image by bayesian combination of edgelet part detectors. In: IEEE International Conference on Computer Vision, IEEE, vol. 1, pp. 90–97 (2005)
88. Yang, Y., Liu, J., Shah, M.: Video scene understanding using multi-scale analysis. In: IEEE International Conference on Computer Vision, IEEE (2009)
89. Yuan, L., Wen, F., Liu, C., Shum, H.Y.: Synthesizing dynamic textures with closed-loop linear dynamic systems. In: European Conference on Computer Vision, pp. 603–616. Springer LNCS (2004)
90. Zhao, T., Nevatia, R.: Bayesian human segmentation in crowded situations. In: IEEE Conference on Computer Vision and Pattern Recognition, IEEE, vol. 2, pp. 459–466 (2003)
91. Zhong, J., Sclaroff, S.: Segmenting foreground objects from a dynamic textured background via a robust Kalman filter. In: IEEE ICCV, IEEE (2003)

Chapter 13
Inferring Leadership from Group Dynamics Using Markov Chain Monte Carlo Methods

Avishy Y. Carmi, Lyudmila Mihaylova, François Septier, Sze Kim Pang, Pini Gurfil, and Simon J. Godsill

Abstract This chapter presents a novel framework for identifying and tracking dominant agents in groups. The proposed approach relies on a causality detection scheme that is capable of ranking agents with respect to their contribution in recognizing the system's collective behavior based exclusively on the agents' observed trajectories. Further, the reasoning paradigm is made robust to multiple emissions and clutter by employing a class of recently introduced Markov chain Monte Carlo-based group tracking methods. Examples are provided that demonstrate the strong potential of the proposed scheme in identifying actual leaders in swarms of interacting agents and moving crowds.

A.Y. Carmi (✉)
Department of Mechanical and Aerospace Engineering, Nanyang Technological University, Singapore
e-mail: acarmi@ntu.edu.sg

L. Mihaylova
School of Computing and Communications, Lancaster University, Lancaster, UK
e-mail: mila.mihaylova@lancaster.ac.uk

F. Septier
Signal Processing and Information Theory Group, TELECOM Lille 1,
Villeneuve d'Ascq Cedex, France
e-mail: francois.septier@telecom-lille1.eu

S.K. Pang
DSO National Laboratories, Singapore
e-mail: pszekim@dso.org.sg

P. Gurfil
Department of Aerospace Engineering, Technion Israel Institute of Technology, Haifa, Israel
e-mail: pgurfil@technion.ac.il

S.J. Godsill
Department of Engineering, University of Cambridge, Cambridge, UK
e-mail: sjg@eng.cam.ac.uk

S. Ali et al. (eds.), *Modeling, Simulation and Visual Analysis of Crowds*, The International
Series in Video Computing 11, DOI 10.1007/978-1-4614-8483-7__13,
© Springer Science+Business Media New York 2013

13.1 Introduction

Tracking interacting objects moving in a coordinated fashion and making inference about the patterns of their behavior has been subject of an increased interest in the last decade. Such problems occur in many areas, especially video surveillance, cell tracking in biomedicine, pollutant clouds monitoring and people rescuing. The common pattern of the whole group is of main interest, not the individual trajectories on their own. In most of the multi-object tracking methods, as opposed to groups tracking methods, tracking of individual objects is the common approach. This is an especially challenging problem when the groups are composed of hundreds or thousands elements and the inference needs to be done quickly, in real time, based on heterogeneous multi-sensor data.

Groups can be considered as structured objects, a term which reflects the interrelationships between their components. These endogenous forces give rise to group hierarchies and are instrumental in producing emergent phenomena. Fortunately, these are exactly the factors essential for maintaining coordination within and between groups, a premise which to some extent allows us to treat them as united entities in a high level tracking paradigm. Any knowledge of existence of such interrelations facilitates sophisticated agent-based behavioral modeling which, in practice, comprises of a set of local interaction rules or mutually interacting processes (e.g., Boids system [31], causality models [17, 30]) – an approach which by itself provides insightful justifications of characteristic behaviors in the fundamental subsystem level and likewise of group hierarchies and emergent social patterns (see [30]).

13.1.1 Reasoning About Behavioral Traits

Being the underlying driving mechanism for evoking emergent phenomena, hierarchies and principal behavior patterns, the ingrained interactions between agents are possibly the most pivotal factors that should be scrutinized in high level scene understanding. Such interrelations can take the form of a causal chain in which an agent's decisions and behavior are affected by its neighbors and likewise have either direct or indirect influence on other agents. The ability to fully represent these interrelations based exclusively on passive observations such as velocity and position, lays the ground for the development of sophisticated reasoning schemes that can potentially be used in applications such as activity detection, intentionality prediction, and artificial awareness.

In this work we demonstrate this concept by developing a causality reasoning framework for ranking agents with respect to their cumulative contribution in shaping the collective behavior of the system. In particular, our framework is able to distinguish leaders and followers based exclusively on their observed trajectories.

13.1.2 Novelties and Contributions

The contribution of this work is twofold. Firstly, a novel *causality* reasoning scheme is derived for ranking agents with respect to their decision-making capabilities (dominance) as substantiated by the observed emergent behavior. Dominant agents in that sense are considered to have a prominent influence on the collective behavior and are experimentally shown to coincide with actual leaders in groups. Secondly, the causality scheme is consolidated with a recently introduced Markov chain Monte Carlo (MCMC)-based particle method [9, 28] for tracking agents and group hierarchies in potentially cluttered environments.

The subsequent Sects. 13.1.3–13.2 provide an overview of existing group tracking schemes with an emphasis on the underlying MCMC-based particle methods.

The remaining part of this chapter is organized in the following way. Section 13.3 develops the causality-driven agent ranking approach. Section 13.4 demonstrates the performance of the causality identification scheme using a few illustrative examples. Finally, concluding remarks and some open issues are discussed in Sect. 13.5.

13.1.3 Multiple Group Tracking

Over the past decade various methods have been developed for group tracking. These can be divided into two broad classes, depending on the underlying complexities: (1) methods for a relatively small number of groups, with a small number of group components [15, 24, 28], and (2) methods for groups comprised of hundreds or thousands of objects (normally referred to as cluster/crowd tracking techniques) [2, 9]. In the second case the whole group is usually considered as an extended object (an ellipse or a circle) which center position is estimated, together with the parameters of the extent.

Different models of groups of objects have been proposed in the literature, such as particle models for flocks of birds [19], and leader-follower models [26]. However, estimating the dynamic evolution of the group structure has not been widely studied in the literature, although there are similarities with methods used in evolving network models [1, 11].

Typically tracking many objects (hundreds or thousands) can be solved by clustering techniques or other methods where the aggregated motion is estimated, as it is in the case of vehicular traffic flow prediction/estimation, with fluid dynamics type of models combined with particle filtering techniques [27]. For thousands of objects forming a group, the only possible solution is to consider them as an extended object. The extended object tracking problem reduces then to joint state and parameter estimation.

Estimation of parameters in general nonlinear non-Gaussian state-space models is a long-standing problem. Since particle filters (PFs) are known with the challenges they face for parameter estimation and for joint state and parameter estimation [4], most solutions in the literature split the problems into two parts: (i) state estimation, followed by (ii) parameter estimation (see e.g., [3]). In [3] an extended object tracking problem is solved when the static parameters are estimated using Monte Carlo methods (data augmentation and particle filtering), whereas the states are estimated with a Mixture Kalman filter or with an interacting multiple model filter.

13.1.3.1 PFs for Tracking in Variable State Dimensions

An extension of the PF technique to a varying number of objects is introduced in [28, 34] and [24]. In [34] a PF implementation of the probability hypothesis density (PHD) filter is derived. This algorithm maintains a representation of the filtering belief mass function using random set realizations (i.e., particles of varying dimensions). The samples are propagated and updated based on a Bayesian recursion consisting of set integrals. Both works of [28] and [24] develop a MCMC PF scheme for tracking varying numbers of interacting objects. The MCMC approach outperforms the conventional PF due to its efficient sampling mechanism. Nevertheless, in its traditional non-sequential form it is inadequate for sequential estimation. The techniques used by Pang et al. [28] and Khan et al. [24] amend the MCMC for sequential filtering (see also [5]). The work in [24] copes with inconsistencies in state dimension by utilizing the reversible jump MCMC method introduced in [18]. In [28], on the other hand the computation of the marginal filtering distribution is avoided as in [5]. The algorithm operates on a fixed dimension state space through indicator variables for labeling of active object states (the two approaches are essentially equivalent).

13.2 Models and Algorithms for Group Tracking

This section briefly reviews the fundamental concepts underlying the MCMC-based group tracking approaches in [28] and [9].

13.2.1 Virtual Leader Model

The idea of group modeling is to adopt a behavioral model in which each member of a group interacts with the other members of the group, typically making its velocity and position more similar to that of others in the same group. In [28], this idea has been conveniently formulated in continuous time through a multivariate stochastic differential equation (SDE) and then derived in discrete time without approximation errors, owing to the assumed linear and Gaussian form of the model.

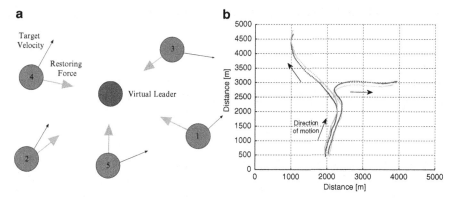

Fig. 13.1 Group model with virtual leader – illustration of the restoring forces (**a**) and of a single realization showing a group of four objects that splits into two groups of two objects (**b**)

In particular, two different models have been proposed. In the first, the basic group model, the group parameter is modeled as a deterministic function of the objects. In the second, the group model with a virtual leader, an additional state variable is introduced in order to model the bulk or group parameter. This second approach is closer in spirit to the bulk velocity model and virtual leader-follower model [26]. Such model provides a more flexible behavior since the virtual leader is no longer a deterministic function of the individual object states. Figure 13.1 gives a graphical illustration of the restoring forces towards the virtual leader for a group of five objects.

The spatio-temporal structure for the ith object in a group, as defined in [28], is given by:

$$d\dot{\mu}_{t,i}^{x} = \left\{ -\alpha[\mu_{t,i}^{x} - v_{t}^{x}] - \gamma_{1}\dot{\mu}_{t,i}^{x} - \beta[\dot{\mu}_{t,i}^{x} - \dot{v}_{t}^{x}] + r_{i} \right\} dt + \sigma_{x}dW_{t,i}^{x} \quad (13.1)$$

$$d\dot{v}_{t}^{x} = -\gamma_{2}\dot{v}_{t}^{x}dt + \sigma_{g}dG_{t}^{x} \quad (13.2)$$

Here $\mu_{t,i}^{x}$ is the Cartesian position in the X direction of the ith object in the group at time t, with $\dot{\mu}_{t,i}^{x}$ the corresponding velocity. v_{t}^{x} and $\dot{v}_{t}^{x}$ represent respectively the Cartesian position and the velocity both in the X direction of the unobserved virtual leader of the group. $W_{t,i}^{x}$ and G_{t}^{x} are two independent standard Brownian motions. $W_{t,i}^{x}$ is assumed to be independently generated for each object i in the group, whereas G_{t}^{x} is a noise component common to all members of a group. The parameters α and β are positive, and reflect the strength of the pull towards the group bulk. The "mean reversion" terms $\gamma_{1}\dot{\mu}_{t,i}^{x}$ and $\gamma_{2}\dot{v}_{t}^{x}$ simply prevent the velocities of the object and the virtual leader drifting up to very large values with time. Finally, in order to reduce or eliminate behavior in which objects become colocated or collide spatially, which are clearly infeasible or highly unlikely in practice, an additional repulsive force r_{i} is introduced in (13.1) when objects become too close.

13.2.2 Modeling Groups of Extended Objects

In practice, objects may produce more than a single emission, and in some cases they may indeed consist of many individual entities moving in a coordinated fashion (i.e., clusters). Such scenarios normally involve additional extent parameters that embody the potentially dynamic physical boundary of an object. In this respect, the fairly simple idea adopted in [9] represents a dynamically evolving group of extended objects, which are otherwise referred to as clusters, by means of a time-varying Gaussian mixture model (i.e., each mixture component corresponds to an individual object). In what follows, we briefly review the essentials of this approach.

Assume that at time k there are l_k clusters, or targets at unknown locations. Each cluster may produce more than one observation yielding the measurement set realization $\mathbf{z}_k = \{\mathbf{y}_k(i)\}_{i=1}^{m_k}$, where typically $m_k >> l_k$. At this point we assume that the observation concentrations (clusters) can be adequately represented by a parametric statistical model.

Letting $\mathbf{z}_{0:k} = \{\mathbf{z}_0, \ldots, \mathbf{z}_k\}$ be the measurement history up to time t_k, the cluster tracking problem may be defined as follows. We are concerned with estimating the posterior distribution of the random set of unknown parameters, i.e. $p(x_k \mid \mathbf{z}_{0:k})$, from which point estimates for x_k and posterior confidence intervals can be extracted.

For reasons of convenience we consider an equivalent formulation of the posterior that is based on existence variables. Thus, following the approach adopted in [9] the random set x_k is replaced by a fixed dimension vector coupled to a set of indicator variables $e_k = \{e_k^j\}_{j=1}^n$ showing the activity status of elements (i.e., $e_k^j = 1$, $j \in [1,n]$ indicates the existence of the jth element where n stands for the total number of elements). To avoid possible confusion, in what follows we maintain the same notation for the descriptive parameter set $\mathbf{x}_k$ which is now of fixed dimension.

In [9], each cluster is modeled via a Gaussian pdf. Following this only the first two moments, namely the mean and covariance, need to be specified for each cluster (under these restrictions, the cluster tracking problem is equivalent to that of tracking an evolving Gaussian mixture model with a variable number of components). It is worth mentioning, that the approach itself does not rely on the Gaussian assumption and other parameterized density functions could equally be adopted in this framework. Thus,

$$\mathbf{x}_k^j = \{\mu_k^j, \dot{\mu}_k^j, \Sigma_k^j, w_k^j, \rho_k^j\}, \quad \mathbf{x}_k = \{\mathbf{x}_k^j\}_{j=1}^n, \tag{13.3}$$

where μ_k^j, $\dot{\mu}_k^j$, Σ_k^j and w_k^j denote the jth cluster's mean, velocity, covariance and associated unnormalized mixture weight at time k, respectively. The additional parameter ρ_k^j denotes the local turning radius of the jth cluster's mean at time k.

13.2.3   Sequential Inference Using MCMC-Based PF

The group tracking problems discussed above can be efficiently solved via the MCMC-based particle method initially proposed for solution of group tracking problems in [28]. This method aims at sequentially approximating the following joint posterior distribution

$$p(\mathbf{x}_k, \mathbf{x}_{k-1} | \mathbf{z}_{0:k}) \propto p(\mathbf{z}_k | \mathbf{x}_k) p(\mathbf{x}_k | \mathbf{x}_{k-1}) p(\mathbf{x}_{k-1} | \mathbf{z}_{0:k-1}) \qquad (13.4)$$

where the state vector $\mathbf{x}_k$ comprises of the objects' instantaneous position, velocity and extent parameters at time t_k. In what follows we would refer to the (discrete) time t_k as simply k.

Since the closed form expression of the distribution $p(\mathbf{x}_{k-1} | \mathbf{z}_{0:k-1})$ is generally unknown, the proposed scheme approximates it by using a set of unweighted particles

$$p(\mathbf{x}_{k-1} | \mathbf{z}_{0:k-1}) \approx \frac{1}{N} \sum_{j=1}^{N} \delta(\mathbf{x}_{k-1} - \mathbf{x}_{k-1}^{(j)}) \qquad (13.5)$$

where N is the number of particles, $\delta(\cdot)$ is the Dirac delta, and (j) is the particle index. Then, by plugging this particle approximation into (13.4), an appropriate MCMC scheme can be used to draw from the joint posterior distribution $p(\mathbf{x}_k, \mathbf{x}_{k-1} | \mathbf{z}_{0:k})$. The converged MCMC outputs are then extracted to give an empirical approximation of the posterior distribution of interest at time k, thus seeding the next step of the filtering at time $k+1$.

At the mth MCMC iteration, the following procedure is performed to obtain samples from $p(\mathbf{x}_k, \mathbf{x}_{k-1} | \mathbf{z}_{0:k})$:

1. Make a joint draw for $\{\mathbf{x}_k, \mathbf{x}_{k-1}\}$ using a Metropolis Hastings step,
2. Update successively some elements in $\mathbf{x}_k$ by using a series of Metropolis Hastings-within-Gibbs.

13.2.3.1   Metropolis Hastings Step for the Cluster Tracking Problem

The Metropolis Hastings (MH) algorithm generates samples from an aperiodic and irreducible Markov chain with a predetermined (possibly unnormalized) stationary distribution. This is a constructive method which specifies the Markov transition kernel by means of acceptance probabilities based on the preceding time outcome. As part of this, a proposal density is used for drawing new samples. In our case, setting the stationary density as the joint filtering pdf of the object states $\mathbf{x}_k, \mathbf{x}_{k-1}$ and the corresponding indicator variables e_k, e_{k-1}, i.e., $p(\mathbf{x}_k, e_k, \mathbf{x}_{k-1}, e_{k-1} \mid \mathbf{z}_{0:k})$ (of which the marginal is the desired filtering pdf), a new set of samples from this distribution can be obtained after the MH burn-in period. This procedure is described next.

Algorithm 3 MCMC particle filtering algorithm

---

1. Given previous time samples $\{\mathbf{x}_{k-1}^{(i)}, e_{k-1}^{(i)}\}_{i=1}^{N}$ perform the following steps
2. for $i = 1, \ldots, N + N_{Burn-in}$
3. for $j = 1, \ldots, n$
4. Cluster evolution: Simulate

$$(\dot{\mu}_{k-1}^{j,(i)}, \mu_{k-1}^{j,(i)}, \Sigma_{k-1}^{j,(i)}, w_{k-1}^{j,(i)}, \rho_{k-1}^{j,(i)}) \longrightarrow (\dot{\mu}_{k}^{j,(i)}, \mu_{k}^{j,(i)}, \Sigma_{k}^{j,(i)}, w_{k}^{j,(i)}, \rho_{k}^{j,(i)})$$

5. end for
6. Perform MCMC move (Algorithm 4).
7. Draw a new set of indicators $e_{k}^{j,(i)}$, $j = 1, \ldots, n$ for the accepted move.
8. Perform Gibbs refinement (Algorithm 5).
9. end for

First, we simulate a sample from the joint propagated pdf $p(\mathbf{x}_k, e_k, \mathbf{x}_{k-1}, e_{k-1} \mid \mathbf{z}_{0:k-1})$ by drawing

$$(\mathbf{x}_k', e_k') \sim p(\mathbf{x}_k, e_k \mid \mathbf{x}_{k-1}', e_{k-1}') \tag{13.6}$$

where $(\mathbf{x}_{k-1}', e_{k-1}')$ is uniformly drawn from the empirical approximation

$$\hat{p}(\mathbf{x}_{k-1}, e_{k-1} \mid \mathbf{z}_{0:k-1}) = N^{-1} \sum_{i=1}^{N} \delta(\mathbf{x}_{k-1}^{(i)} - \mathbf{x}_{k-1}) \delta(e_{k-1}^{(i)} - e_{k-1}) \tag{13.7}$$

This sample is then accepted or rejected using the following Metropolis rule.

Let $(\mathbf{x}_k^{(i)}, e_k^{(i)}, \mathbf{x}_{k-1}^{(i)}, e_{k-1}^{(i)})$ be a sample from the realized chain of which the stationary distribution is the joint filtering pdf. Then the MH algorithm accepts the new candidate $(\mathbf{x}_k', e_k', \mathbf{x}_{k-1}', e_{k-1}')$ as the next realization from the chain with probability

$$\gamma = \min \left\{ 1, \frac{p(\mathbf{z}_k \mid \mathbf{x}_k', e_k', m_k)}{p(\mathbf{z}_k \mid \mathbf{x}_k^{(i)}, e_k^{(i)}, m_k)} \right\} \tag{13.8}$$

where $p(\mathbf{z}_k \mid \mathbf{x}_k', e_k', m_k)$ is the likelihood function. The converged output of this scheme simulates the joint density $p(\mathbf{x}_k, e_k, \mathbf{x}_{k-1}, e_{k-1} \mid \mathbf{z}_{0:k})$ of which the marginal is the desired filtering pdf.

It has already been noted that the above sampling scheme may be inefficient in exploring the sample space as the underlying proposal density of a well behaved system (i.e., of which the process noise is of low intensity) introduces relatively small moves. This drawback is alleviated here by using a secondary Gibbs refinement stage [9].

A single cycle of the basic MCMC cluster tracking algorithm of [9] is summarized in Algorithms 3, 4, and 5.

Algorithm 4 MCMC move

---
1. Compute the MH acceptance probability γ of the new move using (13.8).
2. Draw $u \sim U[0,1]$
3. if $u < \gamma$
4. Accept $\mathbf{s}^{(i)} = (\mathbf{x}_k^{(i)}, e_k^{(i)}, \mathbf{x}_{k-1}^{(i)}, e_{k-1}^{(i)})$ as the next sample of the realized chain.
5. else
6. Retain $\mathbf{s}^{(i)} = \mathbf{s}^{(i-1)}$.
7. end if

Algorithm 5 Particles refinement (Metropolis within Gibbs)

---
1. for $j = 1, \ldots, n$
2. if $e_k^{j,(i)} = 1$
3. for $l = 1, \ldots, N_{\text{MH Steps}}$
4. Propose a move $\bar{\mu}_k^j$.
5. Compute the MH acceptance probability $\bar{\gamma}$ of the new move.
6. Draw $u \sim U[0,1]$
7. if $u < \bar{\gamma}$
8. Accept the new move by setting $\mu_k^{j,(i)} = \bar{\mu}_k^j$.
9. else
10. Retain previous $\mu_k^{j,(i)}$.
11. end if
12. end for
13. end if
14. end for

13.2.3.2 Multiple Chain and Evolutionary MCMC

The theory of multiple chain MCMC grasps that a mixing mechanism for synthesizing samples across chain realizations is necessary for improving robustness to the well known practical problem of quasi-ergodicity otherwise known as poor mixing. Existing multiple chain approaches, such as parallel tempering [13, 14], evolving population particle filters [6–8, 21, 22, 29] and population MCMC [23, 25], utilize exchange mechanisms to expedite convergence. The evolutionary MCMC approach, on the other hand, incorporates an additional structure for generating possibly improved candidates based on convergent chain realizations. This method has been proved successful in high dimensional settings. An evolutionary extension of the basic MCMC filtering scheme is provided in the Appendix part of this work.

13.3 Causality-Driven Agent Ranking

The so-called probabilistic approach to causality, which has reached maturity over the past two decades (see for example Pearl [30], Geffner [12], and Shoam [32] for an extensive overview), establishes a convenient framework for reasoning and inference of causal relations in complex structural models.

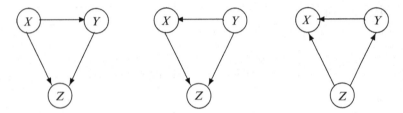

Fig. 13.2 From *left* to *right*: depiction of the causal hierarchies (based on out degrees) (X,Y,Z), (Y,X,Z), and (Z,Y,X). The most influential agents in the causal diagrams from *left* to *right* are X, Y and Z, respectively

Many notions in probabilistic causality rely extensively on structural models and in particular on causal Bayesian networks which are normally referred to as simply causal networks (CN's). A CN is a directed acyclic graph compatible with a probability distribution that admits a Markovian factorization and certain structural restrictions [30].

13.3.1 Causal Hierarchies

In this work the term *causal hierarchies* refers to ranking of agents with respect to their cumulative effect on the actions of the remaining constituents in the system. The word "causal" here reflects the fact that our measure of distinction embodies the intensity of the causal relations between the agent under inspection and its counterparts. Adopting the information-theoretic standpoint, in which the links of a CN are regarded as information channels [10], one could readily deduce that the total effect of an agent is directly related to the local information flow entailed by its corresponding in and out degrees. To be more precise, the total effect of an agent is computed by summing up the associated path coefficients (obtained by any standard Bayesian network learning approach) of either inward or outward links. This concept is further illustrated in Fig. 13.2.

13.3.2 Inferring Causal Hierarchies via PCA

To some extent, causal hierarchies can be inferred using the class of principal component analysis (PCA)-based methods. Probably the most promising one in the context of our problem is the multi-channel singular spectrum analysis (M-SSA), which is otherwise known as extended empirical orthogonal function (EEOF) analysis [16]. The novel approach we suggest has some relations with M-SSA. The relevant details, however, are beyond the scope of this work. A performance evaluation of both our method and M-SSA is provided in the numerical study part in the following sections.

13.3.3 Structural Dynamic Modeling Approach

Structural equation modeling is commonly used for representing the underlying links of a CN [30]. In our case, this formulation assumes a rather dynamic form (i.e., comprising of multiple time series of the agents' observed traits such as velocity and position)

$$\mathbf{x}_k^i = \sum_{j \neq i} \sum_{m=1}^{p} \alpha^{j \to i}(m) \mathbf{x}_{k-m}^j + \varepsilon_k^i, \quad i = 1, .., n \tag{13.9}$$

where $\{\mathbf{x}_k^i\}_{k=0}^{\infty}$ and $\{\varepsilon_k^i\}_{k=0}^{\infty}$ denote the ith random process and a corresponding white noise driving sequence, respectively. The coefficients $\{\alpha^{j \to i}(m)\}_{m=1}^{p}$ quantify the causal influence of the jth process on the ith process. Notice that the Markovian model (13.9) has a finite-time horizon of the order p (also referred to as the wake parameter). In the standard multivariate formulation, the coefficients $\alpha^{j \to i}(m)$ are square matrices of an appropriate dimension. For maintaining a reasonable level of coherency we assume that these coefficients are scalars irrespectively of the dimension of $\mathbf{x}_k^i$. Nevertheless, our arguments throughout this Section can be readily extended to the standard multivariate case.

The methodology underlying the so-called Granger causality [17] considers an F-test of the null hypothesis $\alpha^{j \to i}(m) = 0$, $m = 1, \ldots, p$ for determining whether the jth process G-causes the ith process. The key idea here follows the simple intuitive wisdom that the more significant these coefficients are, the more likely they are to reflect a causal influence. In the framework of CNs the causal coefficients are related to the conditional dependencies within the probabilistic network, which in turn implies that their values can be learned based on the realizations of the time series $\{\mathbf{x}_k^i\}_{k=0}^{\infty}$, $i = 1, \ldots, n$. In what follows, we demonstrate how the knowledge of these coefficients allows us to infer the fundamental role of individual agents within the system. Before proceeding, however, we shall define the following key quantity.

Definition 13.1 (Causation Matrix). The causal influence of the process $\mathbf{x}^j$ on the process $\mathbf{x}^i$ can be quantified by

$$A_{ij} = \sum_m \left[\alpha^{j \to i}(m) \right]^2 \geq 0. \tag{13.10}$$

In the above definition, A_{ij} denotes the coefficient relating the two processes $\mathbf{x}^j$ and $\mathbf{x}^i$ so as to suggest an overall matrix structure that would provide a comprehensive picture of the causal influences among the underlying processes. The matrix $A = [A_{ij}] \in \mathbb{R}^{n \times n}$, termed the *causation matrix*, essentially quantifies the intensity of all possible causal influences within the system (note that according to the definition of a CN, the diagonal entries in A vanish). It can be easily recognized that a single row in this matrix exclusively represents the causal interactions affecting each individual process. Similarly, a specific column in A is comprised of the causal influences of a single corresponding process on the entire system. This premise motivates us to introduce the notion of total causal influence.

Definition 13.2 (Total Causal Influence Measure). The total causal influence (TCI) T_j of the process $\mathbf{x}_k^j$ is obtained as the l_1-norm of the jth column in the causation matrix A, that is

$$T_j = \sum_{i=1}^{n} |A_{ij}| = \sum_{i=1}^{n} A_{ij} \tag{13.11}$$

Having formulated the above concepts we are now ready to elucidate the primary contributions of this work, both of which rely on the TCI measure defined above.

13.3.4 Dominance and Similarity

A rather intuitive, but nonetheless striking, observation about the TCI is that it essentially reflects the dominance of each individual process in producing the underlying emergent behavior. This allows us to decompose any complex act into its prominent behavioral building blocks (processes) using a hierarchical ordering of the form

$$\text{Least dominant } T_{j_1} \leq T_{j_2} \leq \ldots \leq T_{j_n} \text{ Most dominant} \tag{13.12}$$

Equation (13.12) is given an interesting interpretation in the application part of this work, where the underlying processes $\{\mathbf{x}_k^j\}_{j=1}^n$ correspond to the motion of individual agents within a group. In the context of this example, the dominance of an agent is directly related to its leadership capabilities. By using the TCI measure it is therefore possible to distinguish between leaders and followers.

Another interesting implication of the TCI is exemplified in the following argument. Consider the two extreme processes in (13.12), one of which is the most dominant, $\mathbf{x}_k^{j_n}$, while the other is the least dominant, $\mathbf{x}_k^{j_1}$. Now, suppose we are given a new process $\mathbf{x}_k^i$, $i \neq j_1, j_n$ and are asked to assess its dominance based exclusively on the two extremals, with respect to the entire system. Then, a common intuition would suggest to categorize $\mathbf{x}_k^i$ as a dominant process in the system whenever it resembles $\mathbf{x}_k^{j_n}$ more than $\mathbf{x}_k^{j_1}$ in the sense of $|T_{j_n} - T_i| < |T_{j_1} - T_i|$ and vice versa. This idea is summarized below.

Definition 13.3 (Causal Similarity). A process $\mathbf{x}_k^j$ is said to resemble $\mathbf{x}_k^i$ more than $\mathbf{x}_k^l$ if and only if $|T_j - T_i| < |T_j - T_l|$.

In the context of the previously-mentioned example, we expect that dominant agents with high leadership capabilities would possess similar TCIs that would distinguish them from the remaining agents, the followers.

13.3.5 Bayesian MCMC Estimation of $\alpha^{j \to i}$

In typical applications the coefficients $\alpha^{j \to i}(m)$, $m = 1, \ldots, p$ in (13.9) may be unknown. Providing that the realizations of the underlying processes are available it is fairly simple to estimate these coefficients by treating them as regressors. Such an approach by no means guarantees an adequate recovery of the underlying causal structure (see the discussion about the identifiability of path coefficients and a related assertion concerning non-parametric functional modeling in [30] pp. 156–157, both have a clear connotation to the "fundamental problem of causal inference" [20]). Nevertheless, it provides a computationally efficient framework for making inference in systems with exceptionally large number of components. This premise is evident by noting from (13.9) that while fixing i the coefficients $\alpha^{j \to i}(m)$, $\forall j \neq i$, $m = 1, \ldots, p$ are statistically independent of $\alpha^{j \to l}(m)$, $\forall l \neq i$.

In a Bayesian framework we confine the latent causal structure by imposing a prior on the coefficients $\alpha^{j \to i}(m)$. Let p_α^i and $p_\alpha^{j \to i}$ be the priors of $\{\alpha^{j \to i}(m), \forall j \neq i\}$, and $\alpha^{j \to i}(m)$, respectively. Let also p_ε^i be some prescribed (not necessarily Gaussian) probability density of the white noise in (13.9). Then,

$$p(\{\alpha^{j \to i}(m), \forall j \neq i\} \mid \mathbf{x}_{0:k}^{1:n}) \propto$$

$$p_\alpha^i \prod_{t=p}^{k} p(\mathbf{x}_t^i \mid \{\alpha^{j \to i}(m), \mathbf{x}_{t-p:t-1}^j, \forall j \neq i\})$$

$$= p_\alpha^i \prod_{t=p}^{k} p_\varepsilon^i (\mathbf{x}_t^i - \sum_{j \neq i} \sum_{m=1}^{p} \alpha^{j \to i}(m) \mathbf{x}_{t-m}^j), \quad i = 1, \ldots, n \quad (13.13)$$

where $\mathbf{x}_{0:k}^{1:n} = \{\mathbf{x}_0^1, \ldots, \mathbf{x}_0^n, \ldots, \mathbf{x}_k^1, \ldots, \mathbf{x}_k^n\}$, and $\mathbf{x}_{t-p:t-1}^j = \{\mathbf{x}_{t-p}^j, \ldots, \mathbf{x}_{t-1}^j\}$. A viable estimation scheme for $\alpha^{j \to i}(m)$ which works well in most generalized settings is a Metropolis-within-Gibbs sampler that operates either sequentially or concurrently on the conditionals

$$p(\alpha^{j \to i}(m) \mid \mathbf{x}_{0:k}^{1:n}, \{\alpha^{l \to i}, \forall l \neq j, i\}) \propto$$

$$p_\alpha^{j \to i} \prod_{t=p}^{k} p(\mathbf{x}_t^i \mid \{\alpha^{l \to i}(m), \mathbf{x}_{t-p:t-1}^l, \forall l \neq i\}) \quad (13.14)$$

The obtained estimates at time k are then taken as the average of the converged chain (i.e., subsequent to the end of some prescribed burn-in period).

13.3.6 Causal Reasoning in Cluttered Environments

In many practical applications the constituent underlying traits, which are represented here by the processes $\{\mathbf{x}_k^j\}_{j=1}^n$, may not be perfectly known (in the context of our work these could be the object position and velocity, μ_k^j, $\dot{\mu}_k^j$). Hence instead of the actual traits one would be forced to use approximations that might not be consistent estimates of the original quantities (e.g., $\hat{\mu}_k^j$, $\hat{\dot{\mu}}_k^j$). As a consequence, the previously suggested structure might cease being an adequate representation of the latent causal mechanism. A plausible approach for alleviating this problem is to introduce a compensated causal structure that takes into account the exogenous disturbances induced by the possibly inconsistent estimates. Such a model can be readily formulated as a modified version of (13.9), that is

$$\hat{\mu}_k^i = \sum_{j \neq i} \sum_{m=1}^{p} \alpha^{j \to i}(m)\hat{\mu}_{k-m}^j + \varepsilon_k^i + \zeta_k^i, \quad i = 1,..,n, \qquad (13.15)$$

where the additional factor ζ_k^i denotes an exogenous bias. Hence, one can use (13.15) to predict the effects of interventions in ζ_k^i directly from passive observations (which are taken as an output of a tracking algorithm, e.g., $\hat{\mu}_k^j$ or $\hat{\dot{\mu}}_k^j$) without adjusting for confounding factors. See [30] (p. 166) for further elaborations on the subject.

13.4 Illustrative Examples

We demonstrate the performance of our suggested reasoning methodology and some of the previously mentioned concepts using both synthetic and realistic examples. All the scenarios considered here involve a group of dynamic agents, some of which are leaders that behave independently of all others. The leaders themselves may exhibit a highly nonlinear and non-predictive motion pattern which in turn affects the group's emergent behavior. We use a standard CN (13.9) with a predetermined time horizon p for disambiguating leaders from followers based exclusively on their instantaneous TCIs. In all cases the processes $\mathbf{x}_k^i$, $i = 1,\ldots,n$ are taken as either the increment $\dot{\mu}_k^i$ or position μ_k^i of each individual agent in the group. In addition, the unified tracking and reasoning paradigm is demonstrated by replacing the actual position and increment with the corresponding outputs of the MCMC cluster tracking algorithm, $\hat{\mu}_k^i$ and $\hat{\dot{\mu}}_k^i$.

The performance of the causality inference scheme is directly related to its ability to classify leaders based on their TCI values. As leaders are, by definition, more dominant than followers in some measure space, essentially shaping the overall group behavior, we expect that their TCI values would reflect this fact. Furthermore, the hierarchy (13.12) should allow us to disambiguate them from the remaining agents according to the notion of causal similarity which was introduced in Sect. 13.3.4. Following this argument we define a rather distinctive performance measure which allows us to assess the aforementioned qualities.

Let G be a set containing the leaders indices, i.e.,

$$G = \{j \mid x_k^j \text{ is a leader's instantaneous position or velocity}\}.$$

Let also v be a vector containing the agents' ordered indices according to the instantaneous hierarchy at time k

$$T_{j_1} \leq \cdots \leq T_{j_n}, \tag{13.16}$$

i.e., $v = [j_n, \ldots, j_1]^T$. Having stated this we can now define the following *performance index*

$$e = \max\{i \in [1, n] \mid v_i \in G\} \tag{13.17}$$

The above quantity indicates the worst TCI ranking of a leader. As an example, consider a case with, say, five leaders. Then the best performance index we could expect would be five, implying that all leaders have been identified and were properly ranked according to their TCIs. If the performance index yields a value greater than 5, say 10, it implies that all leaders are ranked among the top 10 agents according to their TCIs. The performance index cannot go below the total number of leaders and cannot exceed the total number of agents.

13.4.1 Swarming of Multiple Interacting Agents (Boids)

Our first example pertains to identification of leaders and followers in a dynamical system of multiple interacting agents, collectively performing in a manner usually referred to as *swarming* or *flocking*.

In the current example, Reynolds-inspired flocking [31] is used to create a complex motion pattern of multiple agents. Among these agents, there are leaders, who independently determine their own position and velocity, and followers, who interact among themselves and follow the leader agents.

The inference scheme performance over 100 Monte Carlo runs, in which the agents initial state and velocity were randomly picked, is provided in Fig. 13.3. The synthetic scenario considered consists of 30 agents, 4 of which are actual leaders. The performance index cumulative distribution function (CDF) for this scenario, which is illustrated via the 50, 70 and 90 percentile lines, is shown over the entire time interval in the left panel in this figure. The percentiles indicate how many runs out of 100 yielded a performance index below a certain value. Thus, 50 % of the runs yielded a performance index below the 50 percentile, 70 % of the runs attained values below the 70 percentile, and so on. Following this, it can be readily recognized that from around $k = 150$ the inference scheme is able to accurately identify the actual leaders in 50 % of the runs. A further examination of this figure reveals that the 4 actual leaders are ranked among the top 6 from around $k = 180$ in 90 % of the runs.

A comparison of leaders ranking capabilities of the proposed approach with that of the M-SSA method is provided in the right panel in Fig. 13.3. The instantaneous CDFs of both techniques are shown when using either position or

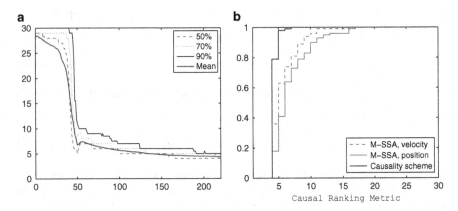

Fig. 13.3 Identification performance over time (abscissa) of the causality scheme (*left*) and the ranking CDF at time $t = 220$ (*right*) of both the causality scheme and the M-SSA method (using either velocity or position data) based on 100 Monte Carlo runs. (**a**) Causal ranking. (**b**) Causal ranking CDF

velocity time series data. This figure clearly demonstrates the superiority of the proposed approach with respect to the M-SSA.

13.4.2 Identifying Extended Leaders in Clutter

In the following example the actual agent tracks are replaced by the output of an MCMC-based tracking approach that was initially derived in [9, 28] and is briefly described in Sect. 13.2. The scenario consists of four agents out of which two are leaders. As before we use the Boids system for simulating the entire system. This time, however, the produced trajectories are contaminated with clutter and additional points representing multiple emissions from possibly the same agent (i.e., agents are assumed to be extended objects). These observations are then used by the MCMC tracking algorithm of which the output is fed to the causality detection scheme, in a fashion similar to the one described in Sect. 13.3.6.

The tracking performance of the MCMC algorithm is demonstrated both in Fig. 13.4 and in the left panel in Fig. 13.5. In Fig. 13.4, the estimated tracks and the cluttered observations are shown for a typical run. The averaged tracking performance of the MCMC approach is further illustrated based on 20 Monte Carlo runs using the Hausdorff distance [9] in Fig. 13.5. From this Figure it can be seen that the mean tracking errors become smaller than 1 after approximately 50 time steps in either cases of cluttered and non-cluttered observations.

The averaged leaders ranking performance in this example is illustrated for three different scenarios in the right panel in Fig. 13.5. Hence, it can be readily recognized

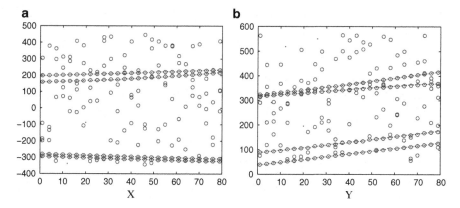

Fig. 13.4 Point observations and estimated tracks over time (abscissa). (**a**) X. (**b**) Y

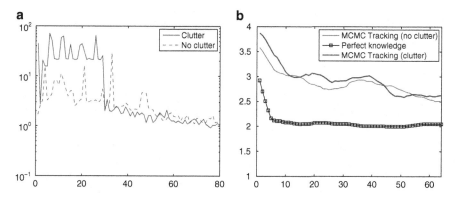

Fig. 13.5 Tracking performance and causality ranking over time (abscissa) averaged over 20 Monte Carlo runs. (**a**) Hausdorff distance. (**b**) Causality ranking

that the two leaders are accurately identified after approximately 10 time steps when the agent positions are perfectly known. As expected, this performance is deteriorated in the presence of clutter and multiple emissions, essentially attaining an averaged ranking metric of nearly 2.5 after 60 time steps.

13.4.3 Identifying Group Leaders from Video Data

Our third, more practical example, deals with the following application. Consider a group of people, among which there are subgroups of leaders and followers. The followers coordinate their paths and motion with the leader. Using video observations only of the group, determine who the group leaders are. To that end,

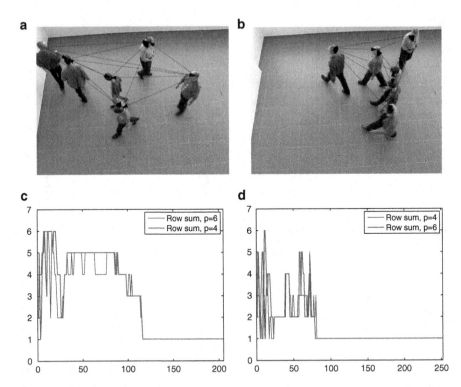

Fig. 13.6 Reconstructed instantaneous causal diagrams shown with the corresponding video frames (*upper panel*), and causality ranking performance over time (*lower panel*). (**a**) Video 1. (**b**) Video 2. (**c**) Video 1. (**d**) Video 2

one must first develop a procedure for estimating the trajectories of n people from a given video sequence. The input to the described procedure is a movie with n moving people, where n is known. The objective is to track each person along the frame sequence, and then feed this information into the CN mechanism for inferring the leaders and followers.

As we are dealing with a rather noiseless and non-cluttered scenario, a simple k-means clustering was used to recover individual person tracks from SIFT (scale-invariant feature transform) features. This approach was applied to two different video sequences in which there were five followers and one leader. Snapshots are shown in the upper panel in Fig. 13.6. In these videos, the actual leader (designated by a red shirt) performs a random trajectory, and the followers loosely follow its motion pattern. The clustering procedure described above is used to estimate the trajectories of the objects (the trajectories were filtered using a simple moving-average procedure to reduce the amount of noise contributed by the k-means clustering method). These trajectories were fed into the causality inference scheme.

The results of this procedure are shown in the bottom panel in Fig. 13.6, which depicts the causality performance index for two values of the finite-time horizon (wake parameter), p. It is clearly seen that from a certain time point the algorithm identifies the *actual* leader in both videos irrespective to the value of p.

13.5 Concluding Remarks

A novel causal reasoning framework has been proposed for ranking agents with respect to their contribution in shaping the collective behavior of the system. The proposed scheme copes with clutter and multiple emissions from extended agents by employing a Markov chain Monte Carlo group tracking method. This approach has been successfully applied for identifying leaders in groups in both synthetic and realistic scenarios.

Appendix

Evolutionary MCMC Implementation

The basic MH scheme can be used to produce several chain realizations each starting from a different (random) state. In that case, the entire population of the converged MH outputs (i.e., subsequent to the burn-in period) approximates the stationary distribution. Using a population of chains enjoys several benefits compared to a single-chain scheme. The multiple-chain approach can dramatically improve the diversity of the produced samples as different chains explore various regions that may not be reached in a reasonable time when using a single chain realization [23, 25]. Furthermore, having a population of chains facilitates the implementation of interaction operators that manipulate information from different realizations for improving the next generation of samples.

Following the approach of [33], the evolutionary MCMC cluster tracking algorithm uses genetic operators to generate new samples. The decoding scheme used here simply transforms the samples into their binary representations.

Let $\mathscr{G}_l = \{\mathbf{x}_k^{(i)}, e_k^{(i)}, \mathbf{x}_{k-1}^{(i)}, e_{k-1}^{(i)}\}_{i=1}^N$ be the lth realization of the converged chain at time k. Define by

$$\mathscr{G} := \{\mathscr{G}_1, \ldots, \mathscr{G}_L\} \tag{13.18}$$

the entire population set consisting of L chain realizations. In order to produce an improved generation of N samples from the joint filtering pdf, members of the population $\mathscr{G}$ undergo two successive genetic operations: crossover and mutation.

Chromosomes and Sub-chromosomes

Any genetic manipulation act on a unique data structure known as a chromosome which usually takes the form of a string. Here, a chromosome refers to a binary representation of a particle $(\mathbf{x}_k^{(i)}, e_k^{(i)})$. Since every particle consists of several clusters endowed with their own individual properties, $\mu_k^{j,(i)}, \dot{\mu}_k^{j,(i)}, \Sigma_k^{j,(i)}, w_k^{j,(i)}$ and $\rho_k^{j,(i)}$, in practice a chromosome consists of several concatenated binary strings each corresponding to a distinct property of a certain cluster. In this work, we term sub-chromosomes, the strings pertaining to individual properties. Assuming there are no more than n clusters for which there are exactly 5 properties yields a chromosome that is built up of $5n$ sub-chromosomes. The active sub-chromosomes within a chromosome are those that belong to active clusters, i.e., clusters for which $e_k^{j,(i)} = 1$, $j = 1, \ldots, n$.

The Crossover Operator

The crossover works by switching genetic material between two parent samples taken from two different chain realizations for producing an offspring. The two parents, $(\mathbf{x}_k, e_k)_1$ and $(\mathbf{x}_k, e_k)_2$ are independently drawn from $\hat{p}(\mathbf{x}_k, e_k \mid z_{0:k})$, i.e., they are picked uniformly at random from the population $\mathscr{G}$. The sub-chromosomes A and B corresponding to the same property in the chosen parents are then manipulated as follows. For every $r \in [1, r_s]$, where r_s denotes the string length of either A or B, the bits A_r and B_r are swapped with some predetermined probability β. The resulting offspring sub-chromosomes are then encoded to produce two new candidates $(\mathbf{x}_k', e_k')_1$ and $(\mathbf{x}_k', e_k')_2$. At this point an additional MH step is performed for deciding whether the new offspring will be a part of the improved population. This step is crucial for maintaining an adequate approximation of the target distribution. In order to ensure that the resulting chain is reversible, on acceptance both new candidates should replace their parents, otherwise both parents should be retained [33].

Following the above argument, it can be easily verified that the acceptance probability of both offspring is [33]

$$\min\left\{1, \left(\frac{1-\beta}{\beta}\right)^a \frac{\hat{p}((\mathbf{x}_k', e_k')_1 \mid z_{0:k})\hat{p}((\mathbf{x}_k', e_k')_2 \mid z_{0:k})}{\hat{p}((\mathbf{x}_k, e_k)_1 \mid z_{0:k})\hat{p}((\mathbf{x}_k, e_k)_2 \mid z_{0:k})}\right\} \tag{13.19}$$

where a denotes the total number of swapped bits.

The Mutation Operator

The mutation operator flips the rth bit within a given chromosome with probability β_m. Let $(\mathbf{x}_k, e_k)$ be a sample drawn from $\hat{p}(\mathbf{x}_k, e_k \mid z_{0:k})$ (i.e., picked uniformly

---

Algorithm 6 Evolutionary MCMC cluster tracking

---

1. Execute Algorithm 3 for L chains in parallel.
2. Subsequent to the end of the burn in period proceed as follows for every chain sample.
3. *Interaction:* Perform genetic operations to obtain an improved offspring

   a. Picking uniformly at random two distinct chain realizations perform crossover between their latest accepted samples.
   b. Compute the acceptance probability of the two offspring using (13.19).
   c. Accept the new offspring accordingly or retain both parent samples.
   d. Mutate the accepted samples of either chains and compute the associated acceptance probability (13.20).
   e. Accept the mutated sample accordingly.

---

at random from the population $\mathscr{G}$). Then, it can be verified that the acceptance probability of a mutated candidate $(\mathbf{x}'_k, e'_k)$ is [33]

$$\min\left\{1, \left(\frac{1-\beta_m}{\beta_m}\right)^a \frac{\hat{p}(\mathbf{x}'_k, e'_k \mid z_{0:k})}{\hat{p}(\mathbf{x}, e_k \mid z_{0:k})}\right\} \tag{13.20}$$

where a denotes the total number of bits changed.

A single cycle of the evolutionary MCMC filtering scheme is summarized in Algorithm 6.

References

1. Albert, R., Barabsi, A.-L.: Statistical mechanics of complex networks. Rev. Mod. Phys. **74**(1), 47–97 (2002)
2. Ali, S., Shah, M.: Floor fields for tracking in high density crowd scenes. In: Computer Vision – ECCV 2008. Volume 5303 of Lecture Notes in Computer Science, pp. 1–14. Springer, Berlin/Heidelberg (2008)
3. Angelova, D., Mihaylova, L.: Extended object tracking using Monte Carlo methods. IEEE Trans. Signal Process. **56**(2), 825–832 (2008)
4. Arulampalam, M., Maskell, S., Gordon, N., Clapp, T.: A tutorial on particle filters for online nonlinear/non-Gaussian Bayesian tracking. IEEE Trans. Signal Process. **50**(2), 174–188 (2002)
5. Berzuini, C., Nicola, G., Gilks, W.R., Larizza, C.: Dynamic conditional independence models and Markov chain Monte Carlo methods. J. Am. Stat. Assoc. **92**(440), 1403–1412 (1997)
6. Bhaskar, H., Mihaylova, L.: Combined data association and evolving population particle filter for tracking of multiple articulated targets. EURASIP J. Image Video Process. **2011**, article ID 642532 (2011)
7. Bhaskar, H., Mihaylova, L., Maskell, S.: Population-based particle filters. In: Proceedings of the from the Institution of Engineering and Technology (IET) Seminar on Target Tracking and Data Fusion: Algorithms and Applications, Birmingham, pp. 31–38 (2008)
8. Cappé, O., Guillin, A., Marin, J.-M., Robert, C.P., Roberty, C.P.: Population Monte Carlo. J. Comput. Gr. Stat. **13**, 907–929 (2004)
9. Carmi, A., Septier, F., Godsill, S.J.: The Gaussian mixture MCMC particle algorithm for dynamic cluster tracking. In: Proceedings of the 12th International Conference on Information Fusion, pp. 1179–1186. Seattle, WA (2009)

10. Cheng, J., Greiner, R., Kelly, J., Bell, D., Liu, W.: Learning Bayesian network from data: an information-theory based approach. Artif. Intell. **137**(1–2), 43–90 (2002)
11. Dorogovtsev, S.N., Mendes, J.F.F.: Evolution of networks. Adv. Phys. **51**, 1079–1187 (2002)
12. Geffner, H.: Default Reasoning: Causal and Conditional Theories. MIT, Cambridge (1992)
13. Geyer, C.: Markov chain maximum likelihood. In: Keramigas, E. (ed.) Computing Science and Statistics: The 23rd Symposium on the Interface. Interface Foundation, Fairfax (1991)
14. Geyer, C., Thompson, E.A.: Annealing Markov chain Monte Carlo with applications to ancestral inference. J. Am. Stat. Assoc. **90**, 909–920 (1995)
15. Gning, A., Mihaylova, L., Maskell, S., Pang, S.K., Godsill, S.: Group object structure and state estimation with evolving networks and Monte Carlo methods. IEEE Trans. Signal Process. **12**(2), 523–536 (2011)
16. Golyandina, N., Nekrutkin, V., Zhigljavsky, A. (eds.): Analysis of Time Series Structure: SSA and Related Techniques. Chapman and Hall, Boca Raton (2001)
17. Granger, C.W.J.: Investigating causal relations by econometric models and cross-spectral methods. Econometrica **37**, 424–438 (1969)
18. Green, P.J.: Reversible jump Markov chain Monte Carlo computation and Bayesian model determination. Biometrika **82**(4), 711–732 (1995)
19. Helbing, D.: Traffic and related self-driven many-particle systems. Rev. Mod. Phys. **73**, 1067–1141 (2002)
20. Holland, P.W.: Statistics and causal inference. J. Am. Stat. Assoc. **81**, 945–960 (1986)
21. Iba, Y.: Population-based Monte Carlo algorithms. J. Comput. Gr. Stat. **13**(4), 175–193 (2000)
22. Iba, Y.: Population Monte Carlo algorithms. Trans. Jpn. Soc. Artif. Intell. **16**, 279 (2000)
23. Jasra, A., Stehphens, D.A., Holmes, C.C.: Population-based reversible jump Markov chain Monte Carlo. Biometrica **94**(4), 787–807 (2007)
24. Khan, Z., Balch, T., Dellaert, F.: MCMC-based particle filtering for tracking a variable number of interacting targets. IEEE Trans. Pattern Anal. Mach. Intell. **27**(11), 1805–1819 (2005)
25. Liu, J.S.: Monte Carlo Strategies in Sceintific Computing. Springer, New York (2001)
26. Mahler, R.: Statistical Multisource-Multitarget Information Fusion. Artech House, Boston (2007)
27. Mihaylova, L., Boel, R., Hegyi, A.: Freeway traffic estimation within recursive Bayesian framework. Automatica **43**(2), 290–300 (2007)
28. Pang, S.K., Li, J., Godsill, S.J.: Detection and tracking of coordinated groups. IEEE Trans. Aerosp. Electron. Syst. **47**(1), 472–502 (2011)
29. Pantrigo, J., Sánchez, A., Gianikellis, K., Monteymayor, A.S.: Combining particle filter and population based metahuristics for visual articulated object tracking. Electron. Lett. Comput. Vis. Image Anal. **5**(3), 68–83 (2005)
30. Pearl, J.: Causality: Models, Reasoning, and Inference. Cambridge University Press, Cambridge, UK (2000)
31. Reynolds, C.W.: Flocks, herds, and schools: a distributed behavioral model. Comput. Gr. **21**, 25–34 (1987)
32. Shoam, Y.: Reasoning About Change: Time and Causation from the Standpoint of Artificial Intelligence. MIT, Cambridge (1988)
33. Strens, M.: Evolutionary MCMC sampling and optimization in discrete spaces. In: Proceedings of the Twentieth International Conference on Machine Learning, Washington, DC (2003)
34. Vo, B., Singh, S., Doucet, A.: Sequential Monte Carlo methods for multi-target filtering with random finite sets. IEEE Trans. Aerosp. Electron. Syst. **41**(4), 1224–1245 (2005)

Chapter 14
Crowd Counting and Profiling: Methodology and Evaluation

Chen Change Loy, Ke Chen, Shaogang Gong, and Tao Xiang

Abstract Video imagery based crowd analysis for population profiling and density estimation in public spaces can be a highly effective tool for establishing global situational awareness. Different strategies such as counting by detection and counting by clustering have been proposed, and more recently counting by regression has also gained considerable interest due to its feasibility in handling relatively more crowded environments. However, the scenarios studied by existing regression-based techniques are rather diverse in terms of both evaluation data and experimental settings. It can be difficult to compare them in order to draw general conclusions on their effectiveness. In addition, contributions of individual components in the processing pipeline such as feature extraction and perspective normalization remain unclear and less well studied. This study describes and compares the state-of-the-art methods for video imagery based crowd counting, and provides a systematic evaluation of different methods using the same protocol. Moreover, we evaluate critically each processing component to identify potential bottlenecks encountered by existing techniques. Extensive evaluation is conducted on three public scene datasets, including a new shopping center environment with labelled ground truth for validation. Our study reveals new insights into solving the problem of crowd analysis for population profiling and density estimation, and considers open questions for future studies.

C.C. Loy (✉)
Department of Information Engineering, The Chinese University of Hong Kong, Shatin, N.T., Hong Kong
e-mail: ccloy@ie.cuhk.edu.hk; ccloy@visionsemantics.com

K. Chen • S. Gong • T. Xiang
Queen Mary University of London, London E1 4NS, UK
e-mail: cory@eecs.qmul.ac.uk; sgg@eecs.qmul.ac.uk; txiang@eecs.qmul.ac.uk

S. Ali et al. (eds.), *Modeling, Simulation and Visual Analysis of Crowds*, The International Series in Video Computing 11, DOI 10.1007/978-1-4614-8483-7__14,
© Springer Science+Business Media New York 2013

14.1 Introduction

The analysis of crowd dynamics and behaviors is a topic of great interest in sociology, psychology, safety, and computer vision. In the context of computer vision, many interesting analyses can be achieved [91], e.g., to learn the crowd flow evolvement and floor fields [3], to track an individual in a crowd [65], to segment a crowd into semantic regions [51,93], to detect salient regions in a crowd [53], or to recognize anomalous crowd patterns [41,60]. A fundamental task in crowd analysis that enjoys wide spectrum of applications is to automatically count the number of people in crowd and profile their behaviors over time in a given region.

One of the key application areas of crowd counting is public safety and security. Tragedies involving large crowds often occur, especially during religious, political, and musical events [35]. For instance, a crowd crush at the 2010 Love Parade music festival in Germany, caused a death of 21 people and many more injured (see Fig. 14.1). And more recently a stampede happened near the Sabarimala Temple, India with death toll crosses 100. These tragedies could be avoided, if a safer site design took place and a more effective crowd control was enforced. Video imagery based crowd counting can be a highly beneficial tool for early detection of over-crowded situations to facilitate more effective crowd control. It also helps in profiling the population movement over time and across spaces for establishing global situational awareness, developing long-term crowd management strategies, and designing evacuation routes of public spaces.

In retail sectors, crowd counting can be an intelligence gathering tool [76] to provide valuable indication about the interest of customers through quantifying the number of individuals browsing a product, the queue lengths, or the percentage of store's visitors at different times of the day. The information gathered can then be used to optimize the staffing need, floor plan, and product display.

Video imagery based crowd counting for population profiling remains a non-trivial problem in crowded scenes. Specifically, frequent occlusion between pedestrians and background clutter render a direct implementation of standard object segmentation and tracking infeasible. The problem is further compounded by visual ambiguities caused by varying individual appearances and body articulations,

Fig. 14.1 Example of surveillance footage frames captured during the Love Parade music festival in Germany, 2010, before the fatalities occurred (Images from www.dokumentation-loveparade. com/)

and group dynamics. External factors such as camera viewing angle, illumination changes, and distance from the region of interest also pose great challenges to the counting problem.

Various approaches for crowd counting have been proposed. A popular method is *counting by detection* [24], which detects instances of pedestrian through scanning the image space using a detector trained with local image features. An alternative approach is *counting by clustering* [7, 63], which assumes a crowd to be composed of individual entities, each of which has unique yet coherent motion patterns that can be clustered to approximate the number of people. Another method is inspired by the capability of human beings, in determining density at a glance without numerating the number of pedestrians in it. This approach is known as *counting by regression* [12, 22], which counts people in crowd by learning a direct mapping from low-level imagery features to crowd density.

In this study, we provide a comprehensive review, comparative evaluation, and critical analysis on computer vision techniques for crowd counting, also known as crowd density estimation, and discuss crowd counting as a tool for population profiling. We first present a structured critical overview of different approaches to crowd counting reported in the literature, including pedestrian detection, coherent motion clustering, and regression-based learning. In particular, we focus on the regression-based techniques that have gain considerable interest lately due to their effectiveness in handling more crowded scenes. We then provide analysis of different regression-based approaches to crowd counting by systematic comparative evaluation, which gives new insights into contributions of key constituent components and potential bottlenecks in algorithm design. To facilitate our experiments, we also introduce a new shopping mall dataset of over 60,000 pedestrians labelled in 2000 video frames, i.e., the largest dataset to date in terms of the number of pedestrian instances captured in realistic crowded public space scenario for crowd counting and profiling research.

14.2 Survey of the State of the Art

The taxonomy of crowd counting algorithms can be generally grouped into three paradigms, namely counting by detection, clustering, and regression. In this section, we provide an overview on each of the paradigms, with a particular focus on the counting by regression strategy that has shown to be effective on more crowded environments.

14.2.1 Counting by Detection

The following is a concise account of pedestrian detection with emphasize on counting application. A more detailed treatment on this topic can be found in [24].

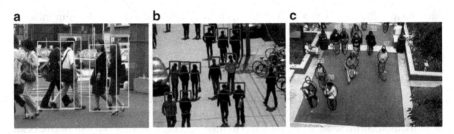

Fig. 14.2 Pedestrian detection results obtained using (**a**) monolithic detection, (**b**) part-based detection, and (**c**) shape matching (Images from [43,46,92])

Monolithic Detection: The most intuitive and direct approach to numerate the number of people in a scene is through detection. A typical pedestrian detection approach is based on monolithic detection [21,43,78], which trains a classifier using the full-body appearance of a set of pedestrian training images (see Fig. 14.2a). Common features to represent the full-body appearance include Haar wavelets [80], gradient-based features such as histogram of oriented gradient (HOG) feature [21], edgelet [85], and shapelets [68]. The choice of classifier imposes significant impact on the speed and quality of detection, often requiring a trade-off between these two. Non-linear classifiers such as RBF Support Vector Machines (SVMs) offer good quality but suffer from low detection speed. Consequently, linear classifiers such as boosting [81], linear SVMs, or Random/Hough Forests [28] are more commonly used. A trained classifier is then applied in a sliding window fashion across the whole image space to detect pedestrian candidates. Less confident candidates are normally discarded using non-maximum suppression, which leads to final detections that suggest the total number of people in a given scene. Whole body monolithic detector can generates reasonable detections in sparse scenes. However, it suffers in crowded scenes where occlusion and scene clutter are inevitable [24].

Part-based Detection: A plausible way to get around the partial occlusion problem to some extent is by adopting a part-based detection method [26, 48, 86]. For instance, one can construct boosted classifiers for specific body parts such as the head and shoulder to estimate the people counts in a monitored area [46] (see Fig. 14.2b). It is found that head region alone is not sufficient for reliable detection due to its shape and appearance variations. Including the shoulder region to form an omega-like shape pattern tends to give better performance in real-world scenarios [46]. The detection performance can be further improved by tracking validation, i.e., associating detections over time and rejecting spurious detections that exhibit coherent motion with the head candidates [62]. In comparison to monolithic detection, part-based detection relaxes the stringent assumption about the visibility of the whole body, it is thus more robust in crowded scenes.

Shape Matching: Zhao et al. [92] define a set of parameterized body shapes composed of ellipses, and employ a stochastic process to estimate the number and shape configuration that best explains a given foreground mask in a scene. Ge and

Collins [29] extend the idea by allowing more flexible and realistic shape prototypes than just simple geometric shapes proposed in [92]. In particular, they learn a mixture model of Bernoulli shapes from a set of training images, which is then employed to search for maximum a posteriori shape configuration of foreground objects, revealing not only the count and location, but also the pose of each person in a scene.

Multi-Sensor Detection: If multiple cameras are available, one can further incorporate multi-view information to resolve visual ambiguities caused by inter-object occlusion. For example, Yang et al. [88] extracted the foreground human silhouettes from a network of cameras to establish bounds on the number and possible locations of people. In the same vein, Ge and Collins [30] estimate the number of people and their spatial locations by leveraging multi-view geometric constraints. The aforementioned methods [30, 88] are restricted since a multi-camera setup with overlapping views is not always available in many cases. Apart from detection accuracy improvement, the speed of detection can benefit from the use of multi-sensors, e.g., the exploitation of geometric context extracted from stereo images [5].

Transfer Learning: Applying a generic pedestrian detector to a new scene cannot guarantee satisfactory cross-dataset generalization [24], whilst training a scene-specific detector for counting is often laborious. Recent studies have been exploring the transfer of generic pedestrian detectors to a new scene without human supervision. The key challenges include the variations of viewpoints, resolutions, illuminations, and backgrounds in the new environment. A solution to the problem is proposed in [82, 83] to exploit multiple cues such as scene structures, spatio-temporal occurrences, and object sizes to select confident positive and negative examples from the target scene to adapt a generic detector iteratively.

14.2.2 Counting by Clustering

The counting by clustering approach relies on the assumption that individual motion field or visual features are relatively uniform, hence coherent feature trajectories can be grouped together to represent independently moving entities. Studies that follow this paradigm include [63], which uses a Kanade-Lucas-Tomasi (KLT) tracker to obtain a rich set of low-level tracked features, and clusters the trajectory to infer the number of people in the scene (see Fig. 14.3a); and [7], which tracks local features and groups them into clusters using Bayesian clustering (see Fig. 14.3b). Another closely related method is [77], which incorporates the idea of feature constancy into a counting by detection framework. The method first generates a set of person hypotheses of a crowd based on head detections. The hypotheses are then refined iteratively by assigning small patches of the crowd to the hypotheses based on the constancy of motion fields and intra-garment color (see Fig. 14.3c).

Fig. 14.3 (Color online) (**a**) and (**b**) show the results of clustering coherent motions using methods proposed in [63] and [7] respectively. (**c**) Shows the pairwise affinity of patches (strong affinity=*magenta*, weak affinity=*blue*) in terms of motion and color constancy; the affinity is used to determine the assignment of patches to person hypotheses [77] (Images from [7, 63, 77])

The aforementioned methods [7, 63] avoid supervised learning or explicit modelling of appearance features as in the counting by detection paradigm. Nevertheless, the paradigm assumes motion coherency, hence false estimation may arise when people remaining static in a scene, exhibiting sustained articulations, or two objects sharing common feature trajectories over time. Note that counting by clustering only works with continuous image frames, not static images whilst the counting by detection and regression do not have this restriction.

14.2.3 Counting by Regression

Despite the substantial progress being made in object detection [24] and tracking [90] in recent years, performing either in isolation or both reliably in a crowded environment remains a non-trivial problem. Counting by regression deliberately avoids actual segregation of individual or tracking of features but estimate the crowd density based on holistic and collective description of crowd patterns. Since neither explicit segmentation nor tracking of individual are involved, counting by regression becomes a feasible method for crowded environments where detection and tracking are severely limited intrinsically.

One of the earliest attempts in exploring the use of regression method for crowd density estimation is by Davies et al. [22]. They first extract low-level features such as foreground pixels and edge features from each video frame. Holistic properties such as foreground area and total edge count are then derived from the raw features. Consequently, a linear regression model is used to establish a direct mapping between the holistic patterns and the actual people counts. Specifically, a function is used to model how the input variable (i.e., the crowd density) changes when the target variables (i.e., holistic patterns) are varied. Given an unseen video frame, conditional expectation of the crowd density can then be predicted

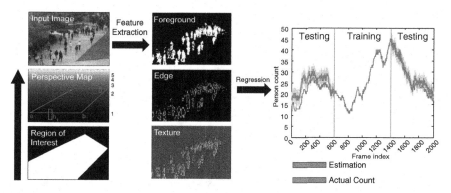

Fig. 14.4 A typical pipeline of counting by regression: first defining the region of interest and finding the perspective normalization map of a scene, then extracting holistic features and training a regressor using the perspective normalized features

given the extracted features from that particular frame. Since the work of Davies et al. [22], various methods have been proposed following the same idea with improved feature sets or more sophisticated regression models, but still sharing a similar processing pipeline as in [22] (see Fig. 14.4). A summary of some of the notable methods is given in Table 14.1. In the following subsections, we are going to have detailed discussion on the main components that constitute the counting by regression pipeline, namely feature representation, geometric correction, and regression modelling.

14.2.3.1   Feature Representation

The question of crowd representation or abstraction must be addressed before a regression function can be established. Feature representation concerns the extraction, selection, and transformation of low-level visual properties in an image or video to construct intermediate input to a regression model. A popular approach is to combine several features with complementary nature to form a large bank of features [13].

Foreground Segment Features: The most common or arguably the most descriptive representation for crowd density estimation is foreground segment, which can be obtained through background subtraction, such as mixture of Gaussians-based technique [73] or mixture of dynamic textures-based method [10]. Various holistic features can be derived from the extracted foreground segment, for example:

* Area – total number of pixels in the segment.
* Perimeter – total number of pixels on the segment perimeter.
* Perimeter-area ratio – ratio between the segment perimeter and area, which measures the complexity of the segment shape.

Table 14.1 A table summarizing existing counting by regression methods. Note that only publicly available datasets are listed in the datasets column

	Year	Segment	Edge	Texture	Shape	Intensity	Gradients	Motion	Others	Learning Regression method	Level	Datasets
Davies et al. [22]	1995	✓	✓	–	–	–	–	–	–	Linear regression	Global	–
Marana et al. [57]	1997	–	–	✓	–	–	–	–	–	Self-organising map neural network	Global	–
Cho et al. [16]	1997	✓	✓	–	–	–	–	–	–	Feedforward neural network	Global	–
Kong et al. [38,39]	2005 2006	✓	✓	–	–	–	–	–	–	Feedforward neural network	Global	–
Dong et al. [25]	2007	–	–	–	✓	–	–	–	–	Shape matching + locally-weighted regression	Segment	USC campus plaza
Chan et al. [12–14]	2008 2009	✓	✓	✓	–	–	–	–	–	Gaussian processes	Global	UCSD pedestrian, PETS 2009
Chan et al. [11]	2009	✓	✓	✓	–	–	–	–	–	Bayesian poisson regression	Global	UCSD pedestrian
Ryan et al. [67]	2009	✓	✓	–	–	–	–	–	–	Feedforward neural network	Segment	UCSD pedestrian
Cong et al. [18]	2009	✓	✓	–	–	–	–	–	–	Polynomial regression	Segment	–
Lempitsky	2010	✓	–	–	–	✓	✓	–	–	Density function minimisation based on maximum excess over subarrays distance	Pixel	UCSD pedestrian
Conte et al. [19]	2010	–	–	–	–	–	–	–	Number of SURF points	Support vector regression	Segment	PETS 2009
Benabbas et al. [4]	2010	✓	–	–	–	–	–	✓	–	Linear regression	Segment	PETS 2009
Li et al. [47]	2011	✓	✓	–	–	–	–	–	–	Pedestrian detector + Linear regression	Segment	CASIA pedestrian [45]
Lin et al. [49]	2011	✓	✓	–	–	–	✓	–	–	Gaussian processes	Segment	UCSD pedestrian, PETS 2009
Ke et al. [15]	2012	✓	✓	✓	–	–	–	–	–	Kernel ridge regression	Segment	UCSD pedestrian, PETS 2009, mall

- Perimeter edge orientation – orientation histogram of the segment perimeter.
- Blob count – the number of connected components with area larger than a predefined threshold, e.g., 20 pixels in size.

Various studies [13, 22, 54] have demonstrated encouraging results using the segment-based features despite its simplicity. Several considerations, however, has to be taken into account during the implementation. Firstly, to reduce spurious foreground segments from other regions, one can confine the analysis within a region of interest (ROI), which can be determined manually or following a foreground accumulation approach [54]. Secondly, different scenarios may demand different background extraction strategies. Specifically, dynamic background subtraction [73] can cope with gradual illumination change but have difficulty in isolating people that are stagnant for a long period of time; static background subtraction [51, 66] is able to segment static objects from the background but is susceptible to lighting change.

Finally, poor estimation is expected if one employs only foreground area due to inter-object occlusion, as it is possible to insert another person into the mixture and end up with the same foreground area. Enriching the representation with other descriptors may solve this problem to certain extent.

Edge Features: While foreground features capture the global properties of the segment, edge features inside the segment carries complementary information about the local and internal patterns [13, 22, 38]. Intuitively, low-density crowds tend to present coarse edges, while segments with dense crowds tend to present complex edges. Edges can be detected using an edge detector such as the Canny edge detector [8]. Note that an edge image is often masked using the foreground segment to discard irrelevant edges. Some common edge-based features are listed as follows

- Total edge pixels – total number of edge pixels.
- Edge orientation – histogram of the edge orientations in the segment.
- Minkowski dimension – the Minkowski fractal dimension or box-counting dimension of the edges [59], which counts how many pre-defined structuring elements are required to fill the edges.

Texture and Gradient Features: Crowd texture and gradient patterns carry strong cues about the number of people in a scene. In particular, high-density crowd region tends to exhibit stronger texture response [54] with distinctive local structure in comparison to low-density region; whilst local intensity gradient map could reveal local object appearance and shape such as human shoulder and head, which are informative for density estimation. Example of texture and gradient features employed for crowd counting include gray-level co-occurrence matrix (GLCM) [34], local binary pattern (LBP) [61], HOG feature [56], and gradient orientation co-occurrence matrix (GOCM) [56]. A comparative studies among the aforementioned texture and gradient features can be found in [56]. Here we provide a brief description on GLCM and LBP, which are used in our evaluation.

Gray-level co-occurrence matrix (GLCM) [34] (See Fig. 14.5) is widely used in various crowd counting studies [13, 56, 58, 87]. For instance, Marana et al. [58] uses GLCM to distinguish five different density levels (very low, low, moderate, high, and very high), and Chan and Vasconcelos [12] employ it as holistic property for Bayesian density regression. To obtain GLCM, a typical process is to first quantize the image into eight gray-levels and masked by the foreground segment. The joint probability or co-occurrence of neighboring pixel values, $p(i, j \mid \theta)$ is then estimated for four orientations, $\theta \in \{0°, 45°, 90°, 135°\}$. After extracting the co-occurrence matrix, a set of features such as homogeneity, energy, and entropy can be derived for each θ

- Homogeneity – texture smoothness, $g_\theta = \sum_{i,j} \frac{p(i,j \mid \theta)}{1+|i-j|}$
- Energy – total sum-squared energy, $e_\theta = \sum_{i,j} p(i,j \mid \theta)^2$
- Entropy – texture randomness, $h_\theta = \sum_{i,j} p(i,j \mid \theta) \log p(i,j \mid \theta)$

An alternative texture descriptor for crowd density estimation [55] is the local binary pattern (LBP) [61]. Local binary pattern has been widely adopted in various

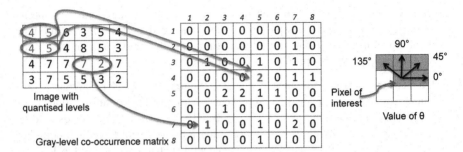

Fig. 14.5 *Gray*-level co-occurrence matrix, with $\theta = 0°$ of a 4-by-6 image. Element (7, 2) in the GLCM contains the value 1 because there is only one instance in the image where two, horizontally adjacent pixels have the values 7 and 2. Element (4, 5) in the GLCM contains the value 2 because there are two instances in the image where two, horizontally adjacent pixels have the values 4 and 5. The value of θ specifies the angle between the pixel of interest and its neighbor

applications such as face recognition [2] and expression analysis [70], due to its high discriminative power, invariance to monotonic gray-level changes, and its computational efficiency.

An illustration of a basic LBP operator is depicted in Fig. 14.6. The LBP operation is governed by a definition of local neighborhood, i.e., the number of sampling point and radius centering the pixel of interest. An example of a circular (8, 1) neighborhood is shown in Fig. 14.6. Following the definition of neighborhood, we sample 8 points at a distance of radius 1 from the pixel of interest and threshold them using the value of the centering pixel. The results are concatenated to form a binary code as the label of the pixel of interest. These steps are repeated over the whole image space and a histogram of labels is constructed as a texture descriptor.

In this study, we employed an extension of the original LBP operator known as *uniform patterns* [61], which frequently correspond to primitive micro-features such as edges and corners. A uniform LBP pattern is binary code with at most two bitwise transitions, e.g., 11110000 (1 transition) and 11100111 (2 transitions) are uniform, whilst 11001001 (4 transitions) is not. In the construction of LBP histogram, we assign a separate bin for every uniform pattern and keep all nonuniform patterns in a single bin, so we have a 58 + 1-dimension texture descriptor.

14.2.3.2 Geometric Correction

A problem commonly encountered in counting by regression framework is perspective distortion, in which far objects appear smaller than those closer to the camera view. As a consequence, features (e.g., segment area) extracted from the same object at different depths of the scene would have huge difference in values. The influence is less critical if one divides the image space into different cells, each of which modelled by a regression function; erroneous results are expected if one only uses a single regression function for the whole image space.

To address this problem geometric correction or perspective normalization is performed to bring perceived size of objects at different depths to the same scale.

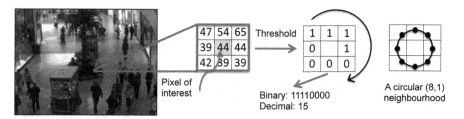

Fig. 14.6 A basic local binary pattern operator [61] and a circular (8, 1) neighborhood

Ma et al. [54] investigate the influence of perspective distortion to people counting and propose a principled way to integrate geometric correction in pixel counting, i.e., to scale each pixel by a weight, with larger weights given to further objects.

A simple and widely adopted perspective normalization method [44, 49, 67] is described in [13]. The method first determines four points in a scene to form a quadrilateral that corresponds to a rectangle (see Fig. 14.7). The lengths of the two horizontal lines of the quadrilateral, $\overline{ab}$ and $\overline{cd}$, are measured as w_1 and w_2 respectively. When a reference pedestrian passes the two extremes, i.e., its bounding box's center touches the $\overline{ab}$ and $\overline{cd}$, its heights are recorded as h_1 and h_2. The weights at $\overline{ab}$ and $\overline{cd}$ are then assigned as 1 and $\frac{h_1 w_1}{h_2 w_2}$ respectively. To determine the remaining weights of the scene, linear interpolation is first performed on the width of the rectangle, and the height of the reference person. A weight at arbitrary image coordinate can then be calculated as $\frac{h_1 w_1}{h' w'}$, where h' and w' representing the interpolants. Here we make an assumption that the horizontal vanishing line to be parallel to the image horizontal scan lines.

When applying the weights to features, it is assumed that the size of foreground segment changes quadratically, whilst the total edge pixels changes linearly with respect to the perspective. Consequently, each foreground segment pixel is weighted using the original weight and the edge features are weighted by square-roots of the weights. Features based on the GLCM are normalized by weighting the occurrence of each pixel pair when accumulating the co-occurrence matrix shown in Fig. 14.5. To obtain perspective-normalized LBP-based features, we multiply the weights to the occurrence of individual LBP labels in the image space prior to the construction of the LBP label histogram.

The aforementioned method [13] requires manual measurement which could be error-prone. There exist approaches to compute camera calibration parameters based on accumulative visual evidence in a scene. For example, a method is proposed in [40] to find the camera parameters by exploiting foot and head location measurements of people trajectories over time. Another more recent method [50] relaxes the requirement of accurate detection and tracking. This method takes noisy foreground segments as input to obtain the calibration data by leveraging the prior knowledge of the height distribution. With a calibrated 3D model, one can also obtain the perspective map as in [14], which moves a virtual person within the 3D world and measures the number of pixels projected onto the 2D image space.

a   b   c

Fig. 14.7 (a) and (b) show a reference person at two extremes of a predefined quadrilateral; (c) a perspective map to scale pixels by their relative size in the three-dimensional scene

14.2.3.3 Regression Models

After feature extraction and perspective normalization, a regression model is trained to predict the count given the normalized features. A regression model may have a broad class of functional forms. In this section we discuss a few popular regression models for crowd density estimation.

Linear Regression: Given a training data comprising N observations $\{\mathbf{x}_n\}$, where $n = 1, \ldots, N$ together with corresponding continuous target values $\{y_n\}$, the goal of regression is to predict the value of y given a new value of $\mathbf{x}$ [6]. The simplest approach is to form of linear regression function $f(\mathbf{x}, \mathbf{w})$ that involves a linear combination of the input variables, i.e.,

$$f(\mathbf{x}, \mathbf{w}) = w_0 + w_1 x_1 + \cdots + w_D x_D, \tag{14.1}$$

where D is the dimension of features, $\mathbf{x} = (x_1, \ldots, x_D)^\mathsf{T}$, and $\mathbf{w} = (w_0, \ldots, w_D)^\mathsf{T}$ are the parameters of the model. This model is often known as *linear regression* (LR), which is a linear function of the parameters $\mathbf{w}$. In addition it is also linear with respect to the input variables $\mathbf{x}$.

In a sparse scene where smaller crowd size and fewer inter-object occlusions are observed, the aforementioned linear regressor [4, 22, 47] may suffice since the mapping between the observations and people count typically presents a linear relationship. Nevertheless, given a more crowded environment with severe inter-object occlusion, one may have to employ a nonlinear regressor to adequately capture the nonlinear trend in the feature space [9].

To relax the linearity assumption, one can take a linear combination of a fixed set of nonlinear functions of the input variables, also known as basis functions $\phi(\mathbf{x})$, to obtain a more expressive class of function. It has the form of

$$f(\mathbf{x}, \mathbf{w}) = \sum_{j=0}^{M-1} w_j \phi_j(\mathbf{x}) = \mathbf{w}^\mathsf{T} \phi(\mathbf{x}), \tag{14.2}$$

where M is the total number of parameters in this model, $\mathbf{w} = (w_0, \ldots, w_{M-1})^\mathsf{T}$, and $\phi = (\phi_0, \ldots, \phi_{M-1})^\mathsf{T}$. The functional form in (14.2) is still known as linear model since it is linear in $\mathbf{w}$, despite the function $f(\mathbf{x}, \mathbf{w})$ is nonlinear with respect to input vector $\mathbf{x}$. A polynomial regression function considered in [18] (see Table 14.1) is a specific example of this model, with the basis functions taking a form of powers of $\mathbf{x}$, that is $\phi_j(\mathbf{x}) = \mathbf{x}^j$. Gaussian basis function and sigmoidal basis function are other possible choices of basis functions.

Parameters in the aforementioned linear model is typically obtained by minimizing the sum of squared errors

$$E(\mathbf{w}) = \frac{1}{2} \sum_{n=1}^{N} \left\{ y_n - \mathbf{w}^\mathsf{T} \phi(\mathbf{x}_n) \right\}^2. \tag{14.3}$$

One of the key limitation of linear model is that the model can get unnecessarily complex give high-dimensional observed data $\mathbf{x}$. Particularly in counting by regression, it is a common practice to exploit high-dimensional features [13]. Some of the elements are not useful for predicting the count. In addition, some of them may be highly co-linear, unstable estimate of parameters may occurs [6], leading to very large magnitude in the parameters and therefore a clear danger of severe overfitting.

Partial Least Squares Regression: A way of addressing the multicollinearity problem is by *partial least squares regression* (PLSR) [31], which projects both input $\mathbf{X} = \{\mathbf{x}_n\}$ and target variables $\mathbf{Y} = \{y_n\}$ to a latent space, with a constraint such that the lower-dimensional latent variables explain as much as possible the covariance between $\mathbf{X}$ and $\mathbf{Y}$. Formally, the PLSR decomposes the input and target variables as

$$\mathbf{X} = \mathbf{T}\mathbf{P}^\mathsf{T} + \varepsilon_x \tag{14.4}$$

$$\mathbf{Y} = \mathbf{U}\mathbf{Q}^\mathsf{T} + \varepsilon_y, \tag{14.5}$$

where $\mathbf{T}$ and $\mathbf{U}$ are known as score matrices, with the column of $\mathbf{T}$ being the latent variables; $\mathbf{P}$ and $\mathbf{Q}$ are known as loading matrices [1]; and ε are the error terms. The decomposition are made so to maximize the covariance of $\mathbf{T}$ and $\mathbf{U}$. There are two typical ways in estimating the score matrices and loading matrices, namely NIPALS and SIMPLS algorithms [1, 89].

Kernel Ridge Regression: Another method of mitigating the multicollinearity problem is through adding a regularization term to the error function in Eq. (14.3). A simple regularization term is given by the sum-of-squares of the parameter vector elements, $\frac{1}{2}\mathbf{w}^\mathsf{T}\mathbf{w}$. The error function becomes

$$E_R(\mathbf{w}) = \frac{1}{2} \sum_{n=1}^{N} \left\{ y_n - \mathbf{w}^\mathsf{T} \phi(\mathbf{x}_n) \right\}^2 + \frac{\lambda}{2} \mathbf{w}^\mathsf{T} \mathbf{w}, \tag{14.6}$$

with λ to control the trade-off between the penalty and the fit. A common way of determining λ is via cross-validation. Using this particular choice of regularization term with $\phi(\mathbf{x}_n) = \mathbf{x}_n$, we will have error function of *ridge regression* [36].

A non-linear version of the ridge regression, known as *kernel ridge regression* (KRR) [69], can be achieved via kernel trick [71], whereby a linear ridge regression model is constructed in higher dimensional feature space induced by a kernel function defining the inner product

$$k(\mathbf{x}, \mathbf{x}') = \phi(\mathbf{x})^\mathsf{T} \phi(\mathbf{x}'). \tag{14.7}$$

For the kernel function, one has typical choices of linear, polynomial, and radial basis function (RBF) kernels. The regression function of KRR is given by

$$f(\mathbf{x}, \alpha) = \sum_{n=1}^{N} \alpha_n k(\mathbf{x}, \mathbf{x}_n), \tag{14.8}$$

where $\alpha = \{\alpha_1, \ldots, \alpha_n\}^\mathsf{T}$ are Lagrange multipliers. This solution is not sparse in the variables α, that is $\alpha_n \neq 0$, $\forall n \in \{1, \ldots N\}$.

Support Vector Regression: *Support vector regression* (SVR) [42, 72] has been used for crowd counting in [87]. In contrast to KRR, the SVR achieves sparseness in α (see Eq. (14.8)) by using the concept of support vectors to determine the solution, which can result in faster testing speed than KRR that sums over the entire training-set [84]. Specifically, the regression function of SVR can be written as

$$f(\mathbf{x}, \alpha) = \sum_{\text{SVs}} (\alpha_n - \alpha_n^*) k(\mathbf{x}, \mathbf{x}_n) + b, \tag{14.9}$$

where α_n and α_n^* represents the Lagrange multipliers, $k(\mathbf{x}, \mathbf{x}_n)$ denotes the kernel, and $b \in \mathbb{R}$. A popular error function for SVR training is ε-insensitive error function [79], which assigns zero error if the absolute difference between the prediction $f(\mathbf{x}, \alpha)$ and the target y is less than $\varepsilon > 0$. *Least-squares support vector regression* (LSSVR) [74] is least squares version of SVR. In LSSVR one finds the solution by solving a set of linear equations instead of a convex quadratic error function as in conventional SVR.

Gaussian Processes Regression: One of the most popular nonlinear methods for crowd counting is *Gaussian processes regression* (GPR) [64]. It has a number of pivotal properties – it allows possibly infinite number of basis functions driven by the data complexity, and it models uncertainty in regression problems elegantly.[1] Formally, we write the regression function as

---

[1] One can also estimate the predictive interval in other kernel methods such as KRR [23].

$$f(\mathbf{x}) \sim \text{GP}(m(\mathbf{x}), k(\mathbf{x}, \mathbf{x}')), \tag{14.10}$$

where Gaussian processes, $\text{GP}(m(\mathbf{x}), k(\mathbf{x}, \mathbf{x}'))$ is specified by its mean function $m(\mathbf{x})$ and covariance function or kernel $k(\mathbf{x}, \mathbf{x}')$

$$m(\mathbf{x}) = \mathbb{E}[f(\mathbf{x})], \tag{14.11}$$

$$k(\mathbf{x}, \mathbf{x}') = \mathbb{E}[(f(\mathbf{x}) - m(\mathbf{x}))(f(\mathbf{x}') - m(\mathbf{x}'))], \tag{14.12}$$

where $\mathbb{E}$ denotes the expectation value.

Apart from the conventional GPR, various extensions of it have been proposed. For instance, Chan and Dong [9] propose a generalised Gaussian process model, which allows different parameterisation of the likelihood function, including a Poisson distribution for predicting discrete counting numbers [11]. Lin et al. [49] employ two GPR in their framework, one for learning the observation-to-count mapping, and another one for reasoning the mismatch between predicted count and actual count due to occlusion. The key weakness of GPR is its poor tractability to large training sets. Various approximation paradigms have been developed to improve its scalability [64].

It is worth pointing out that one of the attractive properties of kernel methods such as KRR, SVR, and GPR is the flexibility of encoding different assumptions about the function we wish to learn. For instance, by combining different covariance functions $k(\mathbf{x}, \mathbf{x}')$, such as linear, Matérn, rational quadratic, and neural network, one has the flexibility to encode different assumptions on the continuity and smoothness of the GP function $f(\mathbf{x})$. This property is exploited in [13], in which linear and a squared-exponential (RBF) covariance functions are combined to capture both the linear trend and local non-linearities in the crowd feature space.

Random Forest Regression: Scalable nonlinear regression modelling can be achieved using *random forest regression* (RFR). A random forest comprises of a collection of randomly trained regression trees, which can achieve better generalisation than a single over-trained tree [20]. Each tree in a forest splits a complex nonlinear regression problem into a set of subproblems, which can be more easily handled by weak learners such as a linear model.[2] To train a forest, one optimizes an energy over a given training set and associated values of target variable. Specifically, parameters θ_j of the weak learner at each split node j are optimized via

$$\theta_j^* = \underset{\theta_j \in \mathcal{T}_j}{\text{argmax}} \, I_j, \tag{14.13}$$

---

[2]There are other weak learners that define the split functions, such as general oriented hyperplane or quadratic function. A more complex splitting function would lead to higher computational complexity.

where $\mathcal{T}_j \subset \mathcal{T}$ is a subset of parameters made available to the j-th node, and I is an objective function that often takes the form of information gain. Given a new observation $\mathbf{x}$, the predictive function is computed by averaging individual posterior distributions of all the trees, i.e.,

$$f(\mathbf{x}) = \frac{1}{T} \sum p_t(y|\mathbf{x}), \tag{14.14}$$

where T is the total number of trees in the forest, $p_t(y|\mathbf{x})$ is the posterior of t-th tree.

The hallmark of random forest is its good performance comparable to state-of-the-art kernel methods (e.g., GPR) but with the advantage of being scalable to large dataset and less sensitive to parameters. In addition, it has the ability of generating variable importance and information about outliers automatically. It is also reported in [20] that forest can yield a more realistic uncertainty in the ambiguous feature region, in comparison to GPR that tends to return largely over-confident prediction.

The weakness of RFR is that it is poor in extrapolating points beyond the value range of target variable within the training data, as we shall explain in more detail in Sect. 14.4.1.

14.2.3.4 Additional Considerations

We have discussed various linear and nonlinear functions for performing crowd density regression. Note that the functional form becomes more critical when one does not have sufficient training set that encompasses all the anticipated densities in a scene. If that is the case, extrapolation outside the training range has to be performed, with increasing room of failure when the extrapolation goes further beyond the existing data range, due to the mismatch between the regression assumption and the actual feature to count mapping.

A closely related consideration is at what level the learning should be performed. Most existing methods (see the 'level' column in Table 14.1) take a global approach by applying a single regression function over the whole image space with input variables being the holistic features of a frame (e.g., total area of foreground segment), and target variable being the total people count in that frame. An obvious limitation of this global approach is that it applies a global regression function over the whole image space, ignoring specific crowd structure in different regions. This can be resolved by dividing the image space up into regions and fitting separate function in each region [56, 87]. The regions can be cells having regular size, or having different resolutions driven by the scene perspective to compensate the distortion [56].

One can also approximate the people count at blob-level [47], i.e., estimates the number of people in each foreground blob and obtains the total people count by summing the blob-level counts. Lempitsky and Zisserman [44] go one step further

a   b

Fig. 14.8 (a) UCSD pedestrian dataset (*ucsd*), (**b**) PETS 2009 benchmark dataset (*pets*)

to model the density at each pixel, casting the problem as that of estimating an image density whose integral over any image region gives the count of objects within that region. The aforementioned segment-and-model strategies facilitate counting at arbitrary locations, which is impossible using a holistic approach. In addition, a potential gain in estimation accuracy may be obtained [44]. This however comes at a price of increased annotation effort. For example, requiring a large amount of dotted annotations on head or pedestrian positions in all training images [44].

14.3   Evaluation Settings

Previous work [12, 44, 54, 56] have independently performed analyses on different components in the crowd counting pipeline such as feature extraction, perspective normalization, and regression modelling. The scenarios studied, however, are rather diverse in terms of both evaluation data and experimental settings. It can be hard to compare them in order to draw general conclusions on their effectiveness. In this study we aim to provide a more exhaustive comparative evaluation to factor out the contributions of different components and identify potential bottlenecks in algorithm design for crowd counting and profile analysis.

14.3.1   Datasets

Two benchmark datasets were used for comparative algorithm evaluation, namely UCSD pedestrian dataset (*ucsd*) and PETS 2009 dataset (*pets*). Example frames are shown in Fig. 14.8. Apart from the two established benchmark datasets, a new and more realistic shopping *mall* dataset is also introduced in this study. This *mall* dataset was collected from a publicly accessible webcam in the course of 2 months

Fig. 14.9 The new shopping mall dataset. The *top-left* figure shows an example of annotated frame

from Feb 2011 to Apr 2011. A portion of 2,000 frames recorded during peak hours were selected for the comparative algorithm evaluation. As can be seen from the sample images in Fig. 14.9, this new dataset is challenging in that it covers crowd densities from sparse to crowded, as well as diverse activity patterns (static and moving crowds), under large range of illumination conditions at different time of the day. Also note that the perspective distortion is more severe than the *ucsd* and *pets* datasets, thus individual objects may experience larger change in size and appearance at different depths of the scene. The details of the three datasets are given in Table 14.2.

For evaluation purpose, we resized the images from the *pets* dataset to 384×288, and the images from the *mall* dataset to 320×240. All color images were converted to grayscale images prior to feature extraction. We annotated the data exhaustively by labelling the head position of every pedestrian in all frames. An example of annotated frame is shown in Fig. 14.9. The ground truth, together with the raw video sequence, extracted features, and the train/test partitions can be downloaded at http://personal.ie.cuhk.edu.hk/~ccloy/.

Table 14.2 Dataset properties: N_f =number of frames, R =resolution, FPS =frame per second, D =density (minimum and maximum number of people in the ROI), and Tp = total number of pedestrian instances

Data	N_f	R	FPS	D	Tp
UCSD [13]	2,000	238×158	10	11–46	49,885
Pets [27]	1,076	384×288	7	0–43	18,289
Mall	2,000	320×240	<2	13–53	62,325

14.3.2   Features and Regression Models

We selected features and regression methods that are both representative and promising in terms of originally reported performance. While we could not evaluate all the available features or methods exhaustively due to unavailability of original codes and practical time and space constraints, we consider that these evaluations giving an accurate portrait of the state-of-the-art.

We extracted segment, edge, GLCM, and LBP features following the methods described in Sect. 14.2.3.1. For both *UCSD* and *pets* datasets, scene lighting were stable so we employed a static background subtraction method based on minimum cuts [17][3] to extract the foreground segments. For the *mall* dataset, gradual illumination change was observed, we therefore adopted a dynamic background modelling method [95].

All features were perspective normalized (see Sect. 14.2.3.2) and a feature vector was formed by concatenating the features, into $\mathbf{x} \in \mathbb{R}^D$, which was used as the input for the regression models. Prior to feeding the features into the regression models, all features were scaled to the [0 1] interval. A list of the regression models and their associated settings is given below

- Linear regression (LR)
- Partial least-squares regression (PLSR) – ten latent components
- Kernel ridge regression (KRR) – linear kernel with four-fold cross-validation for parameter optimization
- Least-squares support vector regression (LSSVR) – linear kernel with four-fold cross-validation for parameter optimization
- Gaussian processes regression (GPR) – linear kernel + RBF kernel as in [13].[4] The parameters are first initialized to random values and optimized using conjugate gradient optimizer.
- Random forest regression (RFR) – 500 trees, the number of parameters made available for node splitting was fixed to square-root of the feature dimension, and the minimum size of terminal nodes was set to 5.

---

[3]Codes available at http://personal.ie.cuhk.edu.hk/~ccloy/.

[4]An interesting aspect not examined in our study is the effect of different kernels and their relations with different kernel methods for crowd regression.

14.3.3 Evaluation Metrics

We employed three metrics in performance evaluation. Two of the metrics are widely used as performance indicators for crowd counting, namely *mean absolute error* and *mean squared error*. Mean absolute error is defined as

$$\varepsilon_{\text{abs}} = \frac{1}{N} \sum_{n=1}^{N} |y_n - \hat{y}_n|. \tag{14.15}$$

Mean squared error is given as

$$\varepsilon_{\text{sqr}} = \frac{1}{N} \sum_{n=1}^{N} (y_n - \hat{y}_n)^2, \tag{14.16}$$

where N is the total number of test frames, y_n is the actual count, and $\hat{y}_n$ is the estimated count of nth frame. Note that as a result of the squaring of each difference, ε_{sqr} effectively penalizes large errors more heavily than small ones. The above two metrics are indicative in quantifying the error of estimation of the crowd count. However, as pointed out by Conte et al. [19], these metrics contain no information about the crowdedness of the region of interest. To that end, [19] proposed another performance metric to take the crowdedness into account – we name it as *mean deviation error*, which is essentially a normalized ε_{abs}

$$x\varepsilon_{\text{dev}} = \frac{1}{N} \sum_{n=1}^{N} \frac{|y_n - \hat{y}_n|}{y_n}. \tag{14.17}$$

14.4 Performance Comparison

In the following we report comparative evaluation results on three aspects, i.e., model choices, feature robustness, and model sensitivity to perspective.

14.4.1 Model Choices

The goals of this experiment are to (1) compare the performance of different regression models under different crowdedness levels, and (2) evaluate their generalization capability to unseen density. These two aspects are somewhat less explicitly studied in existing work. However, they are essential since a regressor may behave differently under different crowdedness levels, and often, it needs to extrapolate outside the anticipated density range in real-world scenarios.

We employed the same segment + edge + LBP features across all regression models. To simulate different crowdedness levels, we divided a dataset into two partitions: one for sparse scenario and another one for crowded scenario, of which the details are provided in Table 14.3.

Table 14.3 Number of frames allocated for the sparse and crowded scenarios. Information inside the brackets contain the definition of crowdedness, together with the training and test set proportions

Data	Sparse scenario (no. frames)	Crowded scenario (no. frames)
ucsd	1,058 ($\leq$23 people, train = 400, test = 658)	942 (>23 people, train = 400, test = 542)
Pets	800 ($\leq$10 people, train = 400, test = 400)	276 (>10 train = 100, test = 176)
Mall	972 ($\leq$30 people, train = 400, test = 572)	1,028 (>30 people, train = 400, test = 628)

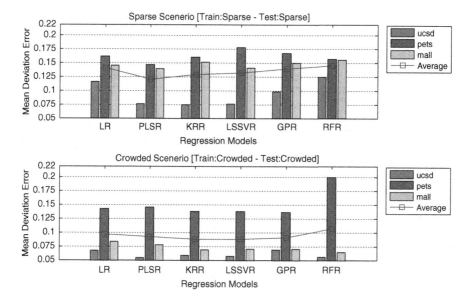

Fig. 14.10 Comparison of mean deviation error (lower is better) between regression models in sparse and crowd scenarios

Model performance under different crowdedness levels: To evaluate a regressor under the sparse scenario, we trained and tested the model using the sparse partition of a dataset. Similar procedures were applied using the crowded partition of a dataset to test a model under crowded scenario. Figure 14.10 shows the performance of the six regression models under the sparse and crowded scenarios. Note that we only presented the mean deviation error since other metrics exhibited similar trends in this experiment.

It is evident that models which can effectively deal with multicollinearity issue, such as LSSVR, PLSR, and KRR, consistently performed better than other models in both the sparse and crowded partitions, as shown in Fig. 14.10. Specifically, over-fitting were less an issue to the aforementioned models, which either add a regularization term[5] into the error function or by projecting the input variables onto a lower-dimensional space.

---

[5]Rasmussen and Williams [64] provide detailed discussion on the regularization approach with the Gaussian process viewpoint.

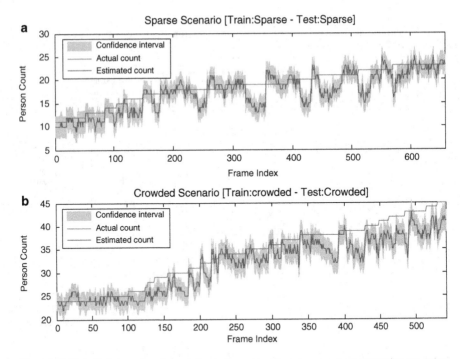

Fig. 14.11 Labelled ground truth vs. estimated count by Gaussian processes regression on sparse and crowded scenarios of *UCSD* dataset. The estimated count is accompanied by $\pm$ two standard deviations corresponding to a 95 % confidence interval

In contrast, LR was ill-conditioned due to highly-correlated features, thus yielding poorer performance as compared to LSSVR, PLSR, and KRR. The performance of GPR was mixed. The error rate of RFR was extremely high in the *pets* crowded partition as the forest structure was too complex given the limited amount of training data. As a result, its generalization capability was compromised due to the over-fitting. In other datasets, RFR showed comparable results to other regression methods.

We found that existing performance metrics including the mean deviation error [19], which is normalized by the actual count (see Sect. 14.3.3), are not appropriate for comparing scenarios with enormous difference in densities. Specifically, our findings were rather counter intuitive in that all regressors performed better in the crowded scenario than the sparse scenario. We note that the lower mean deviation errors in a crowded scene are largely biased by the much larger actual count serving as the denominator in Eq. (14.17). To vindicate our observation, we plotted the performance of GPR on the *UCSD* dataset in Fig. 14.11 and found that the regressor performance did not differ much across sparse and crowded scenarios.

Generalization to unseen density: To evaluate the generalization capability of a regression model to unseen density, we tested it against two scenarios: (1)

generalizing from crowded to sparse environment, and (2) generalizing from sparse to crowded environment. In the first scenario, we trained a regressor with the crowded partition and tested it on the sparse partition. We switched the crowded and sparse partitions in the second scenario. The same data partitions in Table 14.3 were used.

Regression models that worked well within known crowd density may not perform as good given unseen density. In particular, as shown in Table 14.4, simple linear regression models such as LR and PLSR returned surprisingly good performance in both the *UCSD* and *mall* datasets, outperforming their non-linear counterparts. The results suggest that the regression assumption of linear regression models, though simple, could be less susceptible to unseen density and matched closer with the feature-to-density trend in the considered scenarios. The performance of RFR was poorest among the regression models. The results agree with our expectation about its weakness in generalization as discussed in Sect. 14.2.3.3.

It was observed that the generalization performance reported in Table 14.4, were much poorer than those obtained when we trained and tested a regressor using the same density range. In particular, the regressors tend to overestimate or underestimate depending on the extrapolation direction, as shown in Fig. 14.12. In addition, the further the extrapolation goes outside the training range, the larger the error in the estimation due to difference between the learned model and the actual feature-to-density trend. Note that there was no concrete evidence to show that generalizing from crowded to sparse environment was easier than generalizing from sparse to crowded scene.

14.4.2 Feature Robustness

The objective of this experiment is to compare the performance on using different types of features, e.g., segment-based features, edge-based features, texture-based features (in particular GLCM and LBP), as well as their combination, given different crowdedness levels in a scene. As in Sect. 14.4.1, we conducted the evaluation using sparse and crowded partitions. The results are depicted in Figs. 14.13 and 14.14.

Robustness of individual features: It is observed that different features can be more important given different crowdedness levels. In general, the averaged performance suggests that the segment-based features were superior to other features. This is not surprising since the foreground segment carries useful information about the area occupied by objects of interest and it thus intrinsically correlate to the number of pedestrians in a scene. However in the *UCSD* and *mall* datasets, a decrease in performance gap was observed between the edge or texture-based features and the segment-based features when we switched from sparse partition to crowded partition. This observation is intuitive since given a more crowded environment with frequent inter-object occlusion, segment-based features would suffer, whilst edge

Table 14.4 Comparison of generalization capability of different regression models to unseen density. Best performance is highlighted in bold

	Train: crowded – test: sparse			Train: sparse – test: crowded		
	Mean Abs. error	Mean Sq. error	Mean Dev. error	Mean Abs. error	Mean Sq. error	Mean Dev. error
LR	**1.7448**	**4.8034**	**0.1013**	**2.8811**	**13.0382**	**0.0860**
PLSR	2.0208	6.2892	0.1170	4.0934	25.4034	0.1184
KRR	2.0284	6.3176	0.1172	4.1805	26.4459	0.1210
LSSVR	2.0123	6.2202	0.1163	4.2304	27.2070	0.1225
GPR	2.3081	7.6730	0.1330	3.8089	20.6921	0.1119
RFR	6.0851	50.5539	0.3882	9.4671	134.2994	0.2681

(a) *ucsd*

	Train: crowded – test: sparse			Train: sparse – test: crowded		
	Mean Abs. error	Mean Sq. error	Mean Dev. error	Mean Abs. error	Mean Sq. error	Mean Dev. error
LR	1.3137	3.1612	0.2765	2.5833	11.0978	0.1263
PLSR	1.4087	3.6263	0.2835	2.7428	12.3732	0.1337
KRR	**1.2612**	**2.8237**	**0.2643**	**2.5507**	**10.7971**	**0.1248**
LSSVR	1.4737	3.8763	0.3083	2.6051	11.2500	0.1272
GPR	1.4238	3.5463	0.2849	3.3986	20.1159	0.1631
RFR	6.7138	56.4937	1.7037	9.3877	156.5036	0.4279

(b) *pets*

	Train: crowded – test: sparse			Train: sparse – test: crowded		
	Mean Abs. error	Mean Sq. error	Mean Dev. error	Mean Abs. error	Mean Sq. error	Mean Dev. error
LR	5.4959	45.9012	0.2414	**4.5360**	**29.5379**	**0.1225**
PLSR	**4.9877**	**35.0432**	**0.2171**	5.6625	42.8628	0.1499
KRR	5.1070	36.1893	0.2225	5.8006	44.0924	0.1534
LSSVR	5.0216	35.2623	0.2189	5.7704	43.6109	0.1526
GPR	5.4969	39.4660	0.2389	6.9426	59.8687	0.1835
RFR	7.1080	64.0175	0.3127	8.6994	95.4601	0.2276

(l) *mall*

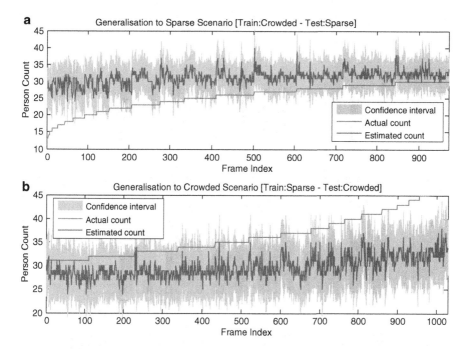

Fig. 14.12 Generalization to unseen density: labelled ground truth vs. estimated count by Gaussian processes regression on *mall* dataset. (**a**) Training on crowded partition and testing on sparse partition results in over-estimation, and (**b**) doing the other way round results in under-estimation. The estimated count is accompanied by ± two standard deviations corresponding to a 95 % confidence interval

and texture that inherently encoded the inter-object boundary and internal patterns would carry more discriminative visual cues for density mapping.

Does combining features help?: From the averaged performance, it is observed that combining different features together could lead to a better performance in general. For instance, when the LBP-based features were used in combination with the segment and edge-based features, the mean deviation error was reduced by 2–14 %. This finding supports the practice of employing a combination of features (see Table 14.1).

Nevertheless, when we examined the performance of individual regression models, it was found that combining all the features did not necessarily produce better performance. For example, using the segment-based features alone in the crowded *mall* partition one would get higher performance; or using the edge features alone with RFR gained more accurate counts in the sparse *UCSD* partition. The results suggests the need for feature selection to discover the suitable set of features given different crowd densities and different regression models.

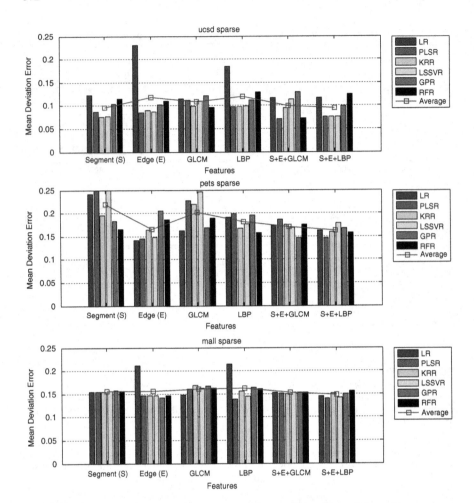

Fig. 14.13 Sparse partition: the mean deviation error (lower is better) vs. different features

14.4.3 Geometric Correction

Geometry correction is critical in crowd counting since objects at different depths of the scene would lead to huge variation in the extracted features. To minimize the influence of perspective distortion, correction is often conducted in existing studies but often without explicit analysis on how its sensitivity would affect the final counting performance. In this experiment, we investigated the sensitivity of crowd counting performance to a widely adopted perspective normalization method described in [13] (see Sect. 14.2.3.2). Evaluation was carried out on the *UCSD* dataset, with 800 frames for training and the remaining 1,200 frames held out for testing following the partitioning scheme suggested in [13].

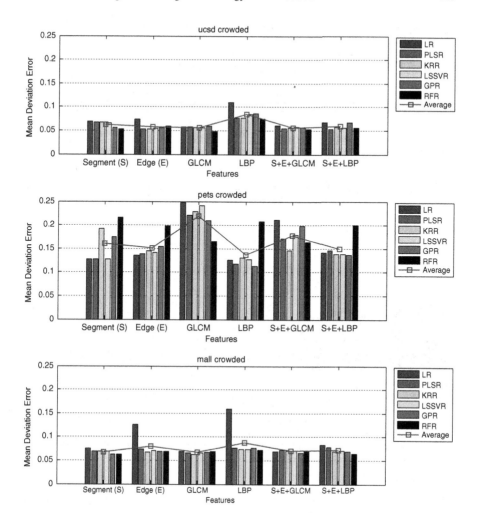

Fig. 14.14 Crowded partition: the mean deviation error (lower is better) vs. different features

Effectiveness of geometric correction: It is evident from Table 14.5 that perspective correction is essential in achieving accurate crowd density estimation. Specifically, depending on different regression models, an improvement of around 20 % was gained in the mean absolute error by applying perspective correction.

Sensitivity to errors in geometric correction: It is interesting to examine how a minor error introduced by manual measurement will propagate through the counting by regression pipeline. We manually measure the heights, denoted as h_1 and h_2, of a reference pedestrian at two extremes of the ground plane rectangle of the *UCSD* dataset (see Fig. 14.15). We varied h_2, the height at the further extreme at $+/-$ 5 pixels with a step size of 1 pixel. Given a frame with resolution of 238×158, this is a reasonable error range that is likely to occur during the manual measurement.

Table 14.5 Comparison of mean absolute error (lower is better) on *UCSD* dataset when crowd density was estimated with and without perspective correction

	With perspective normalisation			Without perspective normalisation		
	Mean Abs.error	Mean Sq.error	Mean Dev.error	Mean Abs.error	Mean Sq.error	Mean Dev.error
LR	2.1608	7.1608	0.1020	2.6308	10.2558	0.1288
PLSR	2.0267	6.6717	0.1007	2.5792	10.0025	0.1271
KRR	2.3433	8.4800	0.1166	2.9167	11.6133	0.1392
LSSVR	2.1100	6.6383	0.1014	2.5825	9.6925	0.1262
GPR	2.1425	7.1358	0.1055	2.7833	10.5200	0.1328
RFR	2.3392	7.9708	0.1129	2.8492	10.8492	0.1332
Average	2.1871	7.3429	0.1065	2.7236	10.4889	0.1312

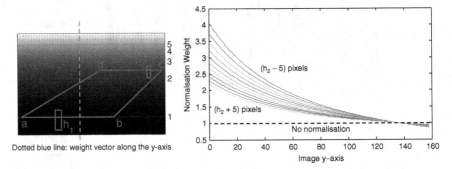

Fig. 14.15 (Color online) (*Top*) Perspective normalization map of *UCSD* dataset. (*Bottom*) Each line in the chart corresponds to a weight vector along the y-axis (e.g., the *dotted blue line*) of each perspective map produced as a result of varying measurement errors in h_2, ranging from -5 pixels to $+5$ pixels with a step size of 1 pixel

Perspective maps within this pixel deviation range were generated, and the crowd counting performances of different models were subsequently recorded.

A minor measurement error in h_2 could result in a great change in perspective map, as shown in Fig. 14.15. Specifically, when h_2 had a smaller value, e.g., $h_2 - 5$ pixels, a steeper slope in the perspective normalization weight vector was observed. On the contrary, given $h_2 + 5$ pixels, the object size at $\overline{cd}$ was larger so the perspective normalization weight vector had a lower slope. Using these different perspective maps we evaluated performances of different regression models.

It is clear from the results depicted in Fig. 14.16 that different perspective maps will lead to drastic difference in estimation performance, e.g., as much as 10 % of difference from that obtained using initial measurement. The results suggest that the initial measurement h_2 may not be accurate, since more accurate counts were obtained at $h_2 - 5$ pixels. A subsequent validation through averaging multiple measurements confirmed that the initial measurement indeed deviated from the accurate value. Hence one should not rely on a single round of measurement, but

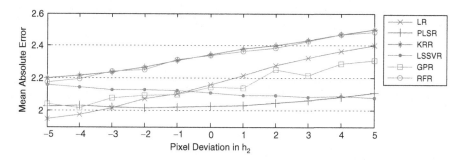

Fig. 14.16 Mean absolute error on *UCSD* dataset as a result of varying measurement errors in h_2: most regression methods experienced drastic performance change as much as over 10 % given just a minor deviation in the manual measurement

to seek for more reliable perspective statistics by averaging measurements obtained across multiple attempts. Note that deviation from the 'exact' perspective map may not necessarily lead to a bad consequence sometimes as the steeper weight slope will counteract the problems of poor segmentation and inter-object occlusion at the back of the scene.

14.5 Crowd Profiling

One of the ultimate goals of crowd counting is to profile the crowd behaviors and density patterns spatially and temporally, e.g., how many people in a region of interest at what time and predicting the trend. The profiling statistic can serve as useful hints for controlling crowd movements, designing evacuation routes, and improving product display strategy to attract more crowds to a shop. An example of such a crowd profiling application is depicted in Fig. 14.17, of which the local density map was generated through learning cell-level counts using separate regressors. A more scalable way based on a single regression model with multiple outputs can also be employed [15].

The top row of Fig. 14.17 shows the footage frames of a shopping mall view overlaid with heat maps, of which the color codes representing the crowd density, with larger crowd represented by red squares and smaller crowd with blue squares. An interesting usage of the crowd density map is to study the crowd movement profile in front of a shop, e.g., the two selected regions (blue and red) in Fig. 14.17. The number of people appear in these areas over time can be profiled as shown in the two plots at the bottom of Fig. 14.17. In addition, activity correlation between these two regions can be computed to examine their crowd flow dependency, as shown in the last plot. Analyzing these local crowd patterns over time and their correlations globally can reveal useful information about the shop visitors, such as their interests towards the product display, walking pace, and intention of buying, without the need for registering individual's identities therefore minimizing privacy violation.

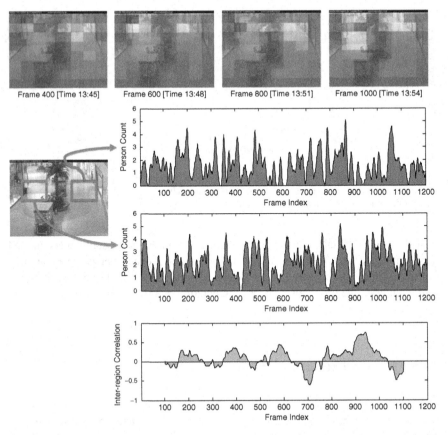

Fig. 14.17 One of the goals of crowd counting is to profile the crowd behaviors and density patterns spatially and temporally, e.g., how many people in a region of interest at what time (see text for details)

The crowd counting application can benefit from extensions such as functional learning of regions [75] (e.g., sitting area, entrance of shops) to better reflect the activity modes at different regions; or combination with cooperative multi-camera network surveillance [33, 52] to model the density and activity correlation in the camera network [94].

14.6 Findings and Analysis

We shall summarize our main findings as follows:

Regression model choices: Our evaluation reveals that regression models that are capable of dealing with multicollinearity among features, e.g., KRR, PLSR, LSSVR generally give better performance than other regression models such as LR and RFR. The aforementioned models, i.e., KRR, PLSR, and LSSVR have not been significantly explored in existing counting by regression literature.

In general, linear model is expected to give poorer performance as its linear property imposes a limitation on the model in capturing only the linear relationship between the people count and low-level features [4, 22, 47]. In most cases especially in crowded environments, the visual observations and people count will not be linearly related. Nonlinear methods in principle allow one to model arbitrary nonlinearities between the mapping from input variables to target people count. In addition, employing a nonlinear method would help in remedying the dimensionality problem since observations typically exhibit strong correlation in a nonlinear manifold, whose intrinsic dimensionality is smaller than the input space [6].

However, our study suggests that the actual performance of a regression model can be quite different from what one may anticipate, subject to the nature of data, especially when it is applied to unseen density. Despite all the evaluated regression techniques suffer poor extrapolation beyond the training data range, simple linear regression models such as LR, is found to be more resistant towards the introduction of unseen density. Its performance can be better than other nonlinear models such as GPR and LSSVR.

We have emphasized that it is impractical to assume the access to all full density range during the training stage, thus the capability of generalizing to unseen density is critical. An unexplored approach of resolving the problem is to transfer the knowledge from other well-annotated datasets that cover wider range of crowd density. This is an open and challenging problem in crowd counting task given different environmental factors of source and target scenes, e.g., variations in lighting conditions and camera orientations.

Features selection: Our results suggest that different features can be more useful given different crowd configurations and densities. In sparse scenes, foreground segment-based features alone can provide sufficient information required for crowd density estimation. However, when a scene becomes crowded with frequent inter-object occlusions, the role of edge-based features and texture-based features becomes increasingly critical. We also found that combining all features do not always help, depending on the dataset and regression model of choice. These findings suggest the importance of feature selection, i.e., selecting optimal feature combinations given different crowd structures and densities, through discarding redundant and irrelevant features. The feature selection problem has been largely ignored in existing crowd counting research.

Perspective correction: The performance of counting by regression can be severely influenced by the accuracy of perspective weight estimation. Perspective map generation based on manual measurement is simple but could be error-prone. We suggest that multiple measurements are necessary to ensure conciseness of the estimation normalization weights. Robust auto-calibration methods such as [40, 50] are also recommended as an alternative to the manual approach.

14.7 Further Reading

Interested readers are referred to the following further readings:

- Gong et al. [33] for a general discussion on applications and advances in automated analysis of human activities for security and surveillance
- Gong and Xiang [32] for a comprehensive treatment of visual analysis of behavior from algorithm-design perspectives
- Jacques et al. [37] for a survey on crowd analysis
- Chan and Vasconcelos [12] for a detailed discussion on using Bayesian techniques for regression-based counting

References

1. Abdi, H.: Partial least square regression (pls regression). In: Salkind, N.J., Rasmussen, K. (eds.) Encyclopedia of Measurement and Statistics, pp. 740–744. SAGE Publications, Thousand Oaks (2007)
2. Ahonen, T., Hadid, A., Pietikainen, M.: Face description with local binary patterns: application to face recognition. IEEE Trans. Pattern Anal. Mach. Intell. **28**(12), 2037–2041 (2006)
3. Ali, S., Shah, M.: Floor fields for tracking in high density crowd scenes. In: European Conference on Computer Vision, Marseille, pp. 1–24 (2008)
4. Benabbas, Y., Ihaddadene, N., Yahiaoui, T., Urruty, T., Djeraba, C.: Spatio-temporal optical flow analysis for people counting. In: IEEE International Conference on Advanced Video and Signal Based Surveillance, Boston, pp. 212–217 (2010)
5. Benenson, R., Mathias, M., Timofte, R., Gool, L.V.: Pedestrian detection at 100 frames per second. In: IEEE Conference Computer Vision and Pattern Recognition, Providence (2012)
6. Bishop, C.M.: Pattern Recognition and Machine Learning. Springer, New York (2007)
7. Brostow, G.J., Cipolla, R.: Unsupervised Bayesian detection of independent motion in crowds. In: IEEE Conference on Computer Vision and Pattern Recognition, pp. 594–601 (2006)
8. Canny, J.: A computational approach to edge detection. IEEE Trans. Pattern Anal. Mach. Intell. **8**(6), 679–698 (1986)
9. Chan A.B., Dong, D.: Generalized Gaussian process models. In: IEEE Conference Computer Vision and Pattern Recognition, Colorado, pp. 2681–2688. IEEE (2011)
10. Chan A.B., Vasconcelos, N.: Modeling, clustering, and segmenting video with mixtures of dynamic textures. IEEE Trans. Pattern Anal. Mach. Intell. **30**(5), 909–926 (2008)
11. Chan A.B., Vasconcelos, N.: Bayesian poisson regression for crowd counting. In: IEEE International Conference on Computer Vision, Kyoto, pp. 545–551. IEEE (2009)
12. Chan A. B., Vasconcelos, N.: Counting people with low-level features and Bayesian regression. IEEE Trans. Image Process. **21**(4), 2160–2177 (2012)
13. Chan, A.B., Liang, Z.S.J., Vasconcelos, N.: Privacy preserving crowd monitoring: counting people without people models or tracking. In: IEEE Conference on Computer Vision and Pattern Recognition, Anchorage, pp. 1–7 (2008)
14. Chan, A.B., Morrow, M., Vasconcelos, N.: Analysis of crowded scenes using holistic properties. In: IEEE International Workshop on Performance Evaluation of Tracking and Surveillance (2009)
15. Chen, K., Loy, C.C., Gong, S., Xiang, T.: Feature mining for localised crowd counting. In: British Machine Vision Conference, Surrey (2012)
16. Cho, S., Chow, T., Leung, C.: A neural-based crowd estimation by hybrid global learning algorithm. IEEE Trans. Syst. Man Cybern. Part B Cybern. **29**(4), 535–541 (1999)

17. Cohen, S.: Background estimation as a labeling problem. In: IEEE International Conference on Computer Vision, Beijing, vol. 2, pp. 1034–1041 (2005)
18. Cong, Y., Gong, H., Zhu, S., Tang, Y.: Flow mosaicking: real-time pedestrian counting without scene-specific learning. In: IEEE Conference Computer Vision and Pattern Recognition, Miami, pp. 1093–1100 (2009)
19. Conte, D., Foggia, P., Percannella, G., Vento, M.: A method based on the indirect approach for counting people in crowded scenes. In: IEEE International Conference on Advanced Video and Signal Based Surveillance, Boston, pp. 111–118. IEEE (2010)
20. Criminisi, A., Shotton, J., Konukoglu, E.: Decision forest for classification, regression, density estimation, manifold learning and semi-supervised learning. Tech. Rep. MSR-TR-2011-114, Microsoft Research (2011)
21. Dalal, N., Triggs, B.: Histograms of oriented gradients for human detection. In: IEEE Conference on Computer Vision and Pattern Recognition, San Diego, pp. 886–893 (2005)
22. Davies, A., Yin, J., Velastin, S.: Crowd monitoring using image processing. Electron. Commun. Eng. J. 7(1), 37–47 (1995)
23. De Brabanter, K., De Brabanter, J., Suykens, J., De Moor, B.: Approximate confidence and prediction intervals for least squares support vector regression. IEEE Trans. Neural Netw. 22(1), 110–120 (2011)
24. Dollar, P., Wojek, C., Schiele, B., Perona, P.: Pedestrian detection: an evaluation of the state of the art. IEEE Trans. Pattern Anal. Mach. Intell. 34(4), 743–761 (2011)
25. Dong, L., Parameswaran, V., Ramesh, V., Zoghlami, I.: Fast crowd segmentation using shape indexing. In: IEEE International Conference on Computer Vision, Rio de Janeiro (2007)
26. Felzenszwalb, P., Girshick, R., McAllester, D., Ramanan, D.: Object detection with discriminatively trained part-based models. IEEE Trans. Pattern Anal. Mach. Intell. 32(9), 1627–1645 (2010)
27. Ferryman, J., Crowley, J., Shahrokni, A.: Pets 2009 benchmark data. http://www.cvg.rdg.ac.uk/WINTERPETS09/a.html
28. Gall, J., Yao, A., Razavi, N., Van Gool, L., Lempitsky, V.: Hough forests for object detection, tracking, and action recognition. IEEE Trans. Pattern Anal. Mach. Intell. 33(11), 2188–2202 (2011)
29. Ge, W., Collins, R.: Marked point processes for crowd counting. In: IEEE Conference on Computer Vision and Pattern Recognition, Miami, pp. 2913–2920 (2009)
30. Ge, W., Collins, R.: Crowd detection with a multiview sampler. In: European Conference on Computer Vision, Heraklion, pp. 324–337 (2010)
31. Geladi, P., Kowalski, B.: Partial least-squares regression: a tutorial. Anal. Chim. Acta 185, 1–17 (1986)
32. Gong, S., Xiang, T.: Visual Analysis of Behaviour: From Pixels to Semantics. Springer, New York (2011)
33. Gong, S., Loy, C.C., Xiang, T.: Security and surveillance. In: Moeslund, T., Hilton, A., Krueger, V., Sigal, L. (eds.) Visual Analysis of Humans: Looking at People, Springer, pp. 455–472 (2011)
34. Haralick, R., Shanmugam, K., Dinstein, I.: Textural features for image classification. IEEE Trans. Syst. Man Cybern. 3(6), 610–621 (1973)
35. Helbing, D., Farkas, I., Molnar, P., Vicsek, T.: Simulation of pedestrian crowds in normal and evacuation situations. In: Schreckenberg, M., Sharma, S.D. (eds.) Pedestrian and Evacuation Dynamics, vol. 21. Springer, Berlin/New York (2002)
36. Hoerl, A., Kennard, R.: Ridge regression: biased estimation for nonorthogonal problems. Technometrics 12, 55–67 (1970)
37. Jacques, J., Jr., Musse, S., Jung, C.: Crowd analysis using computer vision techniques. IEEE Signal Process. Mag. 27(5), 66–77 (2010)
38. Kong, D., Gray, D., Tao, H.: Counting pedestrians in crowds using viewpoint invariant training. In: British Machine Vision Conference, Oxford (2005). Citeseer
39. Kong, D., Gray, D., Tao, H.: A viewpoint invariant approach for crowd counting. In: International Conference on Pattern Recognition, Hong Kong, vol. 3, pp. 1187–1190 (2006)

40. Krahnstoever, N., Mendonca, P.: Bayesian autocalibration for surveillance. In: IEEE International Conference on Computer Vision, Beijing, vol. 2, pp. 1858–1865. IEEE (2005)
41. Kratz, L., Nishino, K.: Anomaly detection in extremely crowded scenes using spatio-temporal motion pattern models. In: IEEE Conference on Computer Vision and Pattern Recognition, Miami, pp. 1446–1453 (2009)
42. Lampert, C.: Kernel Methods in Computer Vision, vol. 4. Now Publishers Inc., Hanover (2009)
43. Leibe, B., Seemann, E., Schiele, B.: Pedestrian detection in crowded scenes. In: IEEE Conference Computer Vision and Pattern Recognition, San Diego, vol. 1, pp. 878–885 (2005)
44. Lempitsky, V., Zisserman, A.: Learning to count objects in images. In: Advances in Neural Information Processing Systems (2010)
45. Li, J., Huang, L., Liu, C.: CASIA pedestrian counting dataset. http://cpcd.vdb.csdb.cn/page/showItem.vpage?id=automation.dataFile/1
46. Li, M., Zhang, Z., Huang, K., Tan, T.: Estimating the number of people in crowded scenes by mid based foreground segmentation and head-shoulder detection. In: International Conference on Pattern Recognition, Tampa, pp. 1–4 (2008)
47. Li, J., Huang, L., Liu, C.: Robust people counting in video surveillance: dataset and system. In: IEEE International Conference on Advanced Video and Signal-Based Surveillance, pp. 54–59. IEEE (2011)
48. Lin, S., Chen, J., Chao, H.: Estimation of number of people in crowded scenes using perspective transformation. IEEE Trans. Syst. Man Cybern. Part A Syst. Hum. 31(6), 645–654 (2001)
49. Lin, T., Lin, Y., Weng, M., Wang, Y., Hsu, Y., Liao, H.: Cross camera people counting with perspective estimation and occlusion handling. In: IEEE International Workshop on Information Forensics and Security (2011)
50. Liu, J., Collins, R.T., Liu, Y.: Surveillance camera autocalibration based on pedestrian height distributions. In: British Machine Vision Conference, Dundee (2011)
51. Loy, C.C., Xiang, T., Gong, S.: Time-delayed correlation analysis for multi-camera activity understanding. Int. J. Comput. Vis. 90(1), 106–129 (2010)
52. Loy, C.C., Xiang, T., Gong, S.: Incremental activity modelling in multiple disjoint cameras. IEEE Trans. Pattern Anal. Mach. Intell. 34(9) 1799–1813 (2011)
53. Loy, C.C., Xiang, T., Gong, S.: Salient motion detection in crowded scenes. In: Special Session on 'Beyond Video Surveillance: Emerging Applications and Open Problems', International Symposium on Communications, Control and Signal Processing, Invited Paper (2012)
54. Ma, R., Li, L., Huang, W., Tian, Q.: On pixel count based crowd density estimation for visual surveillance. In: IEEE Conference on Cybernetics and Intelligent Systems, vol. 1, pp. 170–173. IEEE (2004)
55. Ma, W., Huang, L., Liu, C.: Advanced local binary pattern descriptors for crowd estimation. In: Pacific-Asia Workshop on Computational Intelligence and Industrial Application, vol. 2, pp. 958–962. IEEE (2008)
56. Ma, W., Huang, L., Liu, C.: Crowd density analysis using co-occurrence texture features. In: International Conference on Computer Sciences and Convergence Information Technology, pp. 170–175 (2010)
57. Marana, A., Velastin, S., Costa, L., Lotufo, R.: Estimation of crowd density using image processing. In: Image Processing for Security Applications, pp. 11–1 (1997)
58. Marana, A., Costa, L., Lotufo, R., Velastin, S.: On the efficacy of texture analysis for crowd monitoring. In: International Symposium on Computer Graphics, Image Processing, and Vision, pp. 354–361 (1998)
59. Marana, A., da Fontoura Costa, L., Lotufo, R., Velastin, S.: Estimating crowd density with Minkowski fractal dimension. In: IEEE International Conference on Acoustics, Speech, and Signal Processing, vol. 6, pp. 3521–3524. IEEE (1999)
60. Mehran, R., Oyama, A., Shah, M.: Abnormal crowd behaviour detection using social force model. In: IEEE Conference on Computer Vision and Pattern Recognition, Miami, pp. 935–942 (2009)

61. Ojala, T., Pietikainen, M., Maenpaa, T.: Multiresolution gray-scale and rotation invariant texture classification with local binary patterns. IEEE Trans. Pattern Anal. Mach. Intell. **24**(7), 971–987 (2002)
62. Pätzold, M., Evangelio, R., Sikora, T.: Counting people in crowded environments by fusion of shape and motion information. In: IEEE International Conference on Advanced Video and Signal Based Surveillance, Boston, pp. 157–164. IEEE (2010)
63. Rabaud, V., Belongie, S.: Counting crowded moving objects. In: IEEE Conference on Computer Vision and Pattern Recognition, pp. 705–711 (2006)
64. Rasmussen, C.E., Williams, C.K.I.: Gaussian Process for Machine Learning. MIT, Cambridge (2006)
65. Rodriguez, M., Laptev, I., Sivic, J., Audibert, J.: Density-aware person detection and tracking in crowds. In: IEEE International Conference on Computer Vision, Barcelona (2011)
66. Russell, D., Gong, S.: Minimum cuts of a time-varying background. In: British Machine Vision Conference, Edinburgh, pp. 809–818 (2006)
67. Ryan, D., Denman, S., Fookes, C., Sridharan, S.: Crowd counting using multiple local features. In: Digital Image Computing: Techniques and Applications (2009)
68. Sabzmeydani, P., Mori, G.: Detecting pedestrians by learning shapelet features. In: IEEE Conference on Computer Vision and Pattern Recognition, Minneapolis, pp. 1–8 (2007)
69. Saunders, C., Gammerman, A., Vovk, V.: Ridge regression learning algorithm in dual variables. In: International Conference on Machine Learning, pp. 515–521 (1998)
70. Shan, C., Gong, S., McOwan, P.W.: Facial expression recognition based on local binary patterns: a comprehensive study. Image Vis. Comput. **27**(6), 803–816 (2009)
71. Shawe-Taylor, J., Cristianini, N.: Kernel Methods for Pattern Analysis. Cambridge University Press, Cambridge/New York (2004)
72. Smola, A., Schölkopf, B.: A tutorial on support vector regression. Stat. Comput. **14**(3), 199–222 (2004)
73. Stauffer, C., Grimson, W.E.L.: Learning patterns of activity using real-time tracking. IEEE Trans. Pattern Anal. Mach. Intell. **22**(8), 747–757 (2000)
74. Suykens, J., Vandewalle, J.: Least squares support vector machine classifiers. Neural Process. Lett. **9**(3), 293–300 (1999)
75. Swears, E., Turek, M., Collins, R., Perera, A., Hoogs, A.: Automatic activity profile generation from detected functional regions for video scene analysis. In: Shan, C., Porikli, F., Xiang, T., Gong, S. (eds.) Video Analytics for Business Intelligence, pp. 241–269. Springer, Berlin/New York (2012)
76. Tian, Y., Brown, L., Hampapur, A., Lu, M., Senior, A., Shu, C.: IBM smart surveillance system (s3): event based video surveillance system with an open and extensible framework. Mach. Vis. Appl. **19**(5), 315–327 (2008)
77. Tu, P., Sebastian, T., Doretto, G., Krahnstoever, N., Rittscher, J., Yu, T.: Unified crowd segmentation. In: European Conference on Computer Vision, Marseille (2008)
78. Tuzel, O., Porikli, F., Meer, P.: Pedestrian detection via classification on Riemannian manifolds. IEEE Trans. Pattern Anal. Mach. Intell. **30**(10), 1713–1727 (2008)
79. Vapnik, V.: The Nature of Statistical Learning Theory. Springer, New York (2000)
80. Viola, P., Jones, M.: Robust real-time face detection. Int. J. Comput. Vis. **57**(2), 137–154 (2004)
81. Viola, P., Jones, M., Snow, D.: Detecting pedestrians using patterns of motion and appearance. Int. J. Comput. Vis. **63**(2), 153–161 (2005)
82. Wang, M., Wang, X.: Automatic adaptation of a generic pedestrian detector to a specific traffic scene. In: IEEE Conference on Computer Vision and Pattern Recognition, Colorado Springs, pp. 3401–3408. IEEE (2011)
83. Wang, M., Li, W., Wang, X.: Transferring a generic pedestrian detector towards specific scenes. In: IEEE Conference Computer Vision and Pattern Recognition, Providence (2012)
84. Welling, M.: Support vector regression. Tech. Rep., Department of Computer Science, University of Toronto (2004)

85. Wu, B., Nevatia, R.: Detection of multiple, partially occluded humans in a single image by Bayesian combination of edgelet part detectors. In: Tenth IEEE International Conference on Computer Vision, ICCV 2005, Beijing, vol. 1, pp. 90–97. IEEE (2005)
86. Wu, B., Nevatia, R.: Detection and tracking of multiple, partially occluded humans by Bayesian combination of edgelet based part detectors. Int. J. Comput. Vis. **75**(2), 247–266 (2007)
87. Wu, X., Liang, G., Lee, K., Xu, Y.: Crowd density estimation using texture analysis and learning. In: IEEE International Conference on Robotics and Biomimetics, pp. 214–219. IEEE (2006)
88. Yang, D., González-Baños, H., Guibas, L.: Counting people in crowds with a real-time network of simple image sensors. In: IEEE International Conference on Computer Vision, Nice, pp. 122–129 (2003)
89. Yeniay, O., Goktas, A.: A comparison of partial least squares regression with other prediction methods. Hacet. J. Math. Stat. **31**(99), 111 (2002)
90. Yilmaz, A., Javed, O., Shah, M.: Object tracking: a survey. ACM J. Comput. Surv. **38**(4), 1–45 (2006)
91. Zhan, B., Monekosso, D.N., Remagnino, P., Velastin, S.A., Xu, L.Q.: Crowd analysis: a survey. Mach. Vis. Appl. **19**, 345–357 (2008)
92. Zhao, T., Nevatia, R., Wu, B.: Segmentation and tracking of multiple humans in crowded environments. IEEE Trans. Pattern Anal. Mach. Intell. **30**(7), 1198–1211 (2008)
93. Zhou, B., Wang, X., Tang, X.: Random field topic model for semantic region analysis in crowded scenes from tracklets. In: IEEE Conference Computer Vision and Pattern Recognition, Colorado Springs (2011)
94. Zhu, X., Gong, S., Loy, C.C.: Comparing visual feature coding for learning disjoint camera dependencies. In: British Machine Vision Conference, Surrey (2012)
95. Zivkovic, Z., van der Heijden, F.: Efficient adaptive density estimation per image pixel for the task of background subtraction. Pattern Recognit. Lett. **27**(7), 773–780 (2006)

Chapter 15
Anomaly Detection in Crowded Scenes: A Novel Framework Based on Swarm Optimization and Social Force Modeling

R. Raghavendra, M. Cristani, A. Del Bue, E. Sangineto, and V. Murino

Abstract This chapter presents a novel scheme for analyzing the crowd behavior from visual crowded scenes. The proposed method starts from the assumption that the interaction force, as estimated by the Social Force Model (SFM), is a significant feature to analyze crowd behavior. We step forward this hypothesis by optimizing this force using Particle Swarm Optimization (PSO) to perform the advection of a particle population spread randomly over the image frames. The population of particles is drifted towards the areas of the main image motion, driven by the PSO fitness function aimed at minimizing the interaction force, so as to model the most diffused, normal behavior of the crowd. We then use this proposed particle advection scheme to detect both global and local anomaly events in the crowded scene. A large set of experiments are carried out on public available datasets and results show the consistent higher performances of the proposed method as compared to other state-of-the-art algorithms.

15.1 Introduction

Recently, major research efforts are underway in the computer vision community to develop robust algorithms for understanding the behavior of crowds in video surveillance contexts. Anomaly detection in crowded scenes is an important social problem far from being reliably solved. This is because conventional methods designed for surveillance applications fail drastically for the following reasons: (1) severe overlapping between individual subjects; (2) random variations in the density of people over time; (3) low resolution videos with temporal variations of

R. Raghavendra • M. Cristani • A. Del Bue • E. Sangineto • V. Murino (✉)
Pattern Analysis and Computer Vision (PAVIS), Istituto Italiano di Tecnologia, via Morego 30, 16163 Genova, Italy
e-mail: vittorio.murino@iit.it

S. Ali et al. (eds.), *Modeling, Simulation and Visual Analysis of Crowds*, The International Series in Video Computing 11, DOI 10.1007/978-1-4614-8483-7__15, © Springer Science+Business Media New York 2013

the scene background. Nowadays, crowds are viewed as the very outliers of the social sciences [27]. Such an attitude is reflected by the remarkable paucity of psychological research on crowd processes [27].

The main objective of crowd behavior analysis involves not only modeling of people mass dynamics but also detecting or even predicting possible abnormal or anomalous behaviors in the scene. In particular for surveillance scenarios, this task is of paramount importance since early detection, or even prediction, may reduce the possible dangerous consequences of a threatening event, or may alert a human operator for inspecting more carefully the ongoing situation.

Anomaly detection in crowded scenes can be classified into two types: (1) local abnormal event, indicating that a behavior in a specific local image (or frame) area is different from that of its neighbors in spatio-temporal terms; (2) global abnormal event, indicating that the whole frame is abnormal irrespective of the local regions. In other words, a global abnormal event detection aims at classifying each frame as either abnormal or normal, while in local detection we also want to localize the parts of the given frame which likely contain the abnormal activity.

In this article we present both global [26] and local [25] anomaly detection techniques which have been tested on different real-time scenarios. We developed these techniques based on the assumption that people in the crowd behave in ways like birds (also known as particles) in a swarm. Thus, we try to address crowd behavior analysis by considering the crowd as mutually interacting birds in a swarm.

In general, a crowd can be considered as a collection of mutually interacting people, where random individuals' motion, due to the influence of neighbors, spatial physical structure of the scene, etc., will dominate the dynamics and the flow of the crowd. With this primary idea, we make an attempt to reflect a visual crowd behavior using the concept of Swarm Optimization. Typically, the idea of Swarm Optimization derived from the flight control (defined by a fitness function) of randomly dispersed birds (also referred to as particles) in a given space. In this framework, both local and social behavior among the birds or particles in the swarm is considered. Similarly, we represent people in a crowd as interacting particles following an evolutionary dynamic. These dynamics are driven by a fitness function and they are influenced by the interaction forces among the swarm particles. With this motivation, we propose a novel framework for particle advection using PSO [15] and Social Force Model (SFM) [13]. The proposed method belongs to the class of particle advection schemes and it is based on the assumption that the evolving interaction forces estimated using SFM is a significant feature for analyzing the crowd behavior. Our scheme starts by initializing particles randomly on the initial video frame, which are then optimized and drifted to the main regions of the motion according to a fitness function suitably defined. The aim of the fitness function is to minimize the interaction forces, so as to model the most diffused, normal behavior of the crowd as suggested by behavioral studies. Hence, the anomalies are identified by the particles whose force significantly deviates from the typical force magnitude.

We put forward this framework to detect two different kinds of anomalies namely: global and local anomalies. In order to detect global anomalies, we process the interaction force obtained using the PSO-SFM method by detecting the change in its magnitude. On the other hand, local anomaly detection is carried out by checking if some particles (i.e., their interaction forces) do not fit the estimated "typical" distribution, and this is done using a RANSAC-like method followed by a segmentation algorithm to finely localize the abnormal areas.

There are several characteristics which differentiate our approach with respect to other related works. First, particles are spread randomly over the image and can move in a continuous way according to an optimization criterion, differently from other approaches which constrain the particles in a priori fixed grid. Second, we use PSO for particle advection which considers not only the individual particles motion, but also the global motion of the particles as a whole, i.e., social interactions.

Extensive experiments are carried out on different types of public available video datasets to prove the effectiveness of the proposed scheme. In order to evaluate the global anomaly scheme, we considered four different public available datasets, namely: UMN, PETS 2009, UCF and also a challenging dataset that reflects the prison riots, download by YouTube. In order to evaluate the proposed scheme for local anomaly detection, we consider two different public datasets, namely UCSD and MALL datasets.

The rest of this chapter is organized as follows: Sect. 15.2 shows the state-of-the-art techniques for crowd behavior analysis from the computer vision point of view. Section 15.3 describes the proposed particle advection approach based on the PSO-SFM model and also discusses the global and local anomaly detection schemes. Section 15.4 presents the experimental results. Finally, Sect. 15.5 draws the conclusions.

15.2   Related Work

Several techniques have been proposed for the anomaly detection in visually crowded scenes. State of the art methods can be coarsely classified into two different types: model-based and particle advection-based approaches. Among these two methods, the particle advection based approaches will more naturally represent the holistic view of a crowd and they do not require the segmentation or detection of individuals. On the contrary, the outcome of these algorithms may eventually result in the detection of individuals when they are detected as an anomaly. Here, we first review the literature on model based approaches which is then followed by particle advection schemes.

15.2.1 Model Based Approaches

In [29], a novel unsupervised framework is presented to model the pedestrian activities and interactions in crowded scenes. Here, low level visual features are computed by carrying out the intensity difference between successive frames of a given video. Then, these low level features are labeled using their location and motion direction to form a basic feature set. The features are then quantized into visual words to construct a dictionary. Finally, the activities are classified using two well know classifiers namely: Latent Dirichlet Allocation (LDA) mixture model and Hierarchical Dirichlet Process (HDP) mixture model.

In [20], a dynamic texture model is employed to jointly model the appearance and dynamics of the crowded scene. This method explicitly addresses the detection of both temporal and spatial anomalies. Further, a new dataset of crowded scenes with videos of the walkway of a college campus and crowd with naturally varying densities are made available for the vision community. In [17], steady state motion of the crowd behavior is exploited by analyzing the underlying structure formed by the spatial and temporal variations in the motion. Then, a Hidden Markov Model (HMM) is trained on the motion patterns at each spatial location of the video to predict the motion pattern that is exhibited by the subjects as they transverse through the video. Finally, anomalous activities are detected as low likelihood motion patterns.

In [16], anomaly detection in the crowded scene is carried out using a space-time Markov Random Field (MRF) model. Given a video, a MRF graph is constructed by dividing each frame into a grid of spatio-temporal local regions. Each region corresponds to a single node and neighboring nodes are connected with links. Then, each node is associated with an optical flow observation to learn the atomic motion patterns using a mixture of probabilistic principal component analysis. Finally, inference on the graph is carried out to decide whether each node is normal or abnormal. In [1], a histogram is used to measure the optical flow probability in local patterns of the image and then an ambiguity based threshold is selected to monitor and detect the anomalies in the input videos. Further, a new video dataset with different anomaly scenarios is made available to the vision community. In [3], a new technique based on video parsing is proposed for accurate abnormality detection in the visual crowded scene. Each video frame is parsed by establishing a set of hypotheses that jointly provide information on the entire foreground. Finally, a probabilistic model is employed to localize the abnormality using statistical inference. In [18], dense optical flow fields are computed between two successive frames to obtain the low level motion information in terms of direction and magnitude for each pixel. Then, 2D histograms of motion direction and magnitude for all flow vectors are computed. A symmetry measure is computed by summing the absolute difference between the 2D histogram and a flipped version of itself to determine the anomaly in the scene. Extensive experiments are carried out on the LoveParade 2010 dataset to prove the reliability of the method. In [9], a sparse reconstruction cost is proposed to detect the presence of anomalies in crowded scenes. Here local spatio-temporal patches are used to construct the normal dictionary. Further, to reduce the size of the dictionary, a new selection method is proposed based on sparsity consistency constraints.

15.2.2   Particle Advection Based Approaches

In case of particle advection schemes, a grid of particles is usually considered in each frame which are then advected using the underlying motion data [2,21,22,30]. The assumption here is that each particle is considered as an atomic entity in the mass of people, and the trajectories generated from the particles' advection may portray significant information concerning representative properties of the scene in terms of both characteristics of the physical area and the crowd behavior. The first work using particle advection schemes for crowd behavior analysis was introduced in [2]. Here, the particle flow is computed by moving a grid of particles using the fourth-order Runge-Kutta-Fehlberg algorithm [19] along with the bilinear interpolation of the optical flow field. This method is further extended in [30] using chaotic invariants capable of analyzing both coherent and incoherent scenes. In [22], streaklines are introduced and integrated with a particle advection scheme capable of incorporating the spatial change in the particle flow.

In [21] the social force model (SFM) [13] is exploited to detect abnormal events. After the superposition of a fixed grid of particles on each frame, the SFM is used to estimate the interaction force. In turn, the interaction force is used to describe (abnormal) crowd behavior. So, after estimating the so-called force flow, a bag of words method [4] and a Latent Dirichlet Allocation (LDA) [5] are employed to discriminate between normal and abnormal frames. Possible abnormal areas are localized selecting those regions with the highest force magnitude. In [23] the authors provide an excellent analysis of the above mentioned particle advection schemes in which crowd is dealt with using hydrodynamics principles.

15.2.3   Discussion

In Fig. 15.1a we show the result obtained applying the state-of-the-art people detector of Dalal and Triggs [11] to a crowd image. Only 5 out of 23 persons are correctly detected. Moreover, two false positives (the big rectangles) are also included in the outcome. The situation is even worse in the densely crowded image shown in Fig. 15.1b, where the automatic people detection phase clearly fails in localizing the huge number of persons here represented. These two examples show why approaches based on detection or segmentation of individuals are barely robust when applied to the analysis of non-sparsely crowded scenes.

Conversely, particle advection methods do not rely on people segmentation and assume that a crowd can be represented by a set of particles influenced by the

Fig. 15.1 Examples of common people detector errors on a low-crowded (**a**) and a high-crowded (**b**) scenario. The large number of false positives and false negatives makes the use of people detector-based techniques highly unreliably for crowd analysis

people's movements. The particles' flow is then analyzed trying to detect possible anomalies. In Sect. 15.3.5 we will show that our anomaly detection approach is able to localize an anomaly in the frame shown in Fig. 15.1a (i.e., a man on a bicycle with a velocity higher than the surrounding pedestrians). In fact, we can detect the person(s) in the scene with an anomaly behavior by back-projecting the particle positions corresponding to the localized anomaly into the image.

Before concluding this section, we refer the reader interested in crowd behavior analysis details to recent review papers. In [31], a survey on available techniques for crowd modeling from both the computer vision and the crowd simulation point of view are presented. Emphasis is drawn on discussing the techniques available for crowd modeling using agent based models, nature based models and physical models. In [14] a discussion on the available computer vision techniques for crowd behavior analysis for video surveillance applications is presented. This survey also reports a few computer vision schemes able to address problems like crowd dynamics, crowd analysis and crowd synthesis. In [10] a summary of crowd behavior techniques from a social signal perspective applied to video surveillance is presented.

15.3 Proposed Particle Advection Using PSO-SFM

This section describes our proposed particle advection method using PSO-SFM. In earlier attempts [2, 21], the particle advection is carried out by placing a rectangular grid of particles over each video frames. Then, the velocity for each particle is calculated using fourth-order Runge-Kutta-Fehlberg algorithm [19] along with the bilinear interpolation of the optical flow field. In general, a drawback of this approach is that it assumes that a crowd follows a fluid-dynamical model which is too restrictive when modeling masses of people. The elements of the crowd may also move with unpredictable trajectories that will result in an unstructured flow. Moreover, the use of a rectangular grid for particles is a coarse approximation with respect to the continuous evolution of the social force. To overcome these drawbacks, we propose a novel particle advection scheme using PSO aiming at modeling the crowd behavior. Before presenting the detailed description of our proposed scheme, we first provide a brief introduction on PSO and SFM in the following subsections.

15.3.1 Particle Swarm Optimization

Particle Swarm Optimization is a stochastic, iterative, population-based optimization technique aimed at finding a solution to an optimization problem in a search space [15]. The main objective of PSO is to optimize a given criterion function called fitness function f. PSO is initialized with a population, namely a *swarm*, of N-dimensional particles distributed randomly over the search space (of dimension N too): each particle is so considered as a point in this N-dimensional space and the optimization process manages to move the particles according to the evaluation of the fitness function in an iterative way. More specifically, at each iteration, each particle is updated according to two "best" values, respectively called *pbest_i*, which depends on the *i*-th particle, and *gbest* which is independent from the specific particle. *pbest_i* is the position corresponding to the best (e.g., minimum) fitness value of particle i obtained so far (i.e. taking into account the positions computed from the first iteration to the current one). On the other hand, *gbest* is the best position achieved by the whole swarm:

$$gbest = \arg \min_i f(pbest_i), \tag{15.1}$$

The position change (called "velocity") v_i for the i-th particle is updated according to the following equations [15]:

$$v_i^{new} = I_A \cdot v_i^{old} + C_1 \cdot rand_1 \cdot (pbest_i - x_i^{old})$$
$$+ C_2 \cdot rand_2 \cdot (gbest - x_i^{old}); \tag{15.2}$$
$$x_i^{new} = x_i^{old} + v_i^{new}, \tag{15.3}$$

where I_A is the inertia weight, whose value should be tuned to provide a good balance between global and local explorations, and it may result in fewer iterations on average for finding near optimal results. The scalar values C_1 and C_2 are acceleration parameters used to drive each particle towards *pbest_i* and *gbest*. Low values of C_1 and C_2 allow the particles to roam far from target regions, while high values result in abrupt movements towards the target regions. $rand_1$ and $rand_2$ are random numbers between 0 and 1. Finally, x_i^{old} and x_i^{new} are the current and updated particle positions, respectively, and the same applies for the deviation v_i^{old} and v_i^{new}.

15.3.2 Social Force Model

The SFM [13] provides a mathematical formalization to describe the movement of each individual in a crowd on the basis of its interaction with the environment and other obstacles. The SFM can be written as:

$$m_i \frac{dW_i}{dt} = m_i \left(\frac{W_i^p - W_i}{\tau_i} \right) + F_{int}, \tag{15.4}$$

where m_i denotes the mass of the individual, W_i indicates its *actual velocity* which varies given the presence of obstacles in the scene and τ_i is a relaxing parameter. F_{int} indicates the interaction force experienced by the individual which is defined as the sum of attraction and repulsive forces. Finally, W_i^p is the *desired velocity* of the individual.

Assuming $m_i = 1$ and $\tau_i = 1$, from Eq. (15.4) we obtain:

$$F_{int} = W_i - W_i^p + \frac{dW_i}{dt}. \tag{15.5}$$

Equation (15.5) shows that the higher the difference between the actual and the desired velocities of a particle, the stronger its interaction force. The intuitive idea behind this is that an obstacle (e.g., a person or a group of persons) can make a particle (representing an individual of the analyzed crowd) to deviate from its desired path. The higher this deviation, the stronger the underlying interaction force. Thus, estimating the interaction force of the particle swarm will give us an instrument to assess the total amount of person-to-person interactions in a given frame. Anomalies will be detected as outliers in the interaction force distribution.

In the next section we will see how the optical flow can be used for an operational definition of the velocities involved in Eq. (15.5) and the how the PSO process can be used to simulate the movement of a set of individuals who aim at minimizing their respective interaction forces.

15.3.3 The Proposed Minimization Scheme

The PSO begins with a random initialization of the particles in the first frame. From such initial stage, we obtain a first guess of $pbest_i$, for each particle i, and the global *gbest*. The particles are defined by their 2-D positions corresponding to the pixel coordinates in the frames. At each iteration, the *pbest$_i$* value is updated only if the present position of the particle is better than the previous position according to fitness function evaluated on the model interaction force. Finally, the *gbest* is updated with the position obtained from the best *pbest$_i$* after reaching the maximum number of iterations or if the desired fitness value is achieved. We then use the final particle positions as the initial guess in the next frame and the same iterative process is repeated until the end of the video sequence. Therefore, the movement of the particles is updated according to the fitness function which drives the particles toward the areas of minimum interaction force using SFM.

15.3.3.1 Computing the Fitness Function

The fitness function aims at capturing the interaction force exhibited by each movement in the crowded scene. Each particle is evaluated according to its interaction force calculated using SFM and optical flow [6]. In fact, the Optical Flow (OF) is a good candidate to substitute the pedestrian velocities in the SFM model.

Using OF, we define the actual velocity of particle i as:

$$W_i = O_{avg}(x_i^{new}), \tag{15.6}$$

where $O_{avg}(x_i^{new})$ indicates the average OF at the particle coordinates x_i^{new}, which in turn is estimated using Eq. (15.2). The average is computed over L previous frames. The desired velocity of the particle is defined as:

$$W_i^p = O(x_i^{new}), \tag{15.7}$$

where $O(x_i^{new})$ represents the OF intensity (in the current frame) of the particle i. Both $O()$ and $O_{avg}()$ are computed using interpolation in a small spatial neighborhood to avoid numerical instabilities of the OF. Finally, we calculate the interaction force F_{int} using Eq. (15.5):

$$F_{int}(x_i^{new}) = \frac{dW_i}{dt} - \left(W_i^p - W_i\right), \tag{15.8}$$

where the velocity derivative is approximated as the difference of the OF at the current frame t and $t-1$, that is $\frac{dW_i}{dt} = [O(x_i^{new})|_t - O(x_i^{new})|_{t-1}]$. As above mentioned, the interaction force (Eq. (15.5)) allows an individual to change its movement from the desired path to the actual one. This process is in some way mimicked by the particles which are driven by the OF toward the image areas of larger motion. In this way, the more regular the pedestrians' motion, the less the interaction force, since the people motion flow varies smoothly. So, in a normal crowded scenario the interaction force is expected to stabilize at a certain (low) value

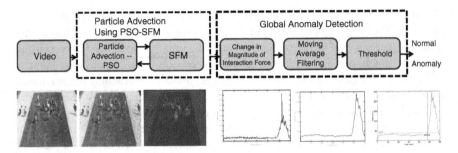

Fig. 15.2 Block diagram of the proposed framework for global anomaly detection

complying with the typical motion flow of the mass of people. It is then reasonable to define a fitness function aimed at minimizing the interaction force and moving particles toward these sinks of small interaction force, thereby allowing particles to simulate a "normal" situation of the crowd.

Hence, we define our fitness function as:

$$f(x_i) = F_{int}(x_i),$$ (15.9)

where x_i denotes the i-th particle's position. With the above definitions we can use the PSO framework presented in Sect. 15.3.1 to minimize $f()$.

15.3.4 Global Anomaly Detection Scheme Using PSO-SFM

In Fig. 15.2 we show the stages of our global anomaly detection system, whose aim is to classify every frame of a given video sequence as either "normal" or "abnormal". In the first stage we estimate the interaction force on each frame using the PSO-SFM scheme described in Sect. 15.3.3. The interaction force associated with each particle is then processed further to identify the global anomaly in the frame.

As an example, Fig. 15.3a–d show the computed interaction force with the proposed particle advection using PSO-SFM for both normal (Fig. 15.3a, b) and anomaly video frames (Fig. 15.3c, d). In these figures, we plotted on the image the magnitude of the interaction forces assigned to every particle. As observed in Fig. 15.3, the presence of the high magnitude interaction force over time can provide useful information about the existence of an anomaly. This allow us to formulate the detection of global anomalies as the detection of the changes in the interaction force magnitude. This process is valid with the proposed particle advection scheme since the presence of global abnormality can be recognized by the presence of high magnitude of the interaction force associated with the particles (see Fig. 15.3). Since all the available test videos contains a certain amount of frames in which normal behavior is assumed, we take advantage of this information in

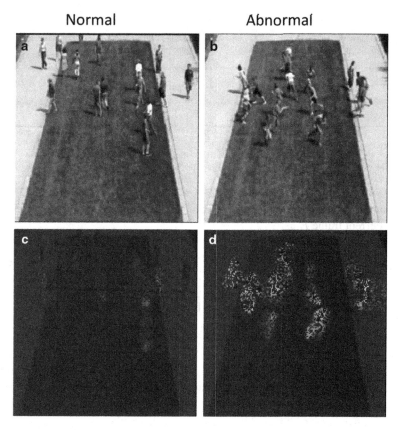

Fig. 15.3 An illustration of the proposed scheme. (**a**) Input normal frame. (**b**) Interaction force corresponding to (**a**). (**c**) Input anomaly frame. (**d**) Interaction force corresponding to (**c**)

the comparison process, like all the other previous algorithms [21]. In practice, we carry out the following steps to decide whether a given frame contains an anomaly or not:

1. First, compute the sum of the interaction forces of a reference frame F_r. This reference frame(s) represents a normal behavior scene in the given video sequence. Actually, all the public datasets considered have an initial (variable, but at least one frame) set of frames representing a normal behavior which can be used as a reference. If k is the number of particles (currently, $k = 15,000$), we obtain F_r as follows:

$$F_r = \sum_{i=1}^{k} F_{int}\left(x_i^{new}\right)\big|_r \tag{15.10}$$

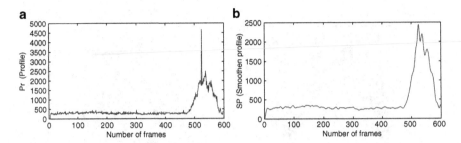

Fig. 15.4 Profile (**a**) before smoothing (**b**) after smoothing

2. Compute the sum of the interaction forces corresponding to all the particles in the current frame F_t as:

$$F_t = \sum_{i=1}^{k} F_{int}(x_i^{new})|_t \tag{15.11}$$

3. Compute the change in the magnitude force at each frame t as:

$$C_t = |F_t - F_r| \tag{15.12}$$

4. Repeat steps 2–3 for all the frames to obtain the profile (values of C_t for all the video frames) corresponding to the change of the force magnitude.

   As an example, Fig. 15.4a shows the profile obtained from a sequence of the UMN dataset after following the above mentioned steps 1–4.
5. Finally, we use the moving average filter to smooth out the short term fluctuations that are present in the obtained profile at the previous step, so to get a smoothed profile C_t^s (see Fig. 15.4b). The moving average is obtained by the simple mean of a few temporally adjacent frames. Once C_t^s is computed, each frame is then classified as either normal or abnormal according to a threshold as follows:

$$L_t = \begin{cases} Abnormal \text{ if } C_t^s > th \\ Normal \text{ otherwise} \end{cases}$$

where C_t^s represents the smoothed profile, th is a threshold value, and L_t holds the final detection result of the given video sequence.

15.3.5   Local Anomaly Detection Scheme Using PSO-SFM

While in the previous section we showed how a frame is classified as either normal or abnormal, the aim of this section is to show how a finer localization of the

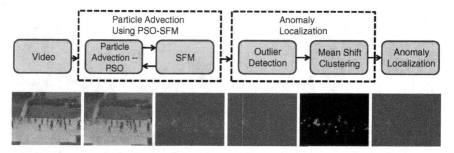

Fig. 15.5 Block diagram of the proposed scheme for anomaly detection and localization

Fig. 15.6 (a) Input frame. (b) Interaction force

anomaly inside the frame is possible. Figure 15.5 summarizes the proposed scheme for accurate localization of the anomaly in a crowd. The first step is the same interaction force optimization approach presented in Sect. 15.3.3 and used for the global case (see Fig. 15.2).

Figure 15.6a–b show the input frame and the corresponding interaction force, respectively. It is interesting to observe that the highest magnitudes of the force are located in the image regions that move differently from the overall image flow (e.g., the man on the bicycle close to the street lamp). Although patterns of high magnitude of the interaction force over a certain period of time can provide useful information about the presence of an anomaly, not necessarily large magnitudes of the force is a direct consequence of the presence of an anomaly. This is due to the fact that particles are not associated to a whole person, but only to person's parts, so, for instance, legs motion can lead to a high interaction force which is obviously not an anomaly. This motivates us to propose a scheme that can capture the high magnitude patterns over a certain period of time and thereby localize the presence of anomalies in the scene. In order to detect structured interaction forces over time, we use an *outlier detection* scheme to eliminate isolated fluctuations of the social force at each time instant. These "outliers" effects are in general due to the approximation

a   b

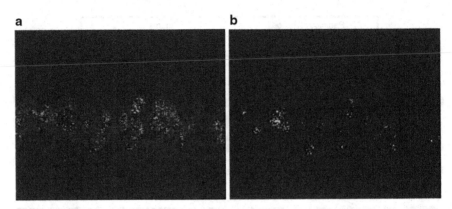

Fig. 15.7 Results of the RANSAC-like algorithm. (**a**) Obtained inliers. (**b**) Corresponding outliers

of the pedestrians velocities with a dense OF computation. For instance, as above observed, we noted that the leg swinging of a walking pedestrian is a cause for false positive (anomaly) detections. This occurs because the local optical flow in this small areas is noisy and may cause some disturbances in the anomaly detection.

The outliers detection process is performed using a custom implementation of the well-known RANdom SAmple Consensus (RANSAC) algorithm [12]. RANSAC is an iterative method used to estimate the parameters of a mathematical model from observed data containing outliers. This algorithm basically assumes that most of the available data consists of inliers whose distribution can be explained by a known parametric model. However, inliers are mixed with outliers which make the direct model parameter estimation inaccurate. Our empirical observations showed that the statistics of the interaction forces associated to a crowd situation in the video datasets can be reasonably well approximated by a Gaussian distribution. Thus, given the interaction force magnitude of the particles at each frame we perform the following steps:

1. Randomly select 5,000 particles (out of 15,000 particles) and their corresponding interaction force magnitude.
2. Estimate the Gaussian distribution using the interaction force magnitude associated with only the selected particles. Let the estimated mean and standard deviation be $\hat{\mu}$ and $\hat{\sigma}$.
3. Consider the remaining particles and evaluate those that are inliers and outliers. Inliers are detected by checking if the particle's force is within the typical $3\hat{\sigma}$ of the estimated model, particles whose force is outside this interval are considered outliers.
4. Repeat the steps 1–3 for R number of iterations, $R = 1,000$ iterations in our case.
5. Finally, choose the Gaussian model with the highest number of inliers.

Figure 15.7a–b show the inliers and outliers obtained using the RANSAC-like algorithm. It is interesting to observe that all high magnitude interaction forces are detected as outliers. In order to achieve a better localization, we perform

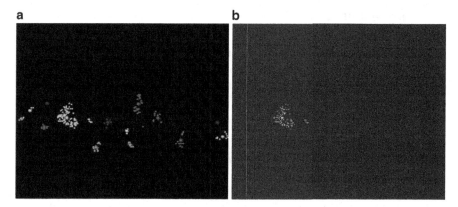

Fig. 15.8 Results of mean-shift clustering. (**a**) Clusters. (**b**) Force magnitude of the largest cluster's particles

Fig. 15.9 An anomaly moving person localized using the positions of the particles in the largest outlier cluster

a spatial clustering of the detected outliers using mean-shift [7, 8] as it works independently on the assumptions regarding the shape of the distribution and the number of modes/clusters. In the end, we finally select the clusters with a number of members larger than a certain threshold, discarding clusters having a small number of particles. This threshold is fixed and kept constant in all the performed experiments; further, assuming that the geometry of the scene is roughly known, this threshold can be set to define the minimal (abnormal) event to be detected.

Figure 15.8a–b show the results of mean-shift clustering and the final anomaly localization obtained after selecting the largest cluster. The positions of the particles of this cluster are plotted on the original input frame in Fig. 15.9. These particles correspond to a moving person on a bicycle, who has been correctly detected as an anomaly because his/her movement does not conform with the movement of the surrounding pedestrians.

15.4   Experiments

In this section we present and discuss the experimental results obtained using the proposed schemes for global and local anomaly detection. We first discuss the results using the global approach and then the experiments performed using the local anomaly scheme.

15.4.1   *Experimental Results and Discussion on the Global Anomaly Scheme*

To validate the performance of the proposed approach for global anomaly detection, we conducted an extensive set of experiments on four different datasets: UMN [28], PETS 2009 [24], UCF [21], and prison riot dataset (collected by us from the web). In the following experiments, all the video frames are resized to a fixed resolution of 200×200 pixels. For the particle advection scheme, the particle density (i.e., the number of particles) is kept constant at 25 % of number of pixels, and the number of iterations is fixed to 100. To detect the changes of the interaction force magnitude, we use the first frame as the reference frame. This is because in all the datasets the initial (roughly) 40 % of the video frames represents the normal behavior which is then followed by the abnormal behavioral frames. Finally, the performance is validated by plotting the ROC curves obtained over all possible values of the threshold th.

15.4.1.1   UMN Dataset

The UMN dataset consists of 11 video sequences acquired in three different crowded scenarios including both indoor and outdoor scenes. All these sequences exhibit an escape panic scenario: they start with the normal behavior frames followed by the abnormal activity. Figure 15.10 illustrates the results of the proposed scheme obtained on the UMN dataset. Figure 15.10a shows two examples of normal and abnormal crowd behavior frames, respectively, and Fig. 15.10b indicates the corresponding interaction force obtained using the proposed PSO-SFM based particle advection approach. From this figure, it can be observed that the presence of high magnitude of the majority of the particles' interaction force is an evidence that an abnormal frame has occurred. Figure 15.10c shows the detection results of the normal and abnormal frames using step 5 of the global anomaly detection algorithm presented in Sect. 15.3.4. Figure 15.11 shows the detection results obtained on two different sequences of the same UMN dataset. Abnormal frames always correspond to a higher interaction force of the particles.

Figures 15.12 and 15.13 show the performance of the proposed scheme on three different scenes of UMN and on the whole dataset, respectively. The quantitative results in Table 15.1 indicate that the proposed scheme obtained the best performance over different available state-of-the-art methods.

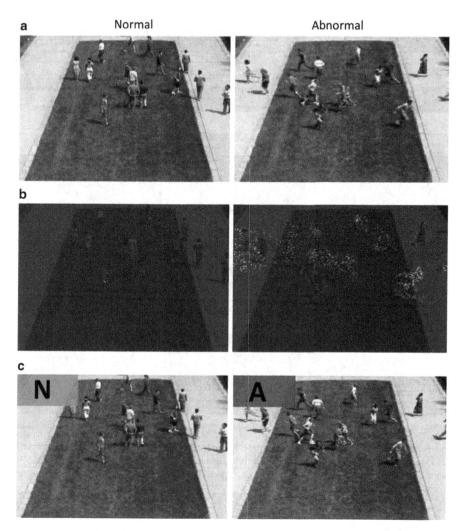

Fig. 15.10 Results on the UMN dataset. (**a**) Input frame. (**b**) Force field. (**c**) Detection (*N* indicates normal and *A* indicates abnormal frame)

15.4.1.2   Prison Riot Dataset

In order to evaluate the proposed method on real applications, we collected a set of real videos from websites such as YouTube and ThoughtEquity.com. The collected video dataset is composed of seven sequences representing riots in prisons that are captured with different angles, resolutions, background and includes abnormality like fighting with each other, clashing, etc. All the collected sequences start with the normal behavior which is then followed by a sequence of abnormal behavior frames. Figure 15.14 shows the interaction force obtained on some of the frames

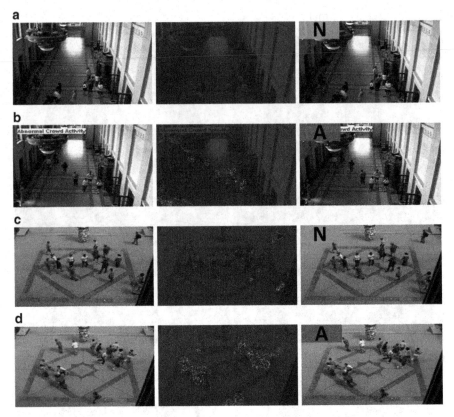

Fig. 15.11 Results of the proposed scheme on other sequences of the UMN dataset. (**a**) Normal behavior in scene 2 with its corresponding interaction force and detection. (**b**) Abnormal behavior in scene 2 with its corresponding interaction force and detection. (**c**) Normal behavior in scene 3 with its corresponding interaction force and detection. (**d**) Abnormal behavior in scene 3 with its corresponding interaction force and detection

of this dataset. Figure 15.15 illustrates the performance of the proposed method on some frames taken from different sequences in this datasets. The ROC curves in Fig. 15.16 demonstrate that the proposed method outperforms the optical flow-based method in distinguishing the abnormal sequences from the normal ones. The quantitative results of this comparison are reported in Table 15.2.

15.4.2 Results on PETS 2009 Dataset

This section describes the results obtained on PETS 2009 'S3' dataset. This dataset is different from the other datasets used in this chapter, in the sense that abnormality begins smoothly and this makes the detection more challenging because of the

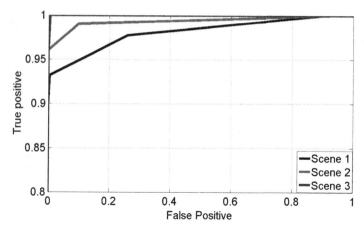

Fig. 15.12 ROC curves of abnormal behavior detection on different scenes in UMN dataset

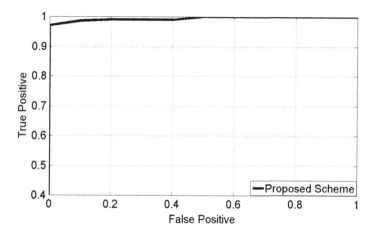

Fig. 15.13 ROC performance on UMN dataset

gradual transaction from normal to abnormal activity. Figure 15.17 shows the interaction force estimated using the proposed scheme on PETS 2009 and Fig. 15.18 shows the corresponding ROC curve. Table 15.3 shows the quantitative results of the comparison, illustrating that the proposed scheme outperforms the optical flow method also with this benchmark.

15.4.2.1 UCF Dataset

Finally, the effectiveness of the proposed algorithm is also evaluated on the UCF dataset [21] composed of 12 video sequences representing normal and abnormal scenes collected from the web. Also in this case, Fig. 15.19 demonstrates that

Table 15.1 Performance of
the proposed scheme on the
UMN dataset

Method	Area under ROC
Optical flow [21]	0.84
Social force [21]	0.96
Chaotic invariants [30]	0.99
NN [9]	0.93
Sparse reconstruction (scene 1) [9]	0.995
Sparse reconstruction (scene 2) [9]	0.975
Sparse reconstruction (scene 3) [9]	0.964
Sparse reconstruction (full dataset) [9]	0.978
Proposed scheme (scene 1)	**0.9961**
Proposed scheme (scene 2)	**0.9932**
Proposed scheme (scene 3)	**0.9991**
Proposed scheme (full dataset)	**0.9961**

the proposed scheme outperforms the optical flow procedure, and this is further
corroborated by the quantitative results reported in Table 15.4 and the qualitative
results reported in Fig. 15.20.

The experiments illustrated so far show that the proposed global anomaly
detection strategy outperforms the available state-of-the-art methods on realistic
datasets like UCF and Prison Riots, other than UMN and PETS 2009 benchmark
datasets. The next section is dedicated to testing the local strategy proposed in
Sect. 15.3.5.

15.4.3 Experimental Results and Discussion on the Local Anomaly Scheme

To evaluate the performances of the local anomaly detection scheme and compare
it with state-of-the-art approaches, we consider two standard datasets used for
abnormal activities detection: UCSD [20] and MALL [1] datasets.

15.4.3.1 UCSD Dataset

The UCSD dataset contains two different sets of surveillance videos called PED1
and PED2. The dataset has a reasonable density of people and anomalies including
bikes, skaters, motor vehicles crossing the scenes. The PED1 has 34 training and
36 testing image sequence and PED2 has 16 training and 12 test image sequences.
These video sequences have two evaluation protocols as presented in [20], namely:
(1) frame-level anomaly detection, and (2) pixel-level anomaly detection. At frame-
level, we verify if the current frame contains a labeled abnormal pixel. In such a
case, the frame is considered containing an abnormal event and compared with
the annotated ground truth status (either normal or abnormal). At pixel-level, the

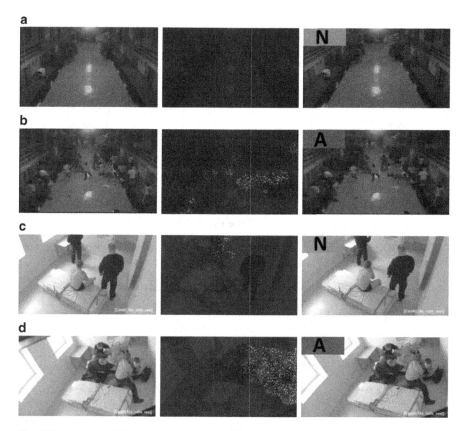

Fig. 15.14 Results of the proposed scheme on the prison dataset. (**a**) A normal behavior frame and its corresponding interaction force and detection result on video sequence 1. (**b**) An abnormal behavior frame and its corresponding interaction force and detection result on sequence 1. (**c**) A normal behavior frame and its corresponding interaction force and detection result on sequence 2. (**d**) An abnormal behavior frame and its corresponding interaction force and detection result on sequence 2

detection of abnormality is compared against the ground truth on a subset of 10 test sequences. If at least 40 % of the detected abnormal pixels match the ground truth pixels, it is presumed that anomaly has been localized otherwise it is treated as a false positive.

Figure 15.21 shows the ROC curve of our method for the frame-level anomaly detection criteria for PED1 and PED2 datasets. We then compare the performance against the state-of-the-art approaches such as the SFM based method [21], MPPCA [16], Adam et al. [1] and Mixture of dynamic textures (MDT) [20]. Table 15.5 shows the quantitative results of the proposed method on frame-level anomaly detection on PED1 and PED2 datasets and Table 15.6 shows the results on anomaly localization. The Equal Error Rate (EER) in Tables 15.5 and 15.6 is defined as the point where false positive rate is equal to false negative rate. Remarkably, the

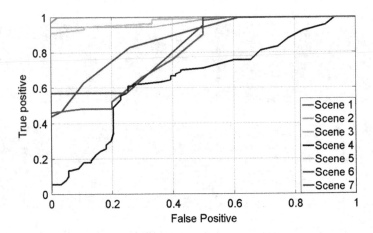

Fig. 15.15 ROC curve of abnormal behavior detection in the different sequences of the prison dataset

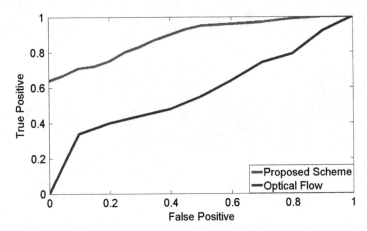

Fig. 15.16 ROC curves showing the comparison of the proposed scheme over the optical flow method on the prison dataset

Table 15.2 Performance of the proposed scheme on the prison dataset

Method	Area under ROC
Optical flow	0.5801
Proposed scheme (full dataset)	**0.8903**

proposed method outperforms all the previous approaches on both frame-level and pixel-level detection, reaching the best performances in the frame-level anomaly detection on the PED2 dataset.

Figure 15.22 shows a few frame samples with anomaly detection and localization for the PED1 and PED2 datasets. It can be observed that the proposed method is capable of detecting anomalies even in the far end of the scene (see Fig. 15.22a, last two frames).

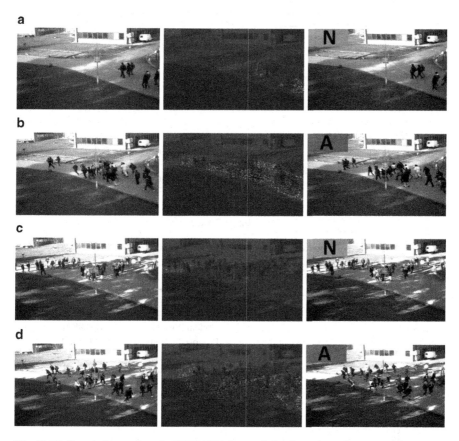

Fig. 15.17 Sample frames from the PETS 2009 dataset (*left column*: input frames, *middle column*: the corresponding interaction force, *right column*: the classification result). (**a**)–(**b**) Sample frames from S3 (14–16). (**c**)–(**d**) sample frames from S3 (14–33)

15.4.3.2 Mall Dataset

The Mall dataset [1] consists of a set of video sequences recorded using three cameras placed in different locations of a shopping mall during working days. The annotated anomalies in such dataset are individuals running erratically in the scene. The evaluation protocol uses only the frame-level anomaly detection criteria. Figure 15.23 shows some frame samples from this dataset in which the anomaly is detected using the proposed method. Table 15.7 shows that the proposed method is extremely accurate in detecting all the frames with an anomaly. Moreover, our approach outperforms the state-of-the-art schemes with respect to the best Rate of Detection (RD) and fewer False Alarm (FA).

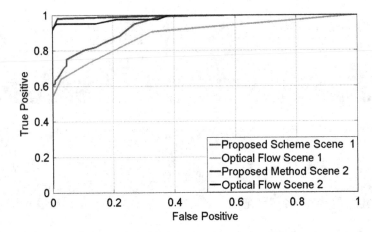

Fig. 15.18 The ROC curves of abnormal behavior detection in the PETS 2009 database

Table 15.3 Performance of the proposed scheme on PETS 2009 dataset

Method	Area under ROC
Optical flow scene 1	0.8834
Proposed scheme scene 1	**0.9414**
Optical flow scene 2	0.9801
Proposed scheme scene 2	**0.9914**

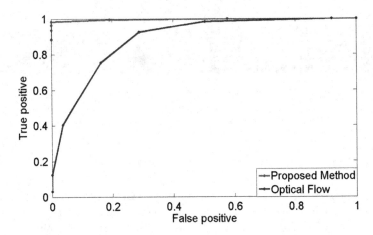

Fig. 15.19 The ROC curves of abnormal behavior detection in the UCF dataset

15.5 Conclusion

We proposed a new particle advection scheme for both global and local anomaly detection in crowded scenes. The main contribution of this work lies in introducing the optimization of the evolving interaction force and performing particle advection

Table 15.4 Performance of the proposed scheme on UCF dataset

Method	Area under ROC
Optical flow	0.884
Proposed scheme (full dataset)	**0.986**

Table 15.5 Equal error rates for frame level anomaly detection on PED1 and PED2 datasets

Approach	SF [21] (%)	MPPCA [16] (%)	Adam et al. [1] (%)	MDT [20] (%)	Proposed method (%)
PED1	31	40	38	25	21
PED2	42	30	42	25	14
Average	37	35	40	25	17

Table 15.6 Anomaly localization: detection rate at the EER

Method	SF [21] (%)	MPPCA [16] (%)	Adam et al. [1] (%)	MDT [20] (%)	Proposed method (%)
Localization	21	18	24	45	**52**

Table 15.7 Performances on the Mall dataset

Dataset	Methods	RD	FA
Mall Cam 1	Adam et al. [1]	95% (19/20)	1
	Proposed method	100% (20/20)	2
Mall Cam 2	Adam et al. [1]	100% (17/17)	6
	Proposed method	100% (17/17)	4
Mall Cam 3	Adam et al. [1]	95% (20/21)	4
	Proposed method	100% (21/21)	3

to capture the optimized interaction force according to the underlying optical flow. The main advantage of the proposed scheme is that the whole anomaly detection/localization process is carried out without any learning phase. This further justifies the applicability of our proposed scheme for real world applications. Finally, empirical results have also indicated that our method is robust and highly performing in detecting abnormal activities on very different types of crowded scenes.

Acknowledgements This article summarizes and incorporates two earlier publications concerning global [26] and local [25] anomaly detection in crowded scenarios.

References

1. Adam, A., Rivlin, E., Shimshoni, I., Reinitz, D.: Robust real-time unusual event detection using multiple fixed-location monitors. IEEE Trans. Pattern Anal. Mach. Intell. **30**(3), 555–560 (2008)
2. Ali, S., Shah, M.: A Lagrangian particle dynamics approach for crowd flow segmentation and stability analysis. In: Proceedings of IEEE Conference on Computer Vision and Pattern Recognition (CVPR), pp. 1–6. Los Alamitos, CA, USA (2007)

Fig. 15.20 Sample frames from UCF dataset indicating normal (*left*) and abnormal (*right*) frames

3. Antic, B., Ommer, B.: Video parsing for abnormality detection. In: Proceedings of IEEE International Conference on Computer Vision (ICCV), pp. 2415–2422. Los Alamitos, CA, USA, (2011)
4. Barnard, K., Duygulu, P., Freitas, D.N., Forsyth, F., Blei, D., Jordan, M.: Matching words and pictures. J. Mach. Learn. Res. 3(1), 1107–1135 (2003)
5. Blei, M.D., Ng, Y.A., Jordan, I.M.: Latent dirichlet allocation. J. Mach. Learn. Res. **34**(1), 993–1022 (1981)
6. Brox, T., Bruhn, A., Papenberg, N., Weickert, J.: High accuracy optical flow estimation based on a theory for warping. In: Proceedings of European Conference on Computer Vision (ECCV), pp. 1–10. Prague (2004)

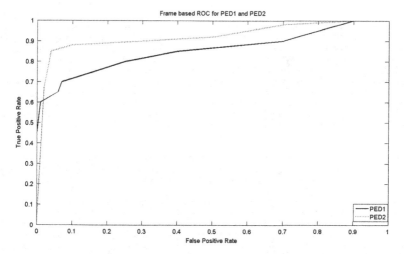

Fig. 15.21 ROC curves obtained on UCSD PED1 and PED2 datasets

Fig. 15.22 (Color online) Examples of anomaly frame detection and localization on PED1 (**a**) and PED2 (**b**) datasets (best viewed in color)

7. Cheng, Y.: Mean shift, mode seeking and clustering. IEEE Trans. PAMI **17**(8), 790–799 (1995)
8. Comaniciu, D., Meer, P.: Mean shift: a robust approach toward feature space analysis. IEEE Trans. PAMI, Colorado Springs, Colorado, USA, **24**(5), 603–619 (2002)
9. Cong, Y., Yuan, J., Liu, J.: Sparse reconstruction cost for abnormal event detection. In: Proceedings of IEEE Conference on Computer Vision and Pattern Recognition (CVPR), pp. 1–10. Colorado Springs, Colorado, USA (2011)
10. Cristani, M., Raghavendra, R., Del Bue, A., Murino, V.: Human behavior analysis in video surveillance: a social signal processing perspective. Neurocomputing, **100**, 86–97 (2013)

a

b

c

Fig. 15.23 (Color online) Examples of anomaly detection on the Mall dataset. (**a**) Mall camera 1. (**b**) Mall camera 2. (**c**) Mall camera 3 (best viewed in color)

11. Dalal, N., Triggs, B.: Histograms of oriented gradients for human detection. In: Proceedings of IEEE Conference on Computer Vision and Pattern Recognition (CVPR), pp. 886–893. San Diego, CA, USA (2005)
12. Fischler, M.A., Bolles, R.C.: Random sample consensus: a paradigm for model fitting with applications to image analysis and automated cartography. Commun. ACM **24**(1), 381–395 (1981)
13. Helbing, D., Molnar, P.: Social force model for pedestrian dynamics. Phys. Rev. E **51**(4), 42–82 (1995)
14. Jacques, J.C.S., Jr., Raupp Musse, S., Jung, C.R.: Crowd analysis using computer vision techniques: a survey. IEEE Signal Process. Mag. **27**(5), 66–77 (2010)
15. Kennedy, J., Eberhart, R.C.: Particle swarm optimization. In: Proceedings of IEEE International Conference on Neural Networks, pp. 1942–1948. Washington, DC, USA (1995)
16. Kim, J., Grauman, K.: Observe locally, infer globally: a space-time MRF for detecting abnormal activities with incremental updates. In: Proceedings of IEEE Conference on Computer Vision and Pattern Recognition (CVPR), pp. 2921–2928. Miami, Florida, USA (2009)
17. Kratz, L., Nishino, K.: Anomaly detection in extremely crowded scenes using spatio-temporal motion pattern models. In: Proceedings of IEEE Conference on Computer Vision and Pattern Recognition (CVPR), pp. 1446–1453. Miami, Florida, USA (2009)

18. Krausz, B., Bauckhage, C.: Automatic detection of dangerous motion behavior in human crowds. In: Proceedings of IEEE International Conference on Advanced Video and Signal-Based Surveillance (AVSS), pp. 224–229. Washington, DC, USA (2011)
19. Lekien, F., Marsden, J.: Tricubic interpolation in three dimensions. J. Numer. Methods Eng. **63**(3), 455–471 (2005)
20. Mahadevan, V., Li, W., Bhalodia, V., Vasconcelos, N.: Anomaly detection in crowded scenes. In: Proceedings of IEEE Conference on Computer Vision and Pattern Recognition (CVPR), pp. 1975–1981. San Francisco (2010)
21. Mehran, R., Oyama, A., Shah, M.: Abnormal crowd behavior detection using social force model. In: Proceedings of IEEE Conference on Computer Vision and Pattern Recognition (CVPR), pp. 935–942. Miami, Florida, USA (2009)
22. Mehran, R., Moore, B., Shah, M.: A streakline representation of flow in crowded scenes. In: Proceedings of European Conference on Computer Vision (ECCV), pp. 1–10. Heraklion, Crete, Greece (2010)
23. Moore, B., Ali, S., Mehran, R., Shah, M.: Visual crowd surveillance through a hydrodynamics lens. Commun. ACM **54**(12), 64–73 (2011)
24. PETS 2009 dataset. http://ftp.cs.rdg.ac.uk/PETS2009/
25. Raghavendra, R., Del Bue, A., Cristani, M., Murino, V.: Abnormal crowd behavior detection by social force optimization. In: Proceedings of Human Behavior Understanding (HBU-2011), pp. 134–145. Amsterdam, The Netherlands (2011)
26. Raghavendra, R., Del Bue, A., Cristani, M., Murino, V.: Optimizing interaction force for global anomaly detection in crowded scenes. In: Proceedings of IEEE Workshop on Modeling, Simulation and Visual Analysis of Large Crowds (MSVLC-2011), pp. 136–143. Barcelona, Spain (2011)
27. Reicher, S.: The Blackwell Handbook of Social Psychology: Group Processes. Blackwell, Oxford (2001)
28. UMN dataset. http://www.mha.cs.umn.edu/movies/crowd-activity-all.avi
29. Wang, X., Ma, X., Grimson, W.E.L.: Unsupervised activity perception in crowded and complicated scenes using hierarchical Bayesian models. IEEE Trans. Pattern Anal. Mach. Intell. **31**(3), 539–555 (2009)
30. Wu, S., Moore, B.E., Shah, M.: Chaotic invariants of Lagrangian particle trajectories for anomaly detection in crowded scenes. In: Proceedings of IEEE Conference on Computer Vision and Pattern Recognition (CVPR) pp. 1–6. San Francisco, CA, USA (2010)
31. Zhan, B., Monekosso, D., Remagnino, P., Velastin, S.A., Xu, L.Q.: Crowd analysis: a survey. Mach. Vis. Appl. **19**(5–6), 345–357 (2008)

Printed in the United States
By Bookmasters